June 24-26, 2013
Como, Italy

Association for
Computing Machinery

Advancing Computing as a Science & Profession

EuroITV'13

Proceedings of the 11th European Conference on Interactive TV and Video

Sponsored by:

Politecnico Milano & Camera di Commercio Como

In Cooperation with:

SIGWEB, SIGCHI, & SIGMM

Supported by:

ContentWise & Moviri

Advancing Computing as a Science & Profession

The Association for Computing Machinery
2 Penn Plaza, Suite 701
New York, New York 10121-0701

Notice to Past Authors of ACM-Published Articles
ACM intends to create a complete electronic archive of all articles and/or other material previously published by ACM. If you have written a work that has been previously published by ACM in any journal or conference proceedings prior to 1978, or any SIG Newsletter at any time, and you do NOT want this work to appear in the ACM Digital Library, please inform permissions@acm.org, stating the title of the work, the author(s), and where and when published.

ISBN: 978-1-4503-1951-5 (Digital)

ISBN: 978-1-4503-2441-0 (Print)

Additional copies may be ordered prepaid from:

ACM Order Department
PO Box 30777
New York, NY 10087-0777, USA

Phone: 1-800-342-6626 (USA and Canada)
+1-212-626-0500 (Global)
Fax: +1-212-944-1318
E-mail: acmhelp@acm.org
Hours of Operation: 8:30 am – 4:30 pm ET

Printed in the USA

Chairs' Welcome

It is our great pleasure to welcome you to the 11th European Interactive TV Conference (EuroITV 2013), held in Como, Italy, on June 24-26.

As EuroITV enters the second decade of life, it has clearly established itself as the premier international venue for research and development in the field of Interactive Television, where leading researchers and practitioners from around the world meet and discuss their latest results and solutions.

The growing maturity of the conference is reflected through several changes in the program as compared to previous years. Authors of short papers have the opportunity to present their work in regular sessions. The tutorial program includes three exciting tutorials and five attractive collocated workshops. We also put a special emphasis on soliciting demonstrations and industrial papers, leading to a record number of 16 demos and 9 industrial papers accepted to the conference, highlighting the strong connection of the conference with real problems and needs. Finally, 4 PhD students have been accepted to participate in the Doctoral Consortium.

In parallel to growing and diversifying participation opportunities, the competitive acceptance rate of papers was maintained. Out of 58 research paper submissions, 21 were accepted (36%) for oral presentation at the conference.

The conference program includes two exceptional keynotes, one from a technological perspective by Sriram Subramanian (University of Bristol) and one from a creative perspective by Chris Noessel (co-author with Nathan Shedroff of the book *Make IT SO*).

Many people have contributed to make EuroITV 2013 a success: we would like to thank the tutorial and workshop chairs Maria da Graça Pimentel and Niloofar Dezfuli; the doctoral consortium chairs David Geerts and Dominik Strohmeier; the demo chairs Jonathan Hook and Diego Martinez-Plasencia; the industry chair Patrick Huber; the track chairs Santosh Basapur, Franca Garzotto, Stefano Tubaro, Shelley Buchinger, Francesca Piredda, Margarete Jahrmann, Falk Heinrich, Margherita Pagani and Elisa Rubegni; the grand challenge chairs Artur Lugmayr, Robert Strzebkowski and Milena Szafir. We would like to thank all the program committee members as well as additional reviewers for their diligent work. Special thanks are dedicated to the local arrangements chair Matteo Picozzi.

We would also like to thank our sponsors, Moviri, ContentWise, Politecnico di Milano, and Camera di Commercio di Como, for their generous support. Finally, we thank our in-cooperation partners, the ACM SIGs (SIGCHI, SIGWEB, and SIGMM).

We hope that you will find this program interesting and thought-provoking and that the conference will provide you with a valuable opportunity to share ideas and have interesting conversations with leading researchers and practitioners in the field. In addition to attending the conference, we hope you will use recommendations from local attendees or past travelers to explore other opportunities to enjoy the lovely city of Como.

Paolo Paolini
General Chair

Paolo Cremonesi
George Lekakos
Conference Chairs

Marianna Obrist
Program Coordinator

Table of Contents

Tutorial Session

Workshop Session

EuroITV 2013 Conference Organization

General Chair: Paolo Paolini *(Politecnico di Milano, Italy)*

Conference Chairs: Paolo Cremonesi *(Politecnico di Milano, Italy)*
George Lekakos *(Athens University of Economics and Business, Greece)*

Program Coordinator: Marianna Obrist *(University of Newcastle, UK)*

Local Arrangements Chair: Matteo Picozzi *(Politecnico di Milano, Italy)*

Steering Committee: Pablo Cesar *(CWI, The Netherlands)*
George Lekakos *(Athens University of Economics and Business, Greece)*
Kostantinos Chorianopoulos *(Ionian University, Greece)*
Artur Lugmayr *(Tampere University of Technology, Finland)*
David Geerts *(CUO/IBBT/K.U. Leuven, Belgium)*
Judith Masthoff *(University of Aberdeen, UK)*
Gunnar Harboe *(University of Zurich, Switzerland)*
Marianna Obrist *(Newcastle University, UK)*
Jens F. Jensen *(Aalborg University, Denmark)*
Lyn Pemberton *(Brighton University, UK)*
Hendrik Knoche *(EPFL, Switzerland)*
Célia Quico *(Universidade Lusófona de Humanidades e Tecnologias, Portugal)*

Track 1 Chairs - Human Computer Interaction: Santosh Basapur *(Motorola Mobility Inc., USA)*
Franca Garzotto *(Politecnico di Milano, Italy)*

Track 2 Chairs - Systems and Enabling Technologies: Stefano Tubaro *(Politecnico di Milano, Italy)*
Shelley Buchinger *(Universität Wien, Austria)*

Track 3 Chairs - Content, Media & Interactive Media Art: Francesca Piredda *(Politecnico di Milano, Italy)*
Margarete Jahrmann *(University of Applied Arts Vienna, Austria and Zurich University of the Arts, Switzerland)*
Falk Heinrich *(Aalborg University, Denmark)*

Track 4 Chairs - Media, Social and Economic Studies: Margherita Pagani *(Università Bocconi, Italy)*
Elisa Rubegni *(Università della Svizzera Italiana, Switzerland)*

Tutorial and Workshop Chairs: Maria da Graça Pimentel *(Universidade de São Paulo, Brazil)*
Niloofar Dezfuli *(Technische Universitäd Darmstadt, Germany)*

Demo Session Chairs: Jonathan Hook *(University of Newcastle, UK)*
Diego Martinez-Plasencia *(Bristol University, UK)*

Doctoral Consortium Chairs: David Geerts *(KU Leuven, Belgium)*
Dominik Strohmeier *(Technische Universität Berlin, Germany)*

iTV in Industry Chair: Patrick Huber *(Sky Germany, Germany)*

Grand Challenge Chairs: Artur Lugmayr *(Tampere University of Technology, Finland)*
Robert Strzebkowski *(Beuth University of Applied Sciences, Berlin)*
Milena Szafir *(Manifesto21.tv)*

Program Committee:

Jorge Abreu
Pedro Almeida
Vera Araújo
Mario Benassi
Frank Bentley
Regina Bernhaupt
Maresa Bertolo
Thomas Bjørner
Monica Borges
Michael Bove
Petter Bae Brandtzaeg
Shelley Buchinger
Brunhild Bushoff
Pablo Cesar
Konstantinos Chorianopoulos
Mariana Ciancia
Manuel José Damásio
Lieven De Marez
Frederic Dufaux
Mycal Elliott
Asbjørn Følstad
Narciso García
David Geerts
Alberto Gil-Solla
Mário Rui Gomes
Gustavo Gonzalez-Sanchez
Rudinei Goularte
Richard Griffiths
Rasmus Grøn
Lone Koefoed Hansen
Jens Jensen
Amela Karahasanovic
Abhijit Karnik
Ian Kegel
Hendrik Knoche
Simone Kriglstein
Rodrigo Laiola Guimarães
Young Lee
Bram Lievens
Rui J. Lopes
Martin Lopez-Nores
Marika Lüders
Ilaria Mariani
Walter Mattana
Alberto Mattiacci
Tuomola Mika
Simone Milani
Marie José Monpetit
Lyndon Nixon
Jose Pazos-Arias
Chengyuan Peng
Jo Pierson
Iacob Prebensen
Celia Quico
Mika Rautiainen
Erika Reponen
Mark Rice
Simone Santini
Raimondo Schettini
Francesco Siliato
Jan Håvard Skjetne
Mijke Slot
Luiz Fernando Gomes Soares
Mark Springett
Roberto Suarez
Marco Tagliasacchi
Matteo Tarantino
Simone Tosoni
Manfred Tscheligi
Tomaz Turk
Chris J Van Aart
Ray Van Brandenburg
Wendy Van Den Broeck
Petri Vuorimaa
Christian Wartena
David Wheatley
Zhiwen Yu
Salvatore Zingale

2013 Sponsors & Supporters

Sponsors:

In Cooperation with:

Supporters:

EuroITV'13 Keynote: Make It So

Christopher Noessel
Cooper
100 First Street
26th Floor, SF CA USA 94107
+01 415 267 3500
chrisnoessel@gmail.com

ABSTRACT
Make It So (Rosenfeld Media, 2012) explores the speculative interfaces seen in science fiction movies and television shows. The authors have developed a model that traces lines of influence between these sci-fi examples and design in the real world, and use this model as a scaffold to investigate how the depiction of technologies evolve over time, how fictional interfaces influence those in the real world, and what lessons interface designers can learn through this process. Only described briefly in the book, this keynote walks the audience through this model in detail, highlighting video clip examples throughout.

Categories and Subject Descriptors
H.5.2 [**User Interfaces**]: General – *Evaluation/methodology*

General Terms
Design

Keywords
Science fiction interfaces, speculative interfaces, apologetics, interaction design

1. THE SURVEY
To build a broad-based understanding of sci-fi interfaces, the authors developed a private online database, cataloging and tagging interfaces from influential and popular sci-fi movies and television shows. After four years of development, the authors had gathered about 10,000 images covering roughly 2,000 interfaces across a variety of properties. We used the resulting tag cloud to prioritize deeper comparative investigations.

2. INDIVIDUAL INFLUENCES
Neither sci-fi nor real world interaction design happen in a vacuum. Shows and interactive systems are created by people, and these people are frequently influenced by things they see in the world, and notably from the other domain.

3. REAL WORLD → SCI-FI
More systemically, the real world influences sci-fi by establishing the paradigms that sci-fi extends. Being a medium of entertainment, sci-fi cannot afford the narrative time to explain very complicated concepts, and so most often only extends cinemagenic aspects of current paradigms.

4. SCI-FI → REAL WORLD
The authors have identified three other ways that sci-fi influences the real world.

EuroITV'13, June 24–26, 2013, Como, Italy.
ACM 978-1-4503-1951-5/13/06.

4.1 Sci-Fi Sets Expectations
Audiences are influenced by the interfaces that they see in cinema and on television when they see useful, plausible, and exciting technologies. Though there is no way, with our current resources and knowledge, for many of these technologies to become a reality, many others of them can become realized. When they enter the market, they benefit from having consumers who are pre-exposed to its desirability and pre-trained in its use.

4.2 Sci-Fi Reminds Us of the Human Context
Technology evolves much more rapidly than biology. Some interfaces in sci-fi illustrate this difference by highlighting the constraints or capabilities of the human actor in speculative systems. Designers must keep these constraints and capabilities in mind as they do real world work, and serve as common touchstones to remind us of these facts.

4.3 Sci-Fi Suggests New Paradigms
Most speculative interfaces focus on new ideas in technology. These are where we can best look for lessons unique to sci-fi. For the purposes of interaction designers, we group these lessons into four categories.

4.3.1 Common Lessons
These interfaces provide lessons that are quite commonly known to practitioners, and being aware of them provides a pop-culture shared reference.

4.3.2 Undistributed Lessons
Other interfaces illustrate less commonly known lessons, things that are under development at academic institutions or in the research and development departments in industry.

4.3.3 Applicable Lessons
Some speculative interfaces provide us with lessons for our real world work, mostly through abstracting what "works" in the sci-fi context.

4.3.4 New Lessons
Other speculative interfaces provide us with lessons that will apply only to the speculative technology when it is available to the real world.

5. CONCLUSION
Sci-fi is the second most popular genre of box-office revenue. It is deeply beloved and influential in ways that reach beyond just the individual. Before makers in the real world simply accept this influence, we must take a deliberate and critical eye to understand what they could mean realistically to us working in the real world. The speaker hopes this framework is a useful one for doing just that.

Haptic Feedback and Shape-shifting Handhelds for iTV

Sriram Subramanian
Department of Computer Science
University of Bristol, UK
sriram@cs.bris.ac.uk

ABSTRACT

Handheld devices are now used to view media content on the move as well as serving as a second screen augmenting content on a large screen. This use of handheld devices provides an opportunity to explore creation and consumption of novel multi-sensory content. The Bristol Interaction and Graphics group has been exploring various approaches to creating devices that support high-fidelity haptic feedback and shape-shifting capabilities. In this talk I will present our approach to creating such systems as well as to provide initial insights into how these systems might be used to enhance the users viewing experience. I will start with our haptic feedback systems before presenting our the actuated, and shape-shifting devices.

Categories and Subject Descriptors

H.5.2 **[Information interfaces and presentation]:** User Interfaces. - Graphical user interfaces;

General Terms

Human Factors.

Keywords

Human-computer Interaction, Haptics, TV, handhelds, shape-shifting.

EuroITV'13, June 24–26, 2013, Como, Italy.
ACM 978-1-4503-1951-5/13/06.

Viewer Behaviors and Practices in the (New) Television Environment

Jorge Abreu
CETAC.MEDIA
University of Aveiro
3810 – 193 Aveiro
Portugal
jfa@ua.pt

Pedro Almeida
CETAC.MEDIA
University of Aveiro
3810 – 193 Aveiro
Portugal
almeida@ua.pt

Bruno Teles
CETAC.MEDIA
University of Aveiro
3810 – 193 Aveiro
Portugal
bmteles@ua.pt

Márcio Reis
CETAC.MEDIA
University of Aveiro
3810 – 193 Aveiro
Portugal
marcio.reis@ua.pt

ABSTRACT

This paper presents the main results of a study conducted in Portugal with the aim of identifying the current trends of the television ecosystem, with special relevance to understand the cognitive processes of viewers when selecting what to watch on TV. To this end, an online survey targeted mainly to young adults with mid to high academic degrees was designed. The survey gathered more than 600 responses, of which 550 were validated and taken into account in the data analysis. The results provide a detailed insight about the habits, dynamics and behaviors of spectators regarding the actual scenario of television. It also allowed the identification of the most relevant criteria considered by viewers when deciding what to watch on TV. The most important ones include the program genre followed by the state of mind (mood), the people who share the couch with and the time available to watch TV. The results also provide a structuring framework for the design of interactive television applications, targeted to this audience in the areas of personalization, recommendation and search and discovery of TV content.

Categories and Subject Descriptors

H.1.2 [**Models and Principles**]: User/Machine Systems - *Human factors*. J.4 [**Social and Behavioral Sciences**]

General Terms

Measurement, Human Factors

Keywords

TV audiences, viewing practices, interactive television, companion devices, user studies, survey.

1. INTRODUCTION

The television ecosystem, especially when considering the increasing offer (either in terms of content and available platforms) and the related dynamics and viewing habits, is in a state of mutation demanding to be understood by several agents like content producers, television stations, multiplatform players and developers of interactive television (iTV) applications.

Viewers are increasingly "gifted" with hundreds of television channels via digital TV systems (IPTV, cable or satellite), which significantly increase the amount and diversity of content from which each viewer can choose. Furthermore, we are witnessing a television (re)offer enhanced by recording features, locally and in the cloud, which allows new dimensions to time-shift TV. These factors increase the effort needed to find and choose what (TV) content to watch at any given time. As a result, potentially interesting content turns unnoticed to many viewers.

In parallel to this trend we can witness a proliferation of multiplatform content offer available in the usual TV screens or in other complementary devices. In the first case media centers (e.g. Apple TV or Boxee) and Smart TVs with access to services like Hulu, Netflix and many other sources of over-the-top content are the most common offers. The second scenario includes mobile devices (smartphones and tablets) that appear as complementary platforms for watching video content and, secondly, as companion devices of the television experience. By companion devices we refer to the possibility of this equipment allowing a greater involvement with what is being watched on TV, through the integration of social and communication features, through the use of content related apps or just browsing in the web.

In this context, it is important to gather a deeper understanding of the dynamics and current relation of users with their television (ecosystem) that can support the specification of iTV applications, particularly those oriented to support the discovery and selection of television content. In the research outlined in this paper we decided to focus on a target group of users closer to the technological context described aiming to understand their relationship with this (new) television environment.

Therefore, this paper is structured as follows: after the introductory section, chapter 2 presents a contextualization of the habits and trends related to the global television consumption and, more specifically, in the Portuguese context; chapter 3 describes the online survey, which was designed and implemented with the aim of complementing the existent (in literature review) characterization, reporting on the results gathered, and, finally; chapter 4 settles, from the data analysis, a set of guidelines with relevance to different players involved in the current television ecosystem.

2. CONTEXTUALIZATION

2.1 Challenges of the TV ecosystem

As mentioned, several technological developments have been fostering changes in the traditional foundations of television in terms of its social and communications dynamics. This phenomenon has been studied in the academia or by consulting institutions due to its increasing economic impact.

When considering TV consumption patterns and the impact of time-shift solutions a Nielsen report, covering the 3rd quarter of 2012 in the U.S., points to an overall decrease in the number of TV viewers, as compared to the same period of 2011, despite an increasing viewing time per spectator. As regards to time-shift viewing, the same study shows a significant increase in the number of users and in the time spent per month (7.1% and 6%, respectively) [31]. Another report developed by Ericsson in 3 different continents shows that despite live content continues to lead the preferences of viewers, their number has been decreasing since 2010, as well as the Digital Video Recorder (DVR) users, losing 16% between 2010 and 2012 [14]. This is in line with the results of a study by Digitalsmiths covering the last period of 2012 that states that more than half of respondents claim to spend less than 10% of their time watching TV from the DVR [11]. In contrast there are the on-demand services with a slight increase in the number of users in the same period [14]. It is also important to highlight that almost 25% of the participants said they *zapped* more than 20 minutes a day and 87% claimed to see the same channels most of the times [11]. Another study from Digitalsmiths reports that only 4% of viewers sit in front of the TV with prior knowledge of what they want to see [10].

The fast evolution of technological devices surrounding the television is changing people's habits. Smartphones and especially tablets brought a new dimension to the television ecosystem, allowing users to do other tasks while watching TV [7], being these often related to what they are seeing at the moment. A Nielsen study shows that 41% of North Americans, in the 2nd quarter of 2012, used tablets and 39% smartphones at least once a day while watching TV [30]. The Ericsson report adds that searching in the Internet, managing social networks and chatting online are the main activities performed while watching TV (or other video source) confirming the growing presence of the complementary devices [14].

The ways to typify the uses of a tablet as a secondary screen depends on the level of correlation between what the user does with it and what he is watching on TV. Thus, Red Bee Media distinguishes 3 types of activities using secondary screens [28]: (i) *dual screening*, which includes any action while watching TV; (ii) *synchronous activity*, activities inspired by what is being shown on TV (e.g. Facebook about it, IMDB, among others), and; (iii) *companion experiences*, including applications developed mostly by broadcasters and TV providers in order to enhance and improve the experience of the viewer with the content being watched (*enhanced TV*) or the programs broadcasted by an operator (e.g. Zeebox [32], The Walking Dead - Walkers Kill Count [15], MEO Remote [24], among others).

As Doughty et al confirmed with a study about the use of Twitter to interact about TV shows, viewers are motivated to socially interact through secondary screens while watching TV either looking for a more communal viewing experience with (remote) friends or to possibly increase their own status or exposure within the network [12].

These studies reflect a plurality in the ways TV content is consumed, particularly given its source (live, time-shifted, DVR and VoD) and the ability to have complementary devices that take advantage of personalization features. The importance of this research track, namely in what relates with content recommendation based in the characteristics and profiles of viewers, is highlighted in the study of Digitalsmiths. It reveals that over 63% of respondents would like that their Electronic Program Guide (EPG) has the ability to show content based on their personal interests and preferences [11]. In line with the results, this desire has been increasingly exploited, either by business corporations (e.g. Red Bee Media [29], ContentWise [8] or Orca Interactive [23]) or by academic teams/research projects (e.g. TV Predictor [16], BBC R&D [17] or NoTube [3]) generating a taxonomy of recommendation methods categorized in: (i) *content-based recommendations* (e.g. IMDB, Bär et al Mobile TV Recommender [4]), recommendations related to the programs previously seen by the viewer; (ii) *collaborative recommendations* (e.g. Amazon, Gregory-Clarke and Newell Prototype [18]), recommendations based on what other viewers with similar profiles liked, and; (iii) *hybrid models* that combine the two previous techniques (e.g. Netflix, Bueno et al Hybrid Recommender [6], AIMED[19]) [1].

These methods have been implemented for several years. Recently and to improve the personalization of recommendations social networks have been used as sources, allowing users to get recommendations based on what their friends, family and other social groups like [27]. According to a study by Ericsson ConsumerLab, suggestions from friends or family is the most widely used method of recommendation and considered the most relevant for these purposes [13]. Another way to enhance personalization relates with the enrichment of content metadata, for example by adding *moods* to content (e.g. Jinni [20], iFelt [22]), *semantics* (e.g. AVATAR [5], NoTube [3]).

In a broad analysis, the challenge in finding/selecting TV content more adapted to a given context is also related with the way the EPG, DVR and other search features or applications are designed and developed. The study from Digitalsmiths [10] reports that almost 62% of the sample does not use text search to find content on TV, however, when participants were asked if they would use this type of search if it were easier to use, more than 57% answered "Yes." In this domain, media centers allow to technically provide more attractive and usually easier and more intuitive interfaces than those found on other TV platforms. Mobile devices also allow greater freedom in terms of Graphic User Interfaces (GUI) enabling the development of engaging interactive applications related to TV content, with the added value of freeing the TV screen from information that disturbs the TV viewing.

2.2 The Portuguese case

Prior to the development of the survey supporting this research and based on the available information, a review based on official sources was carried to understand the context of TV consumption in Portugal. What has been found is that, in the Portuguese context, television continues to be a central media device for viewers, since between 2010 and 2011 the daily viewing time increased almost 10 minutes [21]. The same report from Obercom notes that in the same period comedy and information shows experienced the largest increase in viewers, as opposed to fictional content, that despite being the favorite type of shows for Portuguese viewers faced the biggest drop. Considering the age groups and the generalist TV ratings in 2011, the older group (over 64 years) accounted for the higher result with almost 25% and the younger group (from 4 to 14 years) with the lowest TV ratings (10.2%). In pay-TV services TV ratings have less notable differences between different age groups, but still as opposed to the generalist TV the younger age groups are leaders. With regard to gender, in generalist television women lead with more than

55% of the TV audiences. On the other hand, males lead in cable television audiences in 2011, with almost 57%.

According to an Anacom study [2] focusing on the 3rd quarter of 2012, pay-TV has been gaining ground in Portuguese homes with an increase of 241.000 new subscribers, 8,4% more as compared to the previous year. The same study points to an increase in the number of TV channels per home, as well as an increase in the use of the main features available in the set-top box (STB), including the EPG and DVR. The use of both features was up 10% in 1 year, now being part of the routine for almost 40% of pay-TV subscribers.

Apparently, the increase in the number of households with pay-TV may be related with the switch-off of analog television, which occurred in Portugal between January and April 2012 [9]. The conversion to digital television via *Digital Terrestrial Television* (DTT) forced Portuguese inhabitants to choose between buying a DTT decoder; a new DTT ready television; or subscribing to a pay-TV service. If, on the one hand, the subscription to pay-TV provides a very wide range of channels in exchange for a monthly payment, on the other hand, DTT only provides access to 4 channels. For these and other reasons, many people were reluctant to adopt the digital terrestrial service. A study revealed that in November 2010 34.1% of the Portuguese population stated they have no intention to adopt digital television [26].

Considering pay-TV operators, namely the most popular, MEO [25] and ZON [33], both provide a wide range of interactive services and innovative features including interactive applications (e.g. information apps, games, social networking or media portals); pause and restart TV; Video Store, time-shifted TV (last 7 days), mobile applications (for remote viewing and control), among others. ZON Iris customers are able to integrate Facebook to make likes, shares and recommendations to friends and can create different user profiles for each person in the household.

3. THE ONLINE SURVEY

Taking in consideration that the information available in the literature was not enough for a detailed understanding of the dynamics related with television consumption, an online questionnaire was designed and promoted.

3.1 Survey spread and promotion

In September 26, 2012 the online survey was released aimed at gathering data on viewers' habits and the cognitive process inherent to the moment of deciding which TV program to watch. The survey was created with the support of Google Drive, disseminated in social networks (Facebook, Google +, Twitter and LinkedIn), by email (newsletters, mailing lists, coworkers, acquaintances, friends, family) and in person. It remained online for about 2 months reaching 550 valid responses consisting mainly of students, researchers and teachers. The survey includes about 30 questions, mainly closed ones, which are organised into four sections: (1) personal data, (2) receiver equipment, (3) viewing habits and (4) preferences.

3.2 The results

The results from the questionnaire are presented as follows and grouped into 4 categories according to the purpose and content of the questions.

3.2.1 Sample characteristics

The sample is composed mostly by women (57%) with an average age of around 27 years. The professional/academic areas with higher representation are Education/Teaching/Training with 16%, followed by Engineering with 13%, and Multimedia, with 10%. Regarding academic degrees (completed or to be completed) the majority of respondents have Bachelor and Master levels, with 42% and 27%, respectively.

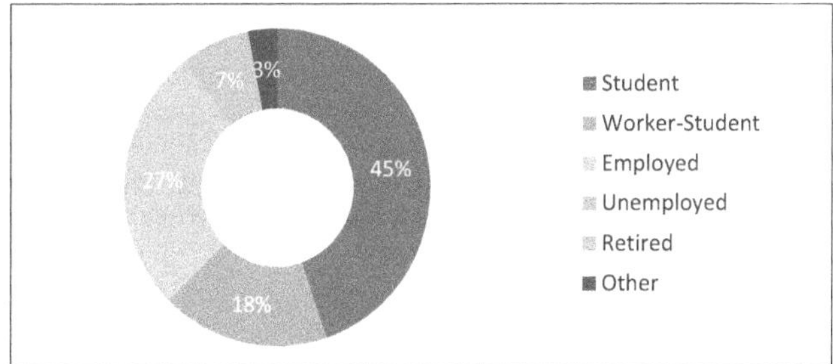

Graphic 1 - The respondents' occupations.

The last two questions in the sample characterisation section inquired respondents about their current occupation and about the people they lived with. As referred above, students are the most dominant group (45%) compared to 27% individuals with professional occupations (Graphic 1). As for household related issues, 382 respondents (70%) state to be living with other adults while only 43 respondents (8%) say they live alone.

As regards to the preferences towards TV genres, it is possible to conclude that for most TV genres, there are not significant differences to point out. Nevertheless, considering the TV genres with lower popularity, "Children and Youth" shows lead with 34% of respondents stating their dislike and 18% their total dislike. On the other hand, "General Knowledge" and "Fictional" shows gather preference from the majority of respondents with 52% likes and 32% strong likes. For these results we need to take in consideration the specificity of this sample with most having a high academic degree.

Considering gender differences, males prefer "General Knowledge" shows in opposition to "Children and Youth" shows. As for females, "Fictional" shows are the preferred ones and "Sport" shows the less interesting.

3.2.2 TV providers

The results show that the vast majority of the sample (83%) uses preferably a digital cable TV service, IPTV or satellite television. The remaining, 17%, refer that they use mainly DTT. Considering the TV operator and taking in account only the respondents with a service other than DTT (456), MEO leads with 41% followed by ZON with 36%. Finally, from those 456 respondents, 407 say that their operator provides a TV guide (EPG), 313 refer having a DVR and 282 a Video Store.

Figure 1- The ZON Iris EPG.

Respondents were also invited to evaluate their EPG. The results show a slight positive opinion regarding the organization of the content and the interface. Of the 407 (87%) respondents who say their TV operator provides an EPG, 43% consider the content layout of the guide as good and 28% point to the mid-level ("Neither good or weak"). Regarding the interface, 39% in the same group classifies it as good and 26% points to the mid-level. The trend keeps up with regard to the information provided in the EPG by the operator, since 37% and 31% answered "Good" and "Neither good or weak". It is important to refer that the highest and lowest levels of the scale are the ones who gathered the smallest number of preferences. On the other hand there are some non-answers to these questions, between 12% and 13%, which may have been caused by the fact that, despite being available in their service, respondents do not use the EPG. Comparing the operators, the MEO customers are the most satisfied with their EPG.

3.2.3 TV consumption dynamics

By examining the habits of respondents as to the frequency of TV viewing along the day, it is very clear, as it could be expected, that the morning and the afternoon are the periods with lowest viewing time, especially on weekdays (Graphic 2). Moreover, it is revealed that the age groups between 22 and 34 years watch more TV during weekend nights, while the youngest age group and the older one prefer the weeknights. We can also notice a direct correlation between the uses of the features in the STB with the increase of the viewing time. This trend is stronger especially on weeknights and weekends in what concerns to the use of the EPG and DVR.

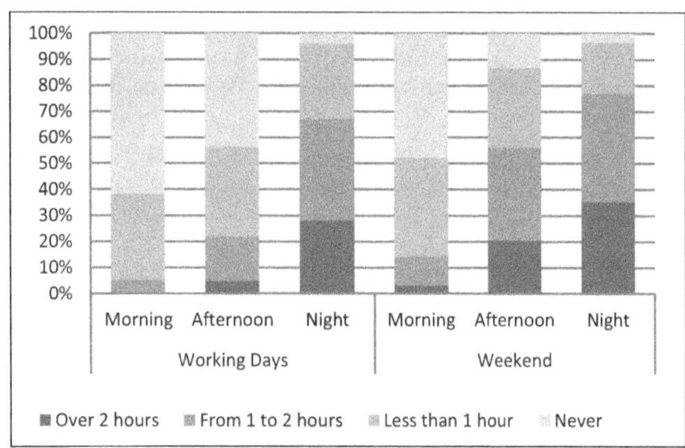

Graphic 2 - TV viewing consumption over the week.

Just over half of the sample (51%) refers having at home 2 to 3 TV sets. The remaining split between one (25%) or more than 3 (24%) TV sets at home. Regarding the places at home where they usually watch TV, the living room is clearly the preferred one for 79% of the individuals. When asked about with whom they watch TV the answers are divided between family viewing (52%) or viewing alone (42%) (Graphic 3).

When watching with others the vast majority of the respondents (75%) refers that the decision regarding what programs to watch is achieved together. When gender is at state, 18% of men, compared to 12% of the opposite sex, refer being the ones to choose what to see on TV.

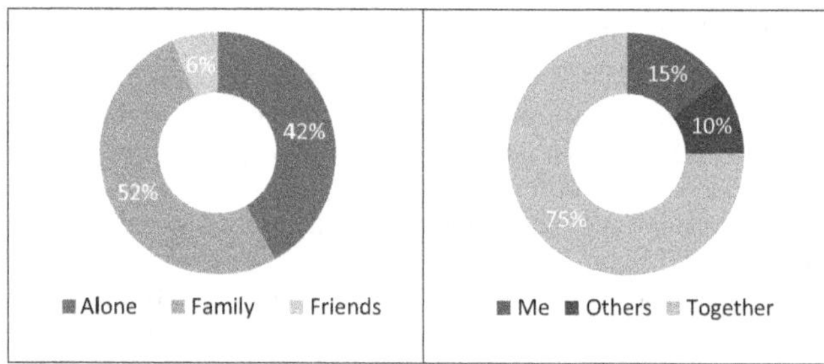

Graphic 3 - With whom viewers often see TV (left) and who decides what program to watch (right).

Another question tried to understand the frequency with which the sample group usually performed other tasks while watching TV. 28% mention they do it more than half of the time, 26% less than half of the time and 24% half of the time. From the ones who claim to perform some type of tasks, 40% say they use mobile devices as complementary information sources to what they are watching on TV (men, younger respondents and workers in the area of Marketing/Advertising appear to be the most frequent users of this type of equipment). From this group, 74% point to Google or other search engines as the type of application or service they use the most on a mobile device related to what they are watching. The second most used type of service, with 71%, is related with websites and apps that provide information about movies and sports (e.g. IMDB), among others (Graphic 4). It is important to notice that sports shows are those that most encourage the use of secondary screens.

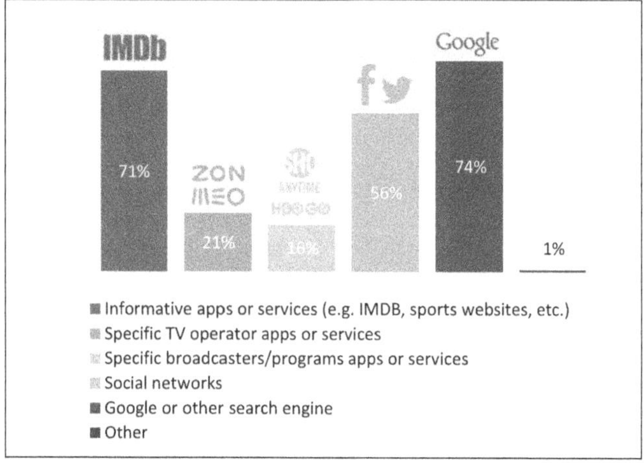

Graphic 4 - Apps or services related to what is being watched.

We also tried to identify the reasons that encourage viewers to watch certain television content. Tips from friends or family are the preferred reasons (54% saying it is relevant and 18% most relevant). The promotion made by each channel to their own content is considered relevant by 43% and most relevant by 7%. With low scores are the highlights in menus or the operator EPG (29% consider it having a low relevance and 27% very low relevance).

As for the most common features of STB, including the EPG and the DVR, a moderate use by respondents is revealed (Graphic 5)[1]. Regarding the EPG, 26% refer using it "two to three times a week", 20% say they never use it and only 14% "more than once a day". Concerning manual DVR recordings, most viewers say they

[1] The results take only in account the number of respondents who, reported having this features in the STB.

do it "two to three times a week" (32%), 16% never do it and 8% "several times a day". Finally, the frequency of use of the Video Store (e.g. Figure 2) or the text search features is significantly lower among the respondents with 69% and 57%, respectively, saying they never use such features.

Figure 2 - MEO Video Store.

Considering gender, although the results are quite close, male are leaders in the use of such features. ZON customers are the ones with a higher frequency of use of the EPG, and MEO customers of the other STB features.

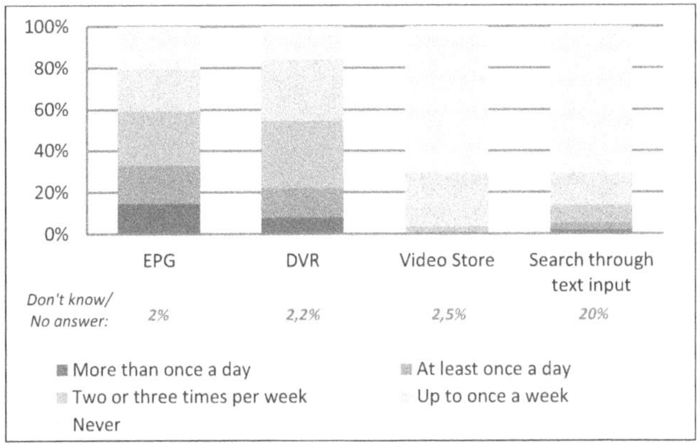

Graphic 5 – The use of STB features.

When asked about the preferred ways to switch between TV channels, 57% of the sample says "By pressing the channel number" and "In a sequential way" (Figure 3). In the opposite direction, the EPG or navigation between channels in the same category, are the least preferred ways, both with 14% of preferences.

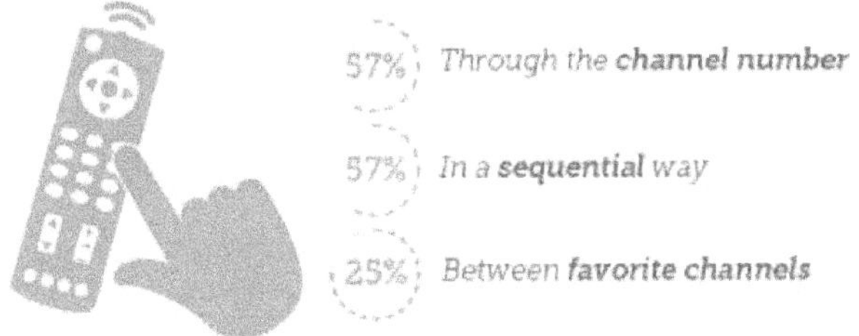

Figure 3 - Most common ways to switch between TV channels.

The next question tried to understand how often viewers know in advance what to watch on TV. 36% of the sample point to a mid-level in a scale from "I never know" to "I always know" (Graphic 6). Women more often declare they know in advance what they want to see. When asked about the time taken, on average, to choose what program to watch, more than half (51%) say they take between "1-3 minutes" and about 31% "less than 1 minute." In these situations, men appear to be faster. Also ZON customers emerge as the fastest finding what they want to see on TV.

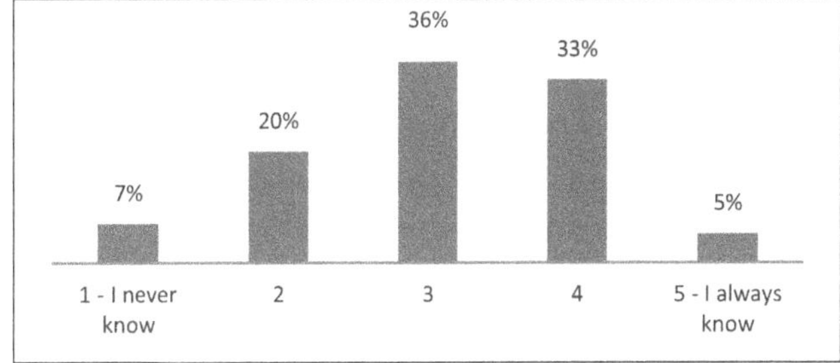

Graphic 6 - How often viewers know in advance what to see on TV.

3.2.4 Decision methods and content selection

Respondents were asked about the possibility of using in their digital television recommendation systems to assist in the process of selecting what to watch based on different methods. The method considered more interesting, by 57%, and very interesting, for 26% of the sample, is the method based on viewing history. The recommendation method based on channels marked as favorite is also rated as interesting by 57% and very interesting by 23% of respondents. Looking at the end of the scale, the method most often regarded as having no interest at all (17%), despite its position on the mid-values, is the one based in information from social networks (Graphic 7). The youngest age group and women tend to be less acquainted with recommendation methods build upon similarities among users with profiles similar to their.

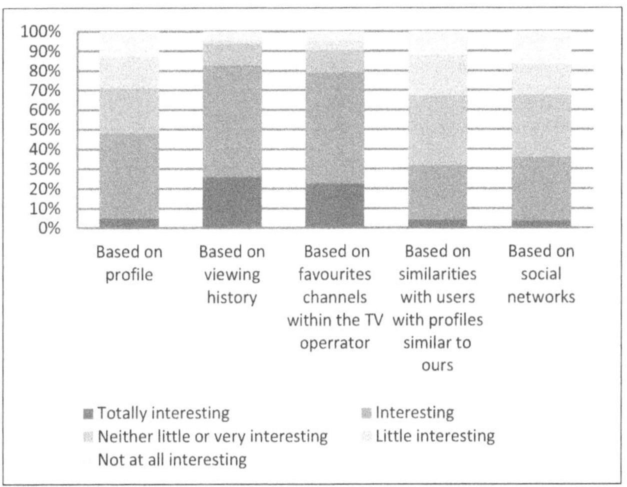

Graphic 7 - Interest for the proposed TV recommendation methods.

The latest survey questions were focused on the criteria viewers have in mind when preparing to choose what to see on TV. With these, the authors intended to build a representation of the cognitive model of viewers when they think about the television program they want to see at a certain moment. Respondents were presented with 8 criteria and where asked to choose up to 5 criteria which they considered to be the most important. In this sense, the three criteria most often pointed by the sample are: the gender of the program (89%), the state of mind at the time (72%) and the companion (being alone or accompanied (60%)). On the other hand, the criterion that is less taken into account by this group of respondents is the language of the program (12%) (Graphic 8).

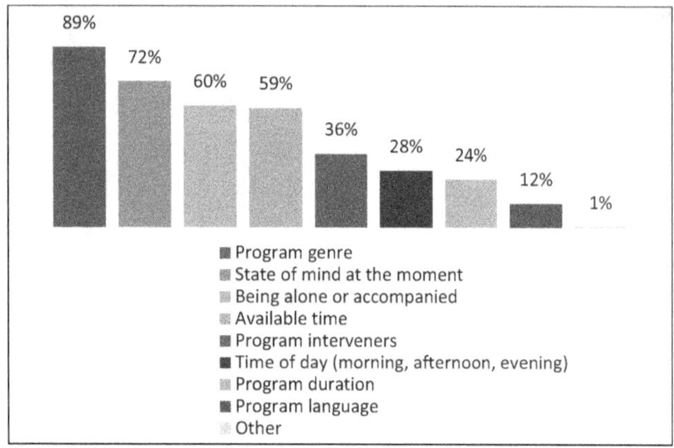

Graphic 8 - Most important criteria when choosing what TV content to see.

At the end they were asked to order the criteria, previously selected, according to their priority. The data analysis shows that the criterion most often considered as the 1st to be considered is the "program genre" with 39% followed by the "state of mind at the moment", or mood, with 20% and finally the "available time" with 18%. In the 2nd priority the "state of mind at the moment" (27%) and "program genre" (24%) change positions followed by "being alone or accompanied". The criterion most often chosen as the 3rd priority is "being alone or accompanied" with about 17% (Table 1).

Table 1 – How viewers order the top five criteria.

	Position				
Criteria	**1º**	**2º**	**3º**	**4º**	**5º**
Available time	17,6%	11,6%	12,7%	**11,3%**	5,1%
Being alone or accompanied	14,7%	14,4%	**17,3%**	8,4%	4,7%
State of mind at the moment	19,8%	**27,1%**	13,8%	7,6%	3,1%
Time of day (morning, afternoon, evening)	4,4%	4,9%	7,6%	6,9%	3,3%
Program genre	**38,7%**	24,0%	15,3%	8,0%	2,4%
Program cast	2,5%	8,0%	10,4%	8,0%	**6,9%**
Program duration	0,2%	4,7%	6,5%	5,6%	6,0%
Program language	1,8%	1,1%	1,8%	2,5%	3,3%
Other	0,2%	0,2%	0,2%	0,0%	0,0%

To establish a ranking of the most relevant criteria a non-parametric statistical test was performed using a Durbin and Skillings-Mack test (Table 2). The test corroborates that there is a significant difference between the presented criteria confirming, with a confidence level higher than 99,99%, that viewers prioritize some features of the content or of the viewing context over others when having to decide what to watch. The test confirms that the "program genre" leads the preferences of viewers, the "state of mind" or mood follows and "being alone or accompanied" and the "available time" split the 3rd place. The other criteria are less relevant for spectators.

Table 2 –Durbin, Skillings-Mack test - criteria mean of ranks.

Criteria	Frequency	Sum of ranks	Mean of ranks
Program genre	485	1099,50	2,27
State of mind	392	1205,00	3,07
Being alone or accompanied	326	1298,00	3,98
Available time	321	1290,00	4,02
Program cast	195	1426,50	7,32
Time of day	149	1357,50	9,11
Program duration	126	1404,50	11,15
Program language	58	1339,50	23,09
Other	3	1302,00	434,00

Doing a cross analysis of these data with the gender of the respondents, we can conclude that the order of the criteria is the same for both genders. Nevertheless we can notice that man have slightly higher preference for the "program genre" and women for the "state of mind". Considering the age groups, we can conclude that for the younger group and the older one the available time is more important than the fact of being alone or accompanied.

With some additional correlations it can be seen that viewers who watch among friends pick the state of mind or mood as the main criterion. Additionally when considering the average time viewers take to select the program it can be seen that those who take more than 10 minutes to choose what to see point to "Being alone or accompanied " as the most important criterion.

4. CONCLUSIONS

This paper focuses on the main results of a study prepared with the aim of characterizing the behaviors and habits of viewers in the current television ecosystem.

The literature review on new TV consumption practices and the technological advances that surround this media device were complemented with an online survey specifically targeted at an audience with higher academic degrees and, therefore, with the potential to have better conditions to experience the current technological landscape. This landscape is characterized not only by the increasing availability of TV content (live, VoD and time-shifted TV), but also by multiplatform offers and convergent technologies that continually embrace television.

The results gathered by this study allow alerting to details of the current ecosystem which should be subject to improvements in order to optimize the viewer experience (as a user of digital TV platforms). On the other hand, these results provide guidelines for the specification of new iTV applications, including those that can support the process of discovery and selection of TV content.

In this context, it is important to highlight some conclusions, such as: i) the substantial TV consumption by the respondents showing that, despite the multiplatform offer and the competition from the Web, television remains a central part on their daily routines; ii) the partial "disenchantment" with the current models of EPG, particularly in terms of its layout, interface and information available, which justifies a continuing effort from the involved players to optimize these type of applications; iii) the importance family context has in the decision of what content to watch together, highlighting the relevance that TV content selection applications need to have to conform not only to the user's profile but also to the surrounding context; iv) in this area is still important not to neglect the factors that encourage viewers to watch a particular television content, including suggestions from friends and family members, who appear as more relevant than the highlights in menus or EPG of the various digital TV platforms; v) moreover, respondents were also favorable to the existence of TV recommendation systems, particularly those based on their viewing history, which is a further contribution to the specification of interactive applications that may be developed in the area of content selection support; vi) the results also show the importance that mobile devices have as companion devices, particularly in the search for detailed information about TV programs, paving the way not only for searching an harmony between the dynamics related with these devices and TV viewing, but also suggesting that the information available in existing EPG is incomplete; vii) finally, the analysis of the cognitive process linked with the selection of TV content allowed to identify the most relevant criteria that must be considered when designing an application with the aim of assisting in this process: television genre, mood, being alone or with companion and the available time.

Therefore the authors hope that this conceptual framework can be useful for different players of the TV ecosystem, including not only the developers of iTV applications, but also content producers, broadcasters and multiplatform players.

These results are also supporting the development of a TV Discovery iTV application by the authors that will soon be available for evaluation with TV viewers.

5. ACKNOWLEDGMENTS

Our thanks to PT Inovação for the financial support granted to this study.

6. REFERENCES

[1] Adomavicius, G., & Tuzhilin, A. (2005). Toward the Next Generation of Recommender Systems: A Survey of the State-of-the-Art and Possible Extensions. *IEEE Trans. on Knowl. and Data Eng.*, *17*(6), 734–749. doi:10.1109/TKDE.2005.99

[2] Anacom. (2012). Serviço de Televisão por subscrição - 3° trimestre 2012. Retrieved January 20, 2013, from http://www.anacom.pt/streaming/3T2012servTVsubscricao.pdf?contentId=1143782&field=ATTACHED_FILE

[3] Aroyo, L., Nixon, L., & Miller, L. (2011). NoTube: The television experience enhanced by online social and semantic data. *2011 IEEE International Conference on Consumer Electronics* (pp. 269–273). Berlin.

[4] Bär, A., Berger, A., Egger, S., & Schatz, R. (2008). A Lightweight Mobile TV Recommender. In M. Tscheligi, M. Obrist, & A. Lugmayr (Eds.), *Changing Television Environments* (Vol. 5066, pp. 143–147). Springer Berlin / Heidelberg.

[5] Blanco, Y., Pazos, J. J., Gil, A., Ramos, M., Fernández, A., Díaz, R. P., López, M., et al. (2005). AVATAR: an approach based on semantic reasoning to recommend personalized TV programs. *Special interest tracks and posters of the 14th international conference on World Wide Web* (pp. 1078–1079). New York, NY, USA: ACM. doi:10.1145/1062745.1062877

[6] Bueno, D., Conejo, R., Martín, D., León, J., & Recuenco, J. (2008). What Can I Watch on TV Tonight? In W. Nejdl, J. Kay, P. Pu, & E. Herder (Eds.), *Adaptive Hypermedia and Adaptive Web-Based Systems* (Vol. 5149, pp. 271–274). Springer Berlin / Heidelberg.

[7] Cesar, P., Bulterman,D., & Jansen, A. (2008). Usages of the Secondary Screen in an Interactive Television Environment: Control, Enrich, Share, and Transfer Television Content. In Tscheligi, M., Obrist, M. & Lugmayr, A. (Eds.), *Proceedings of the 6th European conference on Changing Television Environments* (EUROITV '08). (pp. 168-177) Springer-Verlag, Berlin, Heidelberg,.

[8] ContentWise. (n.d.). Content Discovery, Relevance and Engagement. Retrieved January 16, 2013, from http://www.contentwise.tv/product/discovery/

[9] Denicoli, S. (2012). *A implementação da Televisão Digital Terrestre em Portugal*. Universidade do Minho. Retrieved January 20, 2013, from http://www.lasics.uminho.pt/ojs/index.php/TDT_Portugal/

[10] Digitalsmiths. (2012). *Personalized Video Discovery: Best Practices for Optimizing Revenue Streams*. Retrieved from http://www.digitalsmiths.com/downloads/Tips_to_Optimize_Revenue_with_Personalized_Video_Discovery.pdf

[11] Digitalsmiths. (2012). *Q4 2012 Video Discovery Trends Report: Consumer Behavior Across Pay-TV, VOD, OTT and Next-Gen Features*. Retrieved January 14, 2013, from http://www.digitalsmiths.com/downloads/Digitalsmiths_Q4_2012_Video_Discovery_Trends_Report.pdf

[12] Doughty, M., Rowland, D. & Lawson, S. (2012). Who is on your sofa?: TV audience communities and second screening social networks. In *Proceedings of the 10th European conference on Interactive tv and video* (EuroiTV '12). (pp. 79-86) ACM, New York, NY, USA.

[13] Ericsson ConsumerLab. (2011). *TV & Video 2011: Consumer Trends*. Retrieved January 14, 2013, from http://www.ericsson.com/res/docs/2011/11_1650_RevB_TV_Video_Consumer_Trends_2011_Global_Version.pdf

[14] Ericsson ConsumerLab. (2012). *TV and Video: An analysis of evolving consumer habits*. Retrieved January 14, 2013, from http://www.ericsson.com/res/docs/2012/consumerlab/tv_video_consumerlab_report.pdf

[15] Fox International Channels (UK). (2011). The Walking Dead - Walkers Kill Count. *Apple Store*. Retrieved January 21, 2013, from https://itunes.apple.com/gb/app/walking-dead-walkers-kill/id498529661?mt=8

[16] Fraunhofer FOKUS. (2012). TV Predictor: Personalized Program Recommendations to be displayed on Hybrid TVs. *International Broadcasting Fair (IFA)*.

[17] Gregory-Clarke, R. (2012). Subjective Assessment of a User-controlled Interface for a TV Recommender System. In *EuroITV 2012 – Adjunct Proceedings* (pp. 76–79).

[18] Gregory-Clarke, R., & Newell, C. (2012). TV Recommender System Field Trial Using Dynamic Collaborative Filtering. In *EuroITV 2012 – Adjunct Proceedings* (pp. 123–126).

[19] Hsu, S., Wen, M.-H., Lin, H.-C., Lee, C.-C., & Lee, C.-H. (2007). AIMED - A Personalized TV Recommendation System. In P. Cesar, K. Chorianopoulos, & J. Jensen (Eds.), *Interactive TV: a Shared Experience* (Vol. 4471, pp. 166–174). Springer Berlin / Heidelberg.

[20] Jinni. (n.d.). Jinni. Retrieved January 12, 2013, from http://www.jinni.com/

[21] Obercom. (2011). *Anuário da Comunicação 2010-2011*. Retrieved January 20, 2013, from http://www.obercom.pt/client/?newsId=28&fileName=anuario1011.pdf

[22] Oliveira, E., Martins, P. & Chambel, T. (2011). Ifelt: accessing movies through our emotions. In *Proceddings of the 9th international interactive conference on Interactive television* (EuroITV '11). (pp. 105-114) ACM, New York, NY, USA.

[23] Orca Interactive. (2011). *Recommendations 360°: A Holistic Approach to Content Discovery*. Retrieved January 18, 2013, from http://www.orcainteractive.com/images/stories/pdf/ORCA_WP_Recommendations_360.pdf

[24] PT Comunicações. (2011). MEO Remote. *Apple Store*. Retrieved January 21, 2013, from https://itunes.apple.com/pt/app/meo-remote/id400895046?mt=8

[25] PT Comunicações. (2013). Vantagens TV Fibra e ADSL - MEO - É Outra Vida. Retrieved January 22, 2013, from http://meo.pt/conhecer/tv/vantagens-fibra-e-adsl

[26] Quico, C., Damásio, M. J., & Henriques, S. (2012). Digital TV adopters and non-adopters in the context of the analogue terrestrial TV switchover in Portugal. In *Proceedings of the 10th European conference on Interactive tv and video - EuroiTV '12* (p. 213).

[27] Red Bee Media. (2012). *An Integrated Approach to TV & VOD Recommendations*. Retrieved January 12, 2013, from http://www.redbeemedia.com/sites/all/files/downloads/an_integrated_approach_to_tv_vod_recommendations__white_paper_red_bee_media.pdf

[28] Red Bee Media. (2012). *Second Screen Series - Paper 1: Setting The Scene*. Retrieved January 12, 2013, from http://www.redbeemedia.com/sites/all/files/downloads/second_screen_series_paper_1_whitepaper_red_bee_media.pdf

[29] Red Bee Media. (2012). The Content Discovery Revolution. Retrieved January 12, 2013, from http://www.redbeemedia.com/insights/content-discovery-revolution

[30] The Nielsen Company. (2012). *State of the Media: The Cross-Platform Report Q2 2012 - US*. Retrieved January 15, 2013, from http://www.nielsen.com/content/dam/corporate/us/en/reports-downloads/2012-Reports/Nielsen-Cross-Platform-Report-Q2-2012-final.pdf

[31] The Nielsen Company. (2013). *State of the Media: The Cross-Platform Report Q3 2012 - US*. Retrieved January 15, 2013, fromhttp://nielsen.com/content/dam/corporate/us/en/reports-downloads/2013 Reports/Nielsen Cross Platform Report_Q3_2012.pdf

[32] Zeebox. (2012). Zeebox. Retrieved January 21, 2013, from http://zeebox.com/welcome

[33] ZON Multimedia. (2013). IRIS - A melhor experiência TV do mundo | Residencial. Retrieved January 22, 2013, from http://www.zon.pt/tv/iris/vantagens1/Pages/default.aspx

Tracking and Analyzing TV Content on the Web through Social and Ontological Knowledge

Alessio Antonini
University of Torino
Torino, Italy
antonini@di.unito.it

Ruggero G. Pensa
University of Torino
Torino, Italy
pensa@di.unito.it

Maria Luisa Sapino
University of Torino
Torino, Italy
mlsapino@di.unito.it

Claudio Schifanella
University of Torino
Torino, Italy
schi@di.unito.it

Raffaele Teraoni Prioletti
RAI-CRIT
Torino, Italy
raffaele.teraoni@rai.it

Luca Vignaroli
RAI-CRIT
Torino, Italy
l.vignaroli@rai.it

ABSTRACT

People on the Web talk about television. TV users' social activities implicitly connect the concepts referred to by videos, news, comments, and posts. The strength of such connections may change as the perception of users on the Web changes over time. With the goal of leveraging users' social activities to better understand how TV programs are perceived by the TV public and how the users' interests evolve in time, we introduce a knowledge graph to model the integration of the heterogeneous and dynamic data coming from different information sources, including broadcasters' archives, online newspapers, blogs, web encyclopedias, social media platforms, and social networks, which play a role in what we call the "extended life" of TV content. We show how our graph model captures multiple aspects of the television domain, from the semantic characterization of the TV content, to the temporal evolution of its social characterization and of its social perception. Through a real use-case analysis, based on the instance of our knowledge graph extracted from (the analysis of) a set of episodes of an Italian TV talk show, we discuss the involvement of the public of the considered program.

Categories and Subject Descriptors

H.3.3 [**Information Storage and Retrieval**]: Information Search and Retrieval—*clustering, information filtering*; H.5.1 [**Information Interfaces and Presentation**]: Multimedia Information Systems

Keywords

interactive television, social television, social networks, information integration

EuroITV'13, June 24–26, 2013, Como, Italy.

1. INTRODUCTION

In recent years the way users watch television is radically changing. With the introduction of digital television and the growing number of generic and thematic channels the final user tends to use new forms of navigation in the television content space. To enable users' navigation the broadcasters provide new enriched metadata services such as EPGs (Electronic Program Guide) which describe the scheduled programs. Also the home ambient environment in changing, many smart users watch television while using a portable PC or a tablet as secondary screen related or not to the broadcasted programs [7, 5].

At the same time, social networks allow the final user to be immersed in a collaborative environment providing a powerful reflection of the structure and dynamics of the society. User-generated contents are revolutionizing all phases of the value chain of contents: people can very easily produce content, they can distribute their produced material, and they can experience multiple forms of interactions, such as leaving comments, sharing opinions, supporting other users' contributions, or posting fragments extracted from already online material. We observe in particular that a very large number of user-generated content which users share in their social networks include significant portions of content already broadcasted by the TV broadcaster.

In this direction a number of Social TV services are emerging, which provide the final user with tools to support the social interaction while watching the television, or some media content related to a particular TV program. If properly leveraged, these collaborative social environments can be seen as information-rich data sources, indirectly returning to the broadcasters and the content producers some form of implicit feedback from the final users. A number of services, including user behavior profiling, brand reputation, and recommendation systems for contents and advertisements could benefit from the analysis of the social network data flow. In this paper we address the challenge of exploiting the information gathered from the users' activity in their social networks.

1.1 The extended life of television content

Television content evolves in time: after a television content is produced and broadcasted, a copy of it, enriched with its description (in natural language) together with a collection of related metadata is statically stored in the TV archive to be reused if needed. The broadcaster fills the EPG with the description. Big content producers also make their TV contents available in their Internet site. After a TV product is broadcasted, the content producers usually estimate the associated users' satisfaction by means of quantifiable data, such as audience data (in the case of television) or impression/view count in the case of the internet site, thus ending the "broadcasting phase" of the content's life.

Interestingly, for a significant portion of television content the life cycle spans much beyond that point. In fact, TV programs which better capture the users' interest will probably be published online, either entirely or, more often, in part. During this new phase the television content potentially attracts new users in the network. It will be watched, tagged, liked, commented about and shared again and again. The TV content will act as a magnet attracting the users in the network; it will become a "Social Object".

It is a fact that YouTube is (at least in Italy) the first place where people come to search for a television content they recently missed, although this might violate some copyright requirements. Content producers can choose to contrast this form of piracy, or they can choose to exploit it to their benefit as well, being aware that those users who upload TV content in YouTube are in fact conducting (for free) effective dissemination, description and publishing activities. The upload in Internet of a fragment of a television content starts its "extended life". As an example, we observe that it is very easy to find in YouTube segments of TV programs that have been uploaded long time ago and are now very often watched, commented and liked. This is the case for a short video of Roberto Benigni acting the Dante's "Divina Commedia" published by an unknown YouTube user in 2007 and still watched and commented very frequently in 2013. This content would otherwise be just stored in the archives of the broadcaster and it would be inaccessible to users. As time passes and the users' social context changes, the way any specific television content is perceived also evolves. Capturing how can the TV content evolve and detecting which phenomena can emerge from the contents' evolution are among the objectives of our work. For example, a content can attract a new community of users interested in it, or it might change its own meaning because of a new fact happened in the world. If timely discovered these phenomena could be leveraged by the broadcaster: some of the contents already available in the archives could be considered for repurposing and retransmission.

Some programs also undergo a "short term" evolution. This is the case for news talk shows: some uploaded video burst in number of views and comments. This gives to the public the opportunity to express their opinions about the program and the guests of the show. Analyzing the feedback provided by the public is potentially very beneficial for a number of stakeholders including broadcasters, content producers, ads and media companies.

1.2 Contribution

In this paper, we define a model for the integration of the heterogeneous data coming from the knowledge sources (broadcasters archives, EPGs, collected audience data, social networks, etc.) which play a role in the "extended life" of TV content, starting from its production phase, going through the on-air phase, and continuing with the on-line phase. The model highlights the tight interactions between the Web world and the TV world. A key characteristic of our model is that it is designed to be generic, and it enables a uniform treatment for the different information sources. More specifically, the integrated domain is modelled as a knowledge graph (Section 3.2), in which nodes represent the concepts, while edges capture the relationships existing among them. The key idea that we convey in our model is that the meaning of each entity and relationship within the knowledge graph depends on the context in which they are considered. Thus, persons might be considered as, for instance, authors, reviewers, consumers and so on. Context dependent qualification of entities is not limited to people. For example, a video concerning a piece of news may be regarded in different contexts as part of a news broadcast, a political comment, or a comic sketch (because of some anchorman's gaffe). Network actors and interactions are gathered from existing information sources. Users interact with each other using common social network/media platforms, more or less oriented to TV broadcasting (e.g., YouTube, Dailymotion, Facebook, Twitter, Google+). In this paper, we are interested in extraction and analysis of interactions that are related to TV contents, like videos, TV shows produced by some commercial/public broadcaster (such as RAI). Information from these sources is collected using standard search API's, web crawling techniques or, when possible, by means of social applications. Our framework supports the extraction of both metadata associated to the media contents and related information like user comments. Videos are posted by users by both uploading new content on video-sharing websites, or sharing other people's video content. Usually the original source of these videos are personal home recording. In this case, video content recognition algorithms to map videos to the exact part of TV shows they have been extracted from can be applied [3]. Moreover, our framework is able to support more reliable sources of information produced by a dedicated web-tv platform (e.g., Rai.tv) which enables users to extract portions of TV streams and post them on their preferred social/video-sharing platform. Hence, the original source of posted videos is certified and not subjected to errors. Moreover, citations, posts and comments related to this videos may be directly tracked by the TV service provider. Through a case study, we show how our model captures multiple aspects of the considered domain, from the semantic characterization of the TV content, to the temporal dimension of the problem, to the social characterization and the social perception of a TV event. Last but not least, we provide a non trivial cross-domain analysis scenario on real data gathered from YouTube and Twitter, and related to an Italian TV talk show on politics, broadcasted by RAI.

The paper is organized as follows: in Section 2 we briefly explore some related research work. In Section 3 we introduce the general structure of our integration framework and present the social graph which models the social-driven knowledge in the television context. In Section 4 we apply the model to a real use-case on TV-Web integration. We conclude the paper (Section 5) with a discussion highlighting the potential impact of the model, and showing how it can be used in a number of innovative applications.

2. RELATED WORK

Nowadays we can find an increasing number of emerging

services that aim to enhance the TV experience by offering both extra contents and social platforms on second-screen devices like tablets, smartphones and PCs [7, 5]. Among the most widely used we can mention different enterprise-level products:

- Rai.tv[1] gives to registered (Rai.tv, Facebook or Twitter) users Social TV events linked to the broadcasted television main stream in which the user can watch the program also via IP streaming, comments the program, interacts with comments from other connected users, expresses the feeling about the program, sees the liking of the other users, knows the argument in real time using the associated tag cloud, watches extra contents published by the editorial staff during the social event and answers to real time questions linked to the argument of the program. All these features are available in the web site and also on the secondary screen using the dedicated Apps;
- Tok.tv[2] enables friends to interact with each other within a virtual living room while watching American football matches on TV;
- GetGlue[3] lets users check-in to television shows;
- Miso[4] enables the users to create side shows to support user-generated content;
- IntoNow[5] serves contextual stories from the Web based on real-time mentions;
- Zeebox[6] provides an electronic program guide where the media content is weighted based on social network and also enables social engagement during the viewing experience.

A second screen interactive TV experience conducted by Basapur *et al.* [4] evidenced that the application prototype allows users to better connect with their TV shows and have an enriched social life around live as well as time-shifted TV content. This type of service opens a new kind of television usage and creates a new channel of information from the final user back to broadcasters and content producers. In this case the big challenge is: how can we use this new source of information? The trend of the exploitation of this new type of user interaction has one main direction: the exploitation of user activities to help the broadcaster and the content provider for the acquisition of new audiences, to help with the conservation of the audience and to help maximize the revenue produced by the audience. For these reasons a number of analytics tools are emerging in order to analyze the crowd buzz to try to extract information about the user behavior [8]. The idea of building a framework, based on a knowledge graph, able to capture and track the evolution of television content in the network is our attempt to give a novel approach to efficiently and effectively exploit the huge flow of information coming from social media. Our work is also inspired by a number of projects such as the NoTube project[7] which provided an integration framework between TV, Web, and Semantic Web to build services based on the enrichment and the personalization of the TV content [25, 24]. WinaCS [25] (Web-based Information Network Analysis for Computer Science) is another project that incorporates many recent developments in data sciences to construct a Web-based computer science information network and to discover, retrieve, rank, cluster, and analyze such an information network. However, the scope of WinaCS is limited to scientific content and digital bibliographies. From a more television oriented point of view, in the context of TV and social Web integration and in particular in the social media analytics tools, the recently instituted company Bluefin Labs[8] released a suite of tools to explore the social content related to Social TV programs analyzing the data generated by this mapping between social media and TV media, referred to as the TV Genome. This software is largely based on researches on natural language processing, speech to- text and video-entity recognition carried out by the two co-founders [10, 6].

[1] is a short preliminary version of this paper. While [1] is mostly focused on the presentation of the social and ontological knowledge integration framework, in this paper we discuss the use of the integrated data source, through the detailed analysis of use cases. In particular, our framework allows the application of most network analysis algorithms and tools, such as clustering [19], tensor factorization [14], analysis of diffusion and influence in social networks [12], recommender systems [26], and other social network analysis measures and methods [17].

3. A FRAMEWORK FOR SOCIAL MEDIA DATA INTEGRATION AND ANALYSIS

In this section we introduce the framework which enables the integration of various social and non social information sources in a unique knowledge base. The knowledge base, modelled as a knowledge graph integrating domain and general purpose ontologies as well as social interactions among users and social media, can be queried and analyzed as a whole, enabling the discovery of new and interesting cross-domain patterns.

3.1 The integration framework

Figure 1(a) presents an overview of our integration framework. It consists of three main layers: a source processing layer, a knowledge graph layer and a knowledge query and analysis layer.

The **source processing layer** has the role of collecting all the data which will be conveyed in the model. It accesses a number of predefined web/social/media sources (e.g., broadcasters official web sites, social networks, TV channels, etc) and processes them in order to extract those information units which will be represented as nodes in the knowledge graph, as well as those information that support the existence of relationships (modelled as edges in the graph) among them.

The **knowledge graph layer** manages the knowledge graph, which is the core of our proposal. The graph contains essentially three types of nodes: social objects, subjects and

[1]`http://www.rai.tv/`
[2]`http://www.tok.tv/`
[3]`http://getglue.com/`
[4]`http://gomiso.com//`
[5]`http://www.intonow.com`
[6]`http://zeebox.com/tv/home`
[7]`http://www.notube.tv`
[8]`http://bluefinlabs.com/`

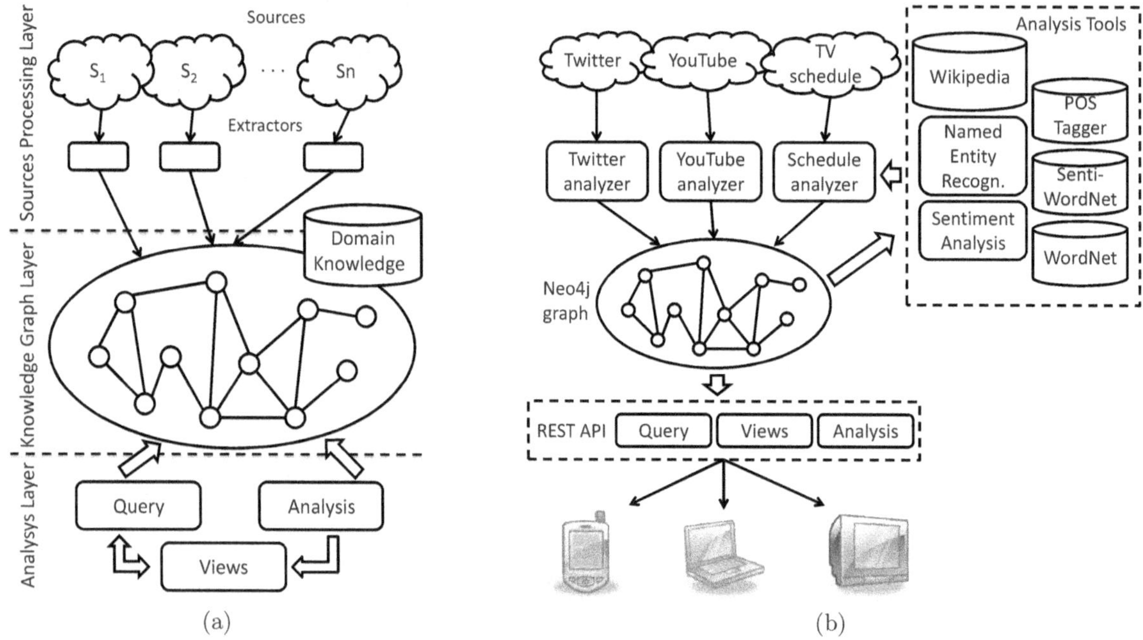

Figure 1: The integration framework and the related system architecture

concepts, and all social representations and structural interactions among them.

The **knowledge query and analysis layer** consists in a set of components for querying, browsing and analyzing the knowledge graph. A query module extracts subgraphs from the knowledge graph based on user's requirements and constraints. Each extracted subgraph can be seen as a "view" over the complete knowledge graph, only containing nodes and edges potentially relevant to the user query. An analysis module provides a set of analysis and data mining tools to obtain models and patterns from the knowledge graph. It can act directly on the knowledge graph, or it can handle the subgraphs extracted from the query module also in terms of matrices or tensors.

The core of our framework is the knowledge network. In particular, we are interested in capturing the dynamic evolution in time of the graph by using temporal nodes associated to social objects and describing their lifecycle.

Notice that in our integration framework a fundamental role is played by a *semantic engine* in two places. First, it is adopted in the source processing layer to provide an interpretation to web/social/media elements taken by the heterogeneous sources. Within this layer, the semantic engine helps understand whether the considered entities should be modelled as a node or an edge in the graph, and helps provide a congruent set of features based on their characteristics. Second, it plays an important role in the graph query and analysis layer, where it is employed to assign a semantic role to each selected node/edge.

Figure 1(b) depicts the actual architecture implementing our framework. In the following sections, we describe each layer in details.

3.2 The knowledge graph

The core of our framework is the knowledge graph that represents the result of public actions of users in social environments, combining different theories from cognitive science [3, 16, 13, 22], language philosophy [20] and social ontology [9, 21]. In this domain we recognize three entities (corresponding to three types of nodes in the knowledge graph): *subjects*, users that act, *social objects*, the result of public acts, and *concepts*, physical and ideal objects referred to by subjects via their public actions. Any act (or set of acts) that can be identified by its trace, and has a recognized social value, is a social object. However, we do not represent single subjects' actions but a unique social object for each group of similar actions. A special subgraph is the one consisting of all concepts, i.e., the forest of the ontology of concepts, or the users' shared conventional knowledge.

We introduce relationships between subjects and social objects and between social objects and concepts as follows: a group of subjects that recognize a social value of an act *supports* the resulting social object (e.g. the contractors *support* the contract); a social object *represents* a social instance of some concepts on a precise context (e.g. a video may represent a volleyball match). Other relationships involve entities of the same type. We call these relationships *structural dependencies*. A social object o_1 is *structural* of another object o_2 if o_1 is part of o_2 (e.g. a comment is part of a video). A subject is *structural* of a group of subjects (e.g. a subscriber is part of playlist subscribers) that performed the same kind of actions on the same social object. A concept may be *structural* of a more general concept (e.g. hilarity is a specialization of joy).

Finally, social objects evolve in time. Hence, as a special case of representation relationship, we consider the *temporal representation* of a social object against a special type of concept called *time objects* (e.g. a video has been posted in a specific time instant, and has been viewed during a specific time period).

The implementation of our knowledge graph is realized

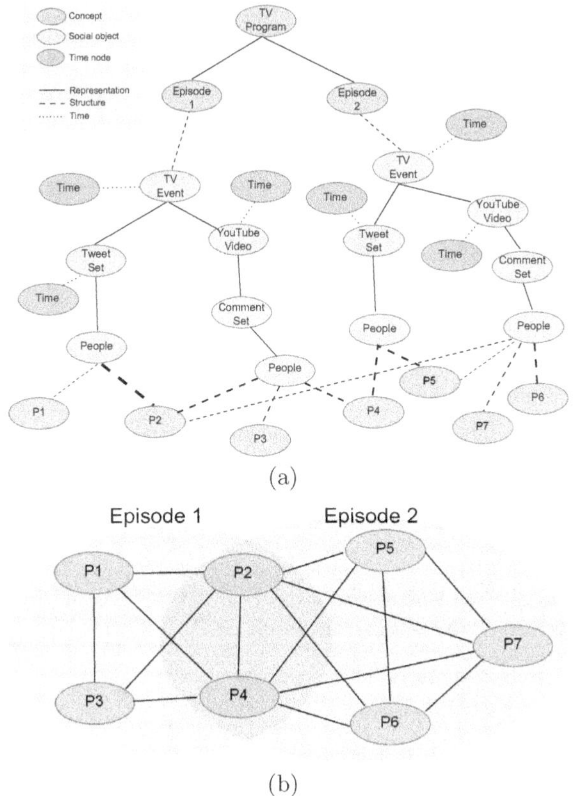

Figure 2: A portion of the knowledge graph (a) and the resulting social network (b).

and stored in Neo4j[9], the well known NoSQL graph database: it offers a comprehensive REST interface, an object-oriented API, and it scales up to billions of nodes and relationships with properties. To populate the knowledge graph, our framework may interact with different and heterogeneous information sources. Each source is first analyzed, then relevant items and relationships are extracted and added to the graph. In the following, we explain how the sources of interest are analyzed.

3.3 Updating the knowledge graph

Our knowledge graph can be fed from any information source. However we distinguish between two kinds of sources: *social sources* and *non social sources*. The first ones consist essentially in social networking platforms, social media platforms and blogs. The second group of sources consists in general purpose or domain ontologies, online newspapers, news feeds, broadcasting websites that are needed to provide a human view on the results of social interactions. In our framework external sources are analyzed in order to extract resources that can be added to the knowledge graph following a set of specific rules.

For each source, we must set an extractor agent that should map each resource into a valid set of social objects, subjects, concepts and relationships among them. To correctly identify each entity, the extractor relies on a set of ontologies. To map each identified entity into a congruent set of vertices and edges in the graph, the extractor leverages a set of rules whose complexity depends on the specific source to be analyzed. In particular, as we mentioned earlier, we use two basic types of extractors: one for social sources, and one for non social sources.

Source Extractors.

Each source (both social and non social) is associated to an analyzer module (the boxes with solid line borders in Figure 1(b)), whose task is to collect the data from the sources and extract concepts, subjects, social objects and their relations through the combined use of different shared modules (the boxes with dotted line borders). The knowledge base extracted by each analyzer will be used to properly update the graph. More in detail, for each TV program that a Schedule Analyzer inserted in the knowledge graph, the Twitter module collects in real-time all related tweets, grouping them into time dependent slices, called tweet sets, where each slice contains the tweets published from time t to $t + \Delta$. Each tweet set is then processed in order to detect the named entities (people, places and events) trough the use of a NER (Named-Entity Recognition) module, while a Sentiment Analysis module allows to extract the opinions contained in a tweet set. Similarly, at each time slice, the YouTube analyzer looks for new videos or new user comments that belong to previously analyzed media and performs the same type of analysis described for Twitter.

Named-Entity Recognition.

Within the Named-Entity Recognition (NER) module, we can detect two different phases: entity detection and entity disambiguation [15]. Entity detection is performed by a combined use of the Freeling POS Tagger [18] and Wikipedia articles[10] as reference knowledge base. In particular, through the use of the Wikipedia search API, the NER module is able to detect the presence of entities starting from hashtags: for example, the hashtag *#barackobama* will be recognized by Wikipedia as the string "Barack Obama". Nevertheless, the most challenging task in Named Entity Recognition is represented by the entity disambiguation (or resolution) [15]. Since our scenario is characterized by the presence of short and sparse texts (both for Twitter and YouTube comments), many of the existing approaches based on the Bag of Words model will fail: for this reason our NER module tries to leverage additional information provided by the context defined by the TV program in which the resolution process is involved, in order to establish which entity is the best among the set of the candidate real-world entities. In details, the context of a TV program is defined by using the Wikipedia categories it belongs to and the set of all entities contained in the knowledge graph previously associated with the program. In this manner, for each detected entity, the NER module tries to establish an order among all real-world candidates extracted from Wikipedia. For example, if the text "Michael Jordan" is contained in a tweet set related to a TV sports program, it is very likely that the tweeter is referring to the famous basketball player rather than the Berkeley's professor, and this is computed by a comparison between the Wikipedia categories of the candidates and the corresponding categories of the TV program. Moreover, if, for example, Michael Jordan is present within the knowledge graph as a real-world entity recognized and associated

[9]`http://www.neo4j.org`

[10]`http://www.wikipedia.org`

with the considered TV program (i.e. because he is the presenter or a frequent guest), the NER module will choose it among all the possible real-world entity candidates. Finally, our module supports the integration of external knowledge generated by a supervised scenario and it allows for user feedback, using an active learning process. In our application, we filter out infrequent recognized entities with the energy cutoff method.

Sentiment Analysis.

The Sentiment Analysis module is used to extract polarity values and emotions from tweet sets. Concerning the former, a first phase of lemmatization is performed by the Freeling POS tagger, while SentiwordNet [2] is used to extract the polarity values: hence, an aggregation function allows us to enrich each tweet set in the knowledge graph with a degree of positivity, negativity and neutrality. With the same approach, WordNet-Affect[23] is used to extract emotions. Where necessary, MultiwordNet[11] is used for cross-language purposes.

Once the extractor agent has analyzed the source, it provides a set of concepts, subjects and social objects that should now be translated into new or updated vertices and edges in the graph. Thanks to this structure, it becomes possible to extract new cross-domain patterns.

3.4 Knowledge Graph in TV domains

In real applications, the graph will not be instantiated with all possible resources extracted from any social or non social source. The reasons are essentially twofold: on the one hand, the huge amount of information could be untractable in practice; on the other hand, many social sources set a limit to the number of resources that can be retrieved in a time slice. For this reason, the way the knowledge graph is populated is somehow constrained by the specific application. We come back now to our case study.

The instantiation of the framework to the TV domain involves a decision process in which we have to choose and define the social and non-social sources, define the resource prototypes and the policies for the source analyzers and decide the detail level of the representation. This last decision depends on what we can extract from social sources, what we want to know about the domain and what we can know about the users' actions. In the next section, we present our solution for the social TV domain. In our view, a user can create and enrich new social uses of the TV media with new metadata, comments, tags, sharing actions and rates. The objects that we will detect and capture are the new correlations introduced between a canonical description of a television event and other new, possibly surprising, concepts. In a nutshell, we want to audit the evolution of the social perception of TV events. The choice of the non-social sources is also critical for the domain definition because it contributes to form the core of the monitored topics on the social sources.

4. A CASE STUDY ON ITALIAN POLITICS

In this section, we describe a real use-case of our framework on an Italian TV show (Ballarò) dealing with politics and broadcasted by RAI. We focused our analysis on the episodes scheduled from October 2, 2012 to November 27, 2012 (nine episodes). This period is interestingly full of political events for many reasons: the past or future elections in many big Italian regions (Sicily, Lazio and Lombardy); the upcoming Italian general elections; the recession; the rise of the populist extra-parliamentarian group M5S (Movimento

[11]http://multiwordnet.fbk.eu

Table 1: Top betweenness centrality scores of nodes from Twitter social network in Fig. 3(a)

Rank	Person	Betweenness centrality
1	Maurizio Crozza	0.2410
2	Mario Monti	0.1783
3	Giovanni Floris	0.0901
4	Matteo Renzi	0.0597
5	Silvio Berlusconi	0.0235
6	Gianfranco Polillo	0.0207
7	Leoluca Orlando	0.0192
8	Bruno Tabacci	0.0172
9	Pier Luigi Bersani	0.0161
10	Luigi Angeletti	0.0135
11	Beppe Grillo	0.0104
12	Guido Crosetto	0.0102
13	Massimo Giannini	0.0051
14	Roberto Formigoni	0.0029
15	Concita De Gregorio	0.0016

Table 2: Top betweenness centrality scores of nodes from YouTube social network in Fig. 3(b)

Rank	Person	Betweenness centrality
1	Beppe Grillo	0.3413
2	Pier Luigi Bersani	0.1907
3	Matteo Renzi	0.1307
4	Giovanni Floris	0.1271
5	Mario Monti	0.1107
6	Gianfranco Fini	0.0512
7	Pier Ferdinando Casini	0.0484

Table 3: Top betweenness centrality scores of nodes from the combined social network in Fig. 3(c)

Rank	Person	Betweenness centrality
1	Maurizio Crozza	0.1710
2	Beppe Grillo	0.1710
3	Mario Monti	0.1305
4	Giovanni Floris	0.0954
5	Pier Luigi Bersani	0.0868
6	Matteo Renzi	0.0723
7	Gianfranco Polillo	0.0259
8	Silvio Berlusconi	0.0236
9	Bruno Tabacci	0.0234
10	Gianfranco Fini	0.0217
11	Luigi Angeletti	0.0213
12	Leoluca Orlando	0.0202
13	Pier Ferdinando Casini	0.0190
14	Guido Crosetto	0.0188
15	Roberto Formigoni	0.0176
16	Concita De Gregorio	0.0165
17	Massimo Giannini	0.0162
18	Rosario Crocetta	0.0162
19	Alessandro Sallusti	0.0162
20	Gianni Alemanno	0.0162

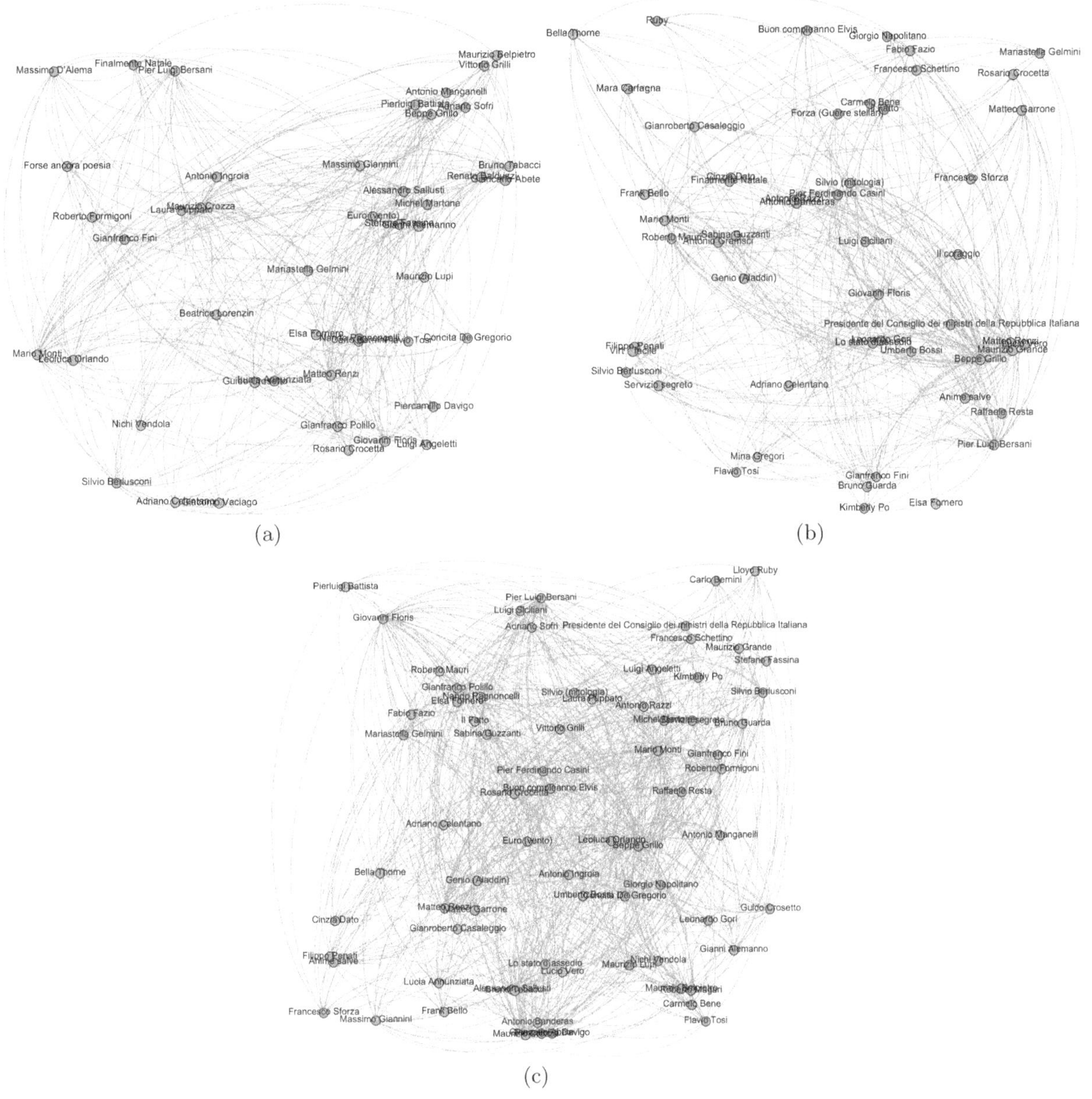

Figure 3: Ballarò social networks extracted from Twitter (a), Youtube (b) and both sources (c)

5 Stelle) that many polling institutes were considering as one of the favorite parties for the next elections in Italy.

We considered two social sources: Twitter and YouTube. For each episode, we collected all tweets containing *#Ballarò* (the official program hashtag) or *@RaiBallaro* (the official program username). YouTube videos were extracted at once by including in the search fields the keyword related to the TV program title ("Ballarò") and the date each episode was broadcasted (e.g., "2-10-2012" or "2 ottobre 2012").

4.1 Social Centrality Study

The first example we consider here concerns the study of the importance (in terms of centrality) of persons (politicians, television people, presenters, hosts), during the observation period. To perform this analysis, we consider all the persons referred by the tweets and videos associated to the nine episodes of the TV show and build the underlying social network. Figure 2 shows how we extracted the social network involving TV people. We add an edge between two person nodes if there exists a path between these two persons, traversing at most one *People* node or at most one *TV Event* node. For instance, following our decision, in Figure 2(a), there is a path between $P4$ and $P6$, but no path between $P1$ and $P7$ exists. Consequently, in Figure 2(b), $P4$ and $P6$ are connected, while $P1$ and $P7$ are not. Notice also that these paths may involve cross-source nodes, i.e., the analysis of an individual source, without our knowledge integration framework, would have led to a different, less precise, social network. Since more than one tweet set and YouTube video may exist during the week associated to each episode, for each episode, all the tweet sets and YouTube videos have been merged to obtain an aggregated episode representation. Each of them is then associated to the set of the most mentioned persons during the considered week.

On our Twitter data, the above described analysis produced the social network presented in Figure 3(a). By computing the betweenness centrality [11, 17] of each node (i.e., the number of shortest paths from all vertices to all others that pass through that node), we obtain the results in Table 1. These results show that Maurizio Crozza is very central for this TV program. He is a satirist that leads a 10 minutes' intervention during each episode of Ballarò TV programs. As such, he usually performs imitations of politicians (like Pierluigi Bersani and Matteo Renzi). Mario Monti, the Italian Prime Minister when these episodes were broadcasted, has been ranked second even if he never participated to the show during the observation period. Among less known politicians, Crosetto (ranked twelfth), had a certain popularity during that period, since he was creating a new political party, in disagreement with Silvio Berlusconi. Among the other top-ranked people, Giovanni Floris is the presenter of Ballarò, while Pier Luigi Bersani and Matteo Renzi were the two main competitors for the leadership of the center-left party, during the observation period.

The same analysis conducted on YouTube data, produced the social network in Fig. 3(b). The betweenness centrality computed for different TV people belonging to this network is reported in Table 2. Interestingly, this analysis shows that the best ranked person is Beppe Grillo. This is probably due to the fact that Grillo's supporters are particularly active in this social media platform. Thus, in this social network, the position of Grillo is more central than in the previous one.

By combining the two information sources (see the social network in Figure 3(c)), we may notice that all the relevant information for both sources are preserved, as shown by the ranked betweenness scores in Table 3. In particular, Grillo and Crozza are equally central, Prime Minister Mario Monti is still in a privileged position, while almost all the most important Italian politics actors are in the first positions of the ranking.

4.2 Popularity Study

The second experiment consists in computing the "episode popularity" of each person. The popularity of a given node is related to the percentage of citations of the associated persons' names in tweets and YouTube comments. Notice that this information is stored in the knowledge graph as the weight of the edge connecting each person to the *People* node (see Figure 2(a)), by the resource extractors. Hence, to conduct this analysis, we only need to aggregate the weights of the out-edges of each person node. Within a single source the aggregation is performed by merging all social objects (tweet set or YouTube video) related to a given episode. Then, each edge weight is multiplied by the total number of occurrences of the concept node *People*. Finally, the cut-off method based on energy is employed to filter out the less important entries. To consider the popularity in both Twitter and YouTube as a whole, we merged the YouTube video nodes and Tweetset nodes associated to each episode. The resulting weight for each person node i is then computed as $w(i)_{all} = \alpha \cdot w(i)_t + (1-\alpha) \cdot w(i)_y$ where $w(i)_t$, $w(i)_y$ and $w(i)_{all}$ are, respectively, the node weights of the edge connecting i to the Tweetset node, the node weights of the edge connecting i to the YouTube video node, and the resulting weight associated to the edge connecting i to the aggregated social object node. In this experiment, we considered all sources with the same weight, i.e., $\alpha = 0.5$.

Figure 4 shows the results for the top-ranked personalities, as computed before. While most popularity values are quite stable during the observation period, the popularity of Matteo Renzi has two peaks, corresponding to the two episodes in which he was hosted in the show. We may also observe that Renzi is more popular on Twitter while Grillo appears to be mentioned more often on YouTube. Berlusconi is almost never mentioned: during the observation period, in fact, he was still not expected to be a key candidate of the center-right party campaign. Monti is mentioned regularly every week, except during the October 30 episode, when Renzi reports the first peak. In those days, in fact, the primary election of the center-left party took place, and Renzi was one of the most observed candidates because he is young and dynamic, and he effectively uses social media and the Web. Bersani, another primary elections candidate, is mentioned regularly but he does not warm the hearts of Web users. The third important candidate is Vendola who seems quite ignored by the Web audience. This may have two explanations: first, his party does not attract many votes and, most interestingly, he is mostly active on Facebook (which we didn't analyze in the discussed use case).

5. DISCUSSION AND CONCLUSIONS

In this paper we have proposed a model for the integration of the heterogeneous data coming from many different knowledge sources, including broadcasters archives, EPGs, ontologies, and social networks. The model highlights the tight interactions between the Web world and the TV world. We have also provided a concrete example of the potential applications of our framework on real data.

We expect the model will have a significant impact on the television production environment. In particular, the ability to track and monitor the second life of Television content will be useful to a number of stakeholders.

- **Broadcasters**: The framework allows the broadcaster to add new references to the static big legacy archive, thus enabling archivists to have a new vision of the evolution of contents that are now frozen inside a huge data base. This new feature makes it easier the personalization of already broadcasted services in which content is customized, adapted to preferences and characteristics of single users or groups of users and provided again to them, exploiting at best the "long tail" phenomena relevant to its own television content. From this viewpoint the broadcaster has the opportunity to reuse materials already exploited by services formerly provided to users, maximizing business logics for these contents that otherwise would be exploitable just in the short term. Furthermore the framework could be useful to provide an alternative approach to the audience analysis of television programs giving more punctual suggestions to optimize the schedule of programs.

- **Service providers** A generic service provider, that only rearranges contents owned by other subjects, will be able to provide new pay services starting from already broadcasted content enforced with a big variety of related content also coming from other media.

- **Final users**: The final user enjoys indirect benefits coming from the use of the model by the broadcaster

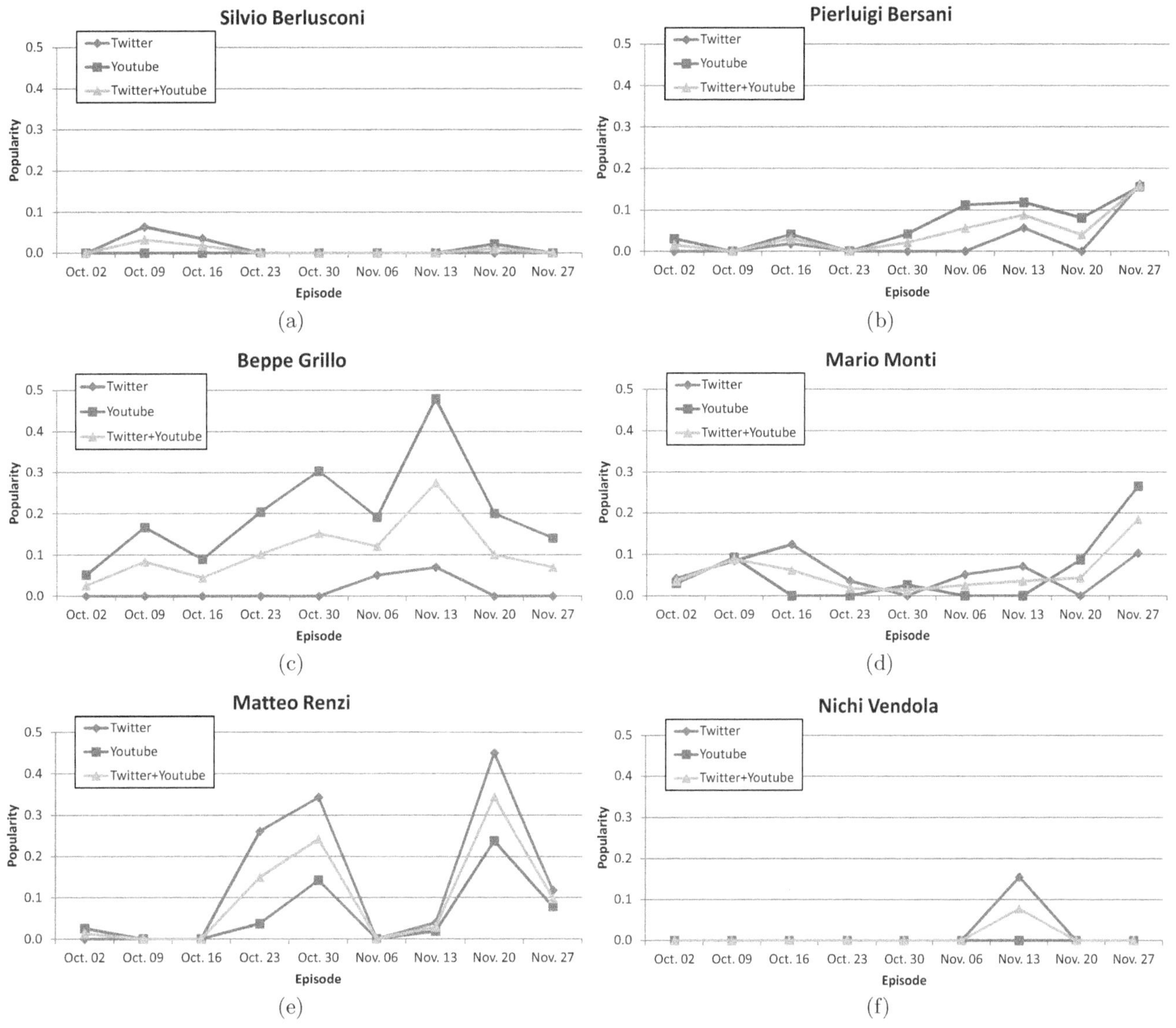

Figure 4: Episode popularity of some cited persons from our knowledge graph

or the service provider. In fact, he/she is able to interact with multimedia material of his own interest in a user suitable format and in time-independent and context-aware modality. Furthermore, the framework could be employed to enhance the interactive TV and "second screen" experiences, by enabling cross-source recommendation techniques, e.g., recommendation of YouTube videos triggered by the usage of particular Twitter hastags [1].

- **Research and industry**: From a research perspective, our framework provides a standard process for data gathering and analysis from different social media sources, thus enabling novel ways to approach sociological studies on the behavior of TV audiences. Moreover, it will relieve data scientists of the ungrateful task of designing ad-hoc data gathering techniques for testing their algorithms and proving their hypotheses. Additionally, from an industrial perspective, the framework could be employed as the underlying architecture for the development of new dedicated services and applications.

Future works will address some limitations of the current architecture. In particular, the correct identification of concepts in the knowledge graph lies in the accuracy of the named-entity recognition module. However, resolution of ambiguities is still an open problem involving information extraction, data mining, natural language processing and other related techniques. We believe that our knowledge graph may be employed to guide the correct identification of persons, places, emotions, events, and other relevant concepts. Hence, we will investigate new active learning techniques for the resolution of ambiguities leveraging the content of the graph, and thus minimizing the intervention of human experts in the named-entity recognition process.

Another weakness of our framework is due to the fact that source analysis is guided by experts that define the correct queries and may possibly adapt them to the new trends and/or needs. We will study self-adaptive strategies to automatically identify emerging keywords and add them to source analyzer queries, while removing obsolete ones.

Finally, since our knowledge graph is able to capture relationships among different types of information, we will investigate new data analysis and mining techniques that take into account the complexity and heterogeneity of the networks.

6. ACKNOWLEDGMENTS

We are grateful to Roberto Del Pero and Fulvio Negro from RAI CRIT for their constructive discussions during the formalization of the integration framework.

7. REFERENCES

[1] A. Antonini, L. Vignaroli, C. Schifanella, R. G. Pensa, and M. L. Sapino. MeSoOnTV: A media and social-driven ontology-based tv knowledge management system. In *Proc. of 24th ACM Conference on Hypertext and Social Media, HT'13, 1-3 May 2013, Paris, France.* ACM, 2013.

[2] S. Baccianella, A. Esuli, and F. Sebastiani. Sentiwordnet 3.0: An enhanced lexical resource for sentiment analysis and opinion mining. In *Proc. of LREC 2010, 17-23 May 2010, Valletta, Malta*, 2010.

[3] B. Bara. *Cognitive Pragmatics : The Mental Processes of Communication.* MIT Press, 2010.

[4] S. Basapur, H. M. Mandalia, S. Chaysinh, Y. S. Lee, N. Venkitaraman, and C. J. Metcalf. Fanfeeds: evaluation of socially generated information feed on second screen as a tv show companion. In *Proc. of 10th European Conference on Interactive TV and Video, EuroITV'12, Berlin, Germany, July 4-6, 2012*, pages 87–96. ACM, 2012.

[5] P. César, D. C. A. Bulterman, and A. J. Jansen. Usages of the secondary screen in an interactive television environment: Control, enrich, share, and transfer television content. In *Proc. of 6th European Conference, EuroITV 2008, Salzburg, Austria, July 3-4, 2008*, volume 5066 of *LNCS*, pages 168–177. Springer, 2008.

[6] P. DeCamp and D. Roy. A human-machine collaborative approach to tracking human movement in multi-camera video. In *Proc. of 8th ACM International Conference on Image and Video Retrieval, CIVR 2009, Santorini Island, Greece, July 8-10, 2009.* ACM, 2009.

[7] M. Doughty, D. Rowland, and S. Lawson. Co-viewing live tv with digital backchannel streams. In *Proc. of 9th European Conference on Interactive TV and Video, EuroITV'11, Lisbon, Portugal, June 29-July 1, 2011*, pages 141–144. ACM, 2011.

[8] M. Doughty, D. Rowland, and S. Lawson. Who is on your sofa?: Tv audience communities and second screening social networks. In *Proc. of 10th European Conference on Interactive TV and Video, EuroITV'12, Berlin, Germany, July 4-6, 2012*, pages 79–86. ACM, 2012.

[9] M. Ferraris. Documentality or why nothing social exists beyond the text. In *Cultures. Conflict - Analysis - Dialogue, Proc. of 29th International Ludwig Wittgenstein-Symposium, Kirchberg, Austria, August 6-12, 2006*, pages 385–401. Austrian Ludwig Wittgenstein Society, 2006.

[10] M. Fleischman and D. Roy. Grounded language modeling for automatic speech recognition of sports video. In *Proc. of 46th Annual Meeting of the Association for Computational Linguistics, ACL 2008, June 15-20, 2008, Columbus, Ohio, USA*, pages 121–129, 2008.

[11] L. C. Freeman. A set of measures of centrality based on betweenness. *Sociometry*, 40(1):pp. 35–41, 1977.

[12] M. Gomez-Rodriguez, J. Leskovec, and A. Krause. Inferring networks of diffusion and influence. *TKDD*, 5(4):21, 2012.

[13] P. N. Johnson-Laird. *Mental Models.* Cambridge University Press, 1983.

[14] T. G. Kolda and B. W. Bader. Tensor decompositions and applications. *SIAM Review*, 51(3):455–500, September 2009.

[15] H. Kopcke and E. Rahm. Frameworks for entity matching: A comparison. *Data and Knowledge Engineering*, 69(2):197 – 210, 2010.

[16] G. Lakoff. *Women, Fire, and Dangerous Things: What Categories Reveal About the Mind.* University of Chicago Press, 1987.

[17] M. Newman. *Networks: An Introduction.* Oxford University Press, 2010.

[18] L. Padró and E. Stanilovsky. Freeling 3.0: Towards wider multilinguality. In *Proc. of LREC 2012, Istanbul, Turkey, May 23-25, 2012*, pages 2473–2479, 2012.

[19] S. E. Schaeffer. Graph clustering. *Computer Science Review*, 1(1):27–64, 2007.

[20] J. R. Searle. *Speech Acts: An Essay in the Philosophy of Language.* Cambridge University Press, 1970.

[21] J. R. Searle. *The Construction of Social Reality.* Free Press, 1997.

[22] R. E. Shaw and J. E. Bransford. *Perceiving, acting, and knowing: Toward an ecological psychology.* Lawrence Erlbaum, 1977.

[23] C. Strapparava and A. Valitutti. Wordnet affect: an affective extension of wordnet. In *Proc. of LREC 2004, May 26-28, 2004, Lisbon, Portugal*, 2004.

[24] L. Vignaroli, R. D. Pero, and F. Negro. Personalized newscasts and social networks: a prototype built over a flexible integration model. In *Proc. of WWW 2012, Lyon, France, April 16-20, 2012 (Companion Volume)*, pages 433–436. ACM, 2012.

[25] T. Weninger, M. Danilevsky, F. Fumarola, J. M. Hailpern, J. Han, T. J. Johnston, S. Kallumadi, H. Kim, Z. Li, D. McCloskey, Y. Sun, N. E. TeGrotenhuis, C. Wang, and X. Yu. Winacs: construction and analysis of web-based computer science information networks. In *Proc. of SIGMOD 2011, Athens, Greece, June 12-16, 2011*, pages 1255–1258. ACM, 2011.

[26] X. Yang, H. Steck, and Y. Liu. Circle-based recommendation in online social networks. In *Proc. of KDD '12, Beijing, China, August 12-16, 2012*, pages 1267–1275. ACM, 2012.

Supporting Interaction and Audience Analysis in Interactive TV Systems

Samuel da C. A. Basílio, Marcelo F. Moreno, Eduardo Barréré
Computer Science Department (DCC/ICE)
Federal University of Juiz de Fora (UFJF)
Rua José Lourenço Kelmer, Juiz de Fora - MG - Brazil
samuelbasilio@ice.ufjf.br, moreno@ice.ufjf.br, eduardo.barrere@ice.ufjf.br

ABSTRACT

Despite of the establishment of digital television services, audience measurement and other media analysis still rely on obsolete and usually expensive techniques that date back to the analog era. Moreover, the introduction of interactive services into TV terminal devices is not adequately considered in such market researches. The analysis of both viewer's audience and interactive behavior can thrive this market, which still didn't come up with killer interactive applications. This paper proposes and compares two approaches for the capture of viewer's audience and interactivity behavior in Interactive TV systems (iTV). The first approach for data capturing relies on new extensions to existing iTV middleware standards, whereas the second one just makes use of software features commonly supported by those standards. Captured data should be sent to aggregation servers that allow multiple reports to be generated and made available online for real time tracking. Aggregation servers and software tools are components of the Interaction and Audience Analysis Service Provider (IAASP), which is also described and implemented in this work.

Categories and Subject Descriptors

H.4.5 [**Hypertext**]: Hypermedia

General Terms

Algorithms, Management, Standardization

Keywords

Digital TV, IPTV, Interactivity, Audience, Measurement

1. INTRODUCTION

Worldwide, most TV broadcasters are private businesses and therefore seek to profit from their services, where the main product sold is the advertisement space and its audience. Even subscription-based content providers obtain a portion of their revenue from selling advertisement space. From the advertisers' standpoint, it is necessary to ensure that they are reaching the target audience, as many people as possible. Hence an effective audience analysis is of common interest of broadcasters and, noticeably, advertisers.

Research institutes that conduct audience measurement on broadcast TV usually adopt two methodologies. The oldest and obsolete methodology (in spite of its wideness) consists in asking viewers to fill a paper form out for a couple of weeks, identifying the programming they watched at small intervals, like 15 minutes. The second methodology uses a device called peoplemeter, which is installed in homes where the audience will be measured. Peoplemeters incur in a high cost of hardware and logistics, forcing research institutes to restrict the number of devices under operation. With such a reduced number of participant viewers, the analysis tend to be concentrated on some regions or cities and may be undesirably biased.

In the peoplemeter methodology, viewers have to be identified before audience data can be collected. The device is usually preconfigured to identify each occupant of the house by a number. This need for the viewer to provide identification before watching TV is debatable. One may say that data analysis may be not accurate if viewers are always reminded that their audience is being measured.

Regarding IPTV audience measurement, despite its inherent ability for data communication, including audience data, the scenario may present more complications. It is an environment where proprietary technology is broadly used for audio/video/data encoding and transmission and likewise each IPTV service provider measures the audience over their services the way it wants. The aggregation of audience data collected from different service providers that use different standards may prove to be something difficult. Moreover, IPTV audience can be measured at several different points in the network besides the client side.

Another issue of current audience measurement systems regards to the data set captured for analysis. Basically, data is limited to the identification of channels, timing and user profile (age, gender, etc.). However, with the establishment of Digital TV and IPTV systems, it becomes obvious that these data do not suffice. For example, broadcasters or advertisers that may be interested in interactive services currently don't expect from measurement systems any feedback on how these services are being used. Undoubtably, an important source of data is being discarded. The analysis of both viewer's audience and interactive behavior could thrive this market, which still didn't come up with killer interactive applications.

This paper presents solutions for the implementation of interaction and audience analysis service providers (IAASP), as a technological evolution that can be adopted by research institutes in this field. Firstly, we propose and compare two approaches for the capture of viewer's audience and interactivity behavior in Interactive TV systems (iTV), including terrestrial, cable and IPTV. Secondly, we present IAASP as a set of tools for interaction and audience analysis (IAA), which allow for the manipulation of such data originated from the interaction between viewers and their iTV terminal devices regardless of the chosen data capture approach. All these proposals are aligned with ITU recommendations for cable, terrestrial and IPTV.

The first approach for data capturing relies on new extensions to existing DTV middleware standards, whereas the second one just makes use of software features commonly supported by those standards. An immediate benefit of these solutions for data capturing is the fact that any viewer possessing an iTV terminal device with an enabled return channel is a potential participant of analyses. Relying on larger samples, analyses may become more comprehensive and dynamically structured. Furthermore, logistics and hardware expenses are replaced by the cost incurred in software development and transmission, which indeed allows for a variety of business models. Finally, the use of resources from their own iTV terminal devices makes latent to viewers the fact that their audience is being measured. Obviously, for privacy reasons, viewers must give their consent in advance to any data capturing.

The main contribution of such a proposal is that iTV terminals, noticeably ones compliant with ITU recommendations, become able to capture user interactions. As far as we know this feature is not supported by other related work in the extent of our approach. Even though many papers deal with the audience measurement question and propose new measurement models, none of them considers the varieties of user interaction in iTV systems.

Moreover, the proposed IAASP architecture and its data structuring allow for the generation of comprehensive, cross-media analyses by seamlessly harmonizing viewers' data from multiple content sources, like terrestrial, cable, satellite, IPTV and Internet TV services.

This paper is structured as follows. Section 2 discusses related work. Section 3 presents an overview of interaction and audience analysis (IAA) workflow. Section 4 proposes two alternative approaches for data capturing in iTV terminal devices. Section 5 describes the Interaction and Audience Analysis Service Provider (IAASP) and its implementation. Section 6 compares the two data capturing approaches and Section 7 is reserved for final remarks.

2. RELATED WORK

Papers that deal with TV audience measurement arise some issues in common, noticeably the lack of standards and the need for new technologies to measure the audience.

The standardization issue mainly exists due to the need to compare audience results from two different content sources. A new way of measurement becomes necessary for several reasons as discussed in [4], where the authors mention that the increasing number of TV channels and their specialization create small groups of audience, which is indeed interesting for advertisers. However, to participate in audience measurement, newer and smaller broadcasters may have to invest as much as larger broadcasters do, something which may be unfeasible for most of them. By not participating in audience measurement, a channel that provides specific content would not attract good advertisers, because there would be no proof of its results. Moreover, a group of viewers would be disregarded in the overall analysis. The authors also mention the shift in control from broadcasters to the audience. Nowadays viewers have a behavior and ways to consume TV that are different from the time when the same methodologies for audience measurement used today were created. A not so long time ago the viewer had no choice of what to view and when, but today, with technologies like Video On Demand (VOD), Digital Video Recorder (DVR), and the Internet itself, there is more control of when and how to watch the preferred TV content in the viewer's hands. In this way, without a proper standardization it will be really hard to measure audience from such a variety of content sources and viewers' behavior.

As a matter of fact, [6] takes this requirement for monitoring multiple media platforms to discuss the need for new methods able to provide a comprehensive audience analysis. Authors argue that a program is not transmitted through a single platform, so ignoring audience from other platforms, like the website of a TV show, makes the analysis fallible. From this point of view, audience measurement should become something dynamically structured, because an audience researcher would choose which elements would be considered in an analysis and what importance each one would have in the result. Our proposal is aligned with such a compositional analysis, by adopting an open architecture for the IAASP and relying on cross-platform technologies for data aggregation and information reporting.

The work described in [3] presents an end-to-end solution to cross-media audience analysis, although their methods depends on additional modules that are not part of TV standards. In certain environments, this may be unfeasible because it would require specialized implementations and these implementations need to be somehow embedded into TV terminal devices. Moreover, functionalities currently available in existing standards that would allow for the creation of simpler solutions are ignored. Despite being capable of covering different platforms, these platforms are limited to the traditional TV consumption and each of them needs a specific module to work. Unlike [3], our proposal focuses on how to measure interactive content consumption in a seamless, harmonized way.

The shortcomings of solutions that depend on non-standardized modules are also reported in [8], where the authors propose a network-based approach to audience measurement for IPTV networks, instead of measuring audience on the client side. Their approach increases the number of users that can participate in audience analyses, however it does not support analysis of users' interaction. The authors claim that logging viewers' behavior may be complex because it is data too sensitive to acquire. The work is limited to IPTV systems.

3. OVERVIEW OF THE IAA WORKFLOW

Figure 1 provides an overview of the IAA workflow, identifying its main actors like the TV service provider, the terminal device and the IAASP. TV service provider is the component responsible for transmitting the content to be consumed, including audio, video and data (applications). Examples of TV service providers include free-to-air, subscription-based and internet broadcasters. The arrow between TV service provider and terminal device represents the distribution network. A terminal device is the viewer's content receiver (e.g. Digital TV receiver, IPTV set-top box, personal computer etc).

The workflow slightly differs between the two proposed approaches for IAA data capturing, which namely are: Extended Middleware (Figure 1(a)) and Capturer Application (Figure 1(b)). These approaches are detailed in Section 4 and may be chosen as alternative or complementary methods, since captured data is uniformly defined for both. The choice between one method over another or even both is a negotiation between IAASP and TV service providers, taking into account their respective business models and the impact over the participant viewers.

Depending on the IAA capturing method, the TV service provider may or may not be required to preprocess interactive applications that are to be transmitted. Conditions that imply the use of the Application Preprocessing Module are also explained in Section 4. This module is responsible for handling interactive applications before they are sent, so that they become able to perform data capture regarding viewers' audience and interactivity behavior.

The same way as the TV service provider is required to perform additional tasks depending on the chosen IAA data capture method, the viewer's terminal device must also be prepared accordingly. From the viewer's point of view, the adaptation to the capture method may be transparent or not. In any case, in the terminal device runs some sort of software in charge of capturing IAA data.

If the adopted method is based on Extended Middleware (see subsection 4.1), the terminal device must have an embedded software component (usually installed by means of a firmware update) that includes modules for IAA data capture, storage and transmission. Otherwise, if the method is based on Capturer Application (see subsection 4.2), broadcasted applications directly transmit to the IAASP all the IAA data. In this case, there is no need for new modules to be installed in terminal devices.

The IAASP is the component that receives, stores and processes IAA data and additionally provides IAA reports. For each of these functions the IAASP has a specific module. The IAASP's Receive Module provides an interface that terminal devices can use to send the IAA data, independently of the capture method. After receiving IAA data from a terminal device, IAASP's Storage Module stores the received data in a way to preserve the original structure of the data, without losing or adding any information. The IAASP's Processing Module prepares and adds external data into the database (e.g.: programming schedules). Some data are useful for some queries, but are not usually available in terminal devices. This kind of data needs to be inserted directly into the database. The same module also transforms the data from the database if it is needed, so that new and optimized queries can be performed. This process is called ETL (Extract, Transform and Load) and is detailed in Subsection 5.1. The IAASP's Access Module provides both the original and transformed data through a WWW interface in form of standard reports or customizable queries.

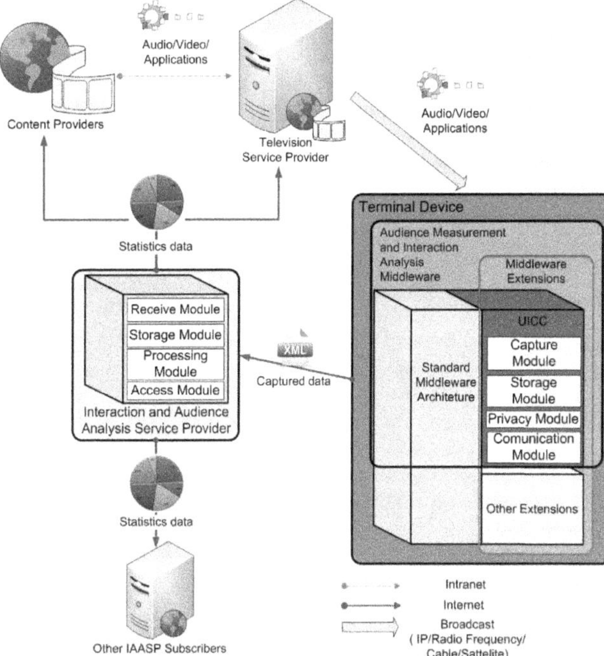

(a) IAA Workflow for Extended Middleware approach

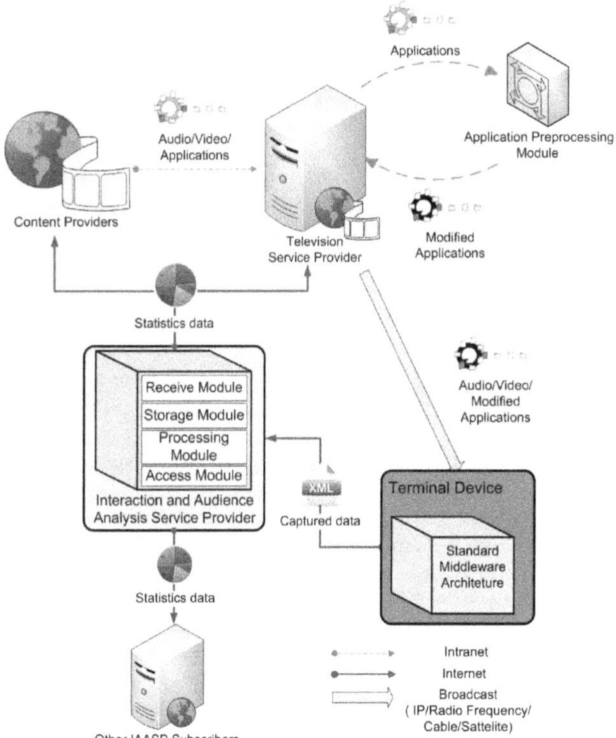

(b) IAA Workflow for Capturer Application approach

Figure 1: IAA Workflow Overview.

4. DATA CAPTURE METHODS

The first step for audience measurement is the capture of relevant data for measuring. According to ITU recommendation H.741.0 [2], audience measurement functions measure the end-user behavior and may optionally collect end-user information. These audience measurement functions may be present at different locations: at the terminal device, at the home network, on the distribution network, at service control and at the content delivery. Among these locations, the ideal place for data capture is the terminal device, as recommended by ITU, that is where we perform the data capture. In this paper we propose two alternative methods for IAA data capture.

4.1 Data Capture based on Extended Middleware

With the development of extensions to the original middleware standard, many new features can be added to terminal devices. The goal of these extensions should be adding new functionalities that were not envisioned at the time the standards were created, without compromising previously existing features.

One of these possible extensions is the User Interaction Capture Component (UICC) as proposed in [7] to be available for interested applications and services. Since the main input method for terminal devices is the remote control, UICC works capturing and storing all of the user interactions with the remote control. Different UICC implementations must be developed to different platforms.

The fact that the UICC is implemented as an extension to a middleware[1] product, it is safe to assume that all data needed for IAA can be properly captured, given the privileges an embedded software running at that level usually have. The UICC's Capture Module (see Figure 2) is responsible for listening the interaction events and pass them to the Storage Module. Undoubtedly, any event generated by the viewer can be captured by an extended middleware, from channel change to volume control, from navigation arrows to colored buttons. The UICC's Storage Module is able to storage all interactions in the terminal device following the desired data structures and formats. The UICC also includes the Privacy Module that asks for the viewer's consent whenever IAA data capture starts. The use of UICC combined with a uniform IAA data structure and format allows the IAASP to process and provide detailed reports.

However, a major limitation of this approach is the need to install a native software in the terminal devices of viewers that will have their interaction and audience analyzed. If the addition of this component in a particular terminal device can not be made, for compatibility reasons, for example, that viewer can not participate in the interaction analysis. In fact, the extended middleware approach may not be feasible in a scenario where the viewer uses his own device to participate in media analysis, specially considering the variety of middleware products, terminal device models and brands. However, this approach can be useful for certain business plans, like for research institutes that may want to replace their peoplemeters by terminal devices with extended middleware. Indeed, this would be a reasonable approach for some institutes, to be adopted in a transition period towards other IAA data capture methods that could coexist.

[1]We refer to middleware as a software that provides all needed interfaces to the terminal device to be able to run audio, video and applications received from the TV service provider. Examples: DTV middleware, IPTV Multimedia Application Frameworks, Web Browsers etc

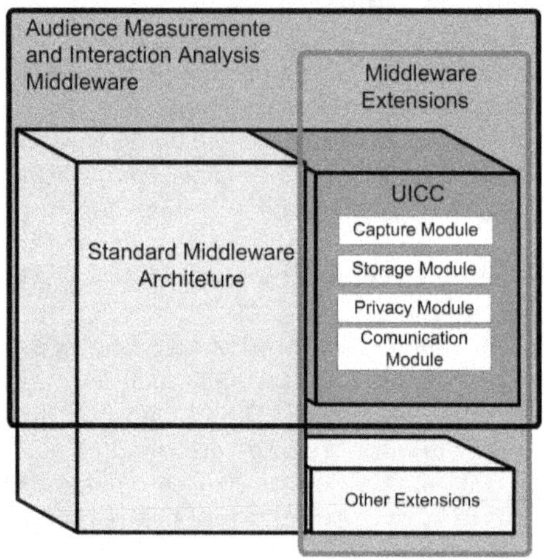

Figure 2: Extended Middleware Modules.

4.2 Data Capture based on Capturer Application

Another IAA data capture method we propose is based on a Capturer Application that eliminates the need for any extensions to the standard. In this approach all IAA data capture is performed by modified applications that uses only standard middleware APIs. Conversely, it represents a shift in audience measurement paradigm, since research institutes will need to establish partnerships with TV service providers interested in having their audience measured. The research institute needs to include in the data stream of each TV service provider a Capturer "base-application"[2]. It is flagged in the data stream as an autostart application.

Figure 3 illustrates a simplified flowchart for the base-appli-cation. When it starts, the base-application firstly looks for an enabled return channel in the terminal device, which is needed to send data in real time to the IAASP. Noticeably, in the interest of a given research institute, it can subsidize the return channel during the sampling period, encouraging the participation of the viewers needed.

If there is no return channel enabled, the base-application finishes. Otherwise, it checks if an IAA configuration file exists in the terminal device's filesystem[3]. If this file does not exist, the viewer is questioned if he wishes to participate in data capture. If the viewer does not agree to participate, the configuration file is created contained information to the base-application that states that the viewer does not want to participate. If the viewer wishes to participate, he is asked to perform the registration. After the registration

[2]Base-application is a small application, added to the broadcaster data stream, responsible for initiating and controlling the session between the terminal device and the IAASP.

[3]The IAA configuration file may be located in the IAASP if terminal devices and users can be uniquely identified

is complete, the configuration file is created, informing the base-application that the user wants to participate. Thus, if the configuration file already exists and it contains an approval for the data capturing, then the base-application can start to capture and send interaction and audience data to the IAASP.

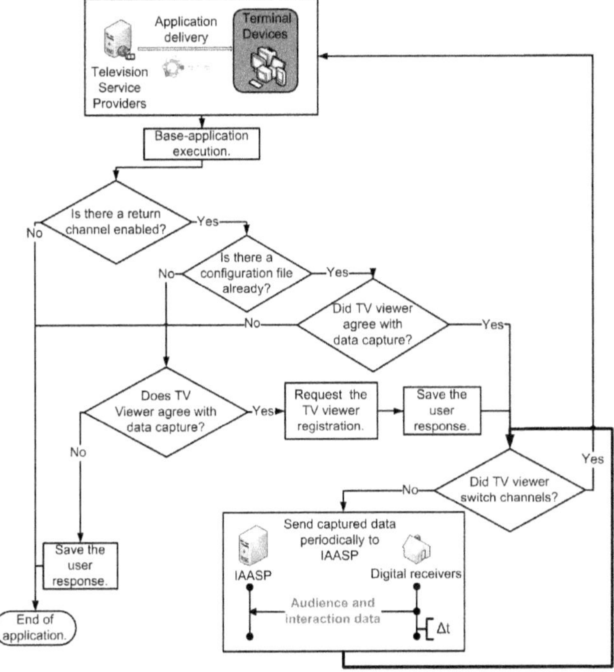

Figure 3: Capturer base-application.

After this handshake, the application will continue to send viewer's IAA data at each Δt time (see Figure 3) while the viewer remains on the same channel. When the user switch channels, the carousel contained in the tuned channel stream is removed and all applications sent by the station are lost, including the base-application. Thus, the base-application will be received again, now at a different channel stream. As aforementioned, all broadcasters who wish to have their audience measured should add to their data carousel the base-application for audience measurement.

On some platforms/standards it may be forbidden for a broadcasted interactive application to store data in the viewer's terminal device. Another issue may arise if applications sent by different broadcasters cannot access shared files in the filesystem. In this case, the configuration file must be saved in the IAASP. However, for some time, the base-application is not aware of which viewer is using the terminal device, and therefore the the configuration files in the IAASP should be written referencing the device via its serial number.

Note that the described procedure only captures audience data, like tuned channel, selected sub-channel, user profile, timing, among others. The capture of viewer's behavior in interactive applications is not in the scope of the base-application.

4.3 Preprocessing Component at the TV Service Provider

As mentioned in Section 3, the TV service provider must include in its workflow a preprocessor for interactive applications. This preprocessor analyzes the code of each application to be sent, searching for points where user action is permitted.

The purpose of the Preprocessing Module is to add to the application code extra actions to be performed as a result of user-generated events, possible to be triggered as found in the application code. This extra action cannot change the normal flow of the original application and usually consists in preparing and sending data about that event to the IAASP. After the preprocessor finishes to modify all user-generated events in an application, the application can be sent on the data stream.

Since most standardized languages used for the development of interactive applications are declarative and for a specific purpose, it is easy to the preprocessor to identify the points where viewers' interactions may happen. The difficulty here is the creation of a specific preprocessor to each language. Some languages like NCL [1] facilitates this task because NCL supports relationships among media objects with multiple cardinality. In other words, an application may specify an event that triggers multiple actions. Other languages such as HTML may require the use of external resources such as javascript, but even so it is simple to be done.

When dealing with general-purpose languages, like Java, the creation of a preprocessor that adds actions to each possible user-generated event found, may not be possible. In this case, it is necessary to manually change the applications to add new actions for IAA data capture.

5. THE INTERACTION AND AUDIENCE ANALYSIS SERVICE PROVIDER

The server side part of the interaction and audience analysis system is the IAASP. This service provider provides interfaces for both terminal devices and clients that want statistics on how users are interacting. Regardless of the chosen method for data capture from the terminal device, the IAASP receives a uniform XML file containing viewers' interaction data (IAA data file).

Another difference between the two data capture approaches resides on the possibility for Extended Middleware approach to create larger IAA data files or buffers and thus IAA captured data does not need to be sent right away. If the UICC, which is responsible for capture, storage and send the viewer's interactions, is set to send the data from the interactions only at certain $\Delta t1$ time intervals (see Figure 5), the IAA data file that will be sent will have all viewer's interactions since the last sending. Otherwise, if the IAA data capture is based on a Capturer Application, the file will contain just one interaction and every time the user performs an interaction, this will be send right away. Figure 4 shows an example of a generated IAA data file containing the interactions.

The IAASP has a specific module for each of its main functions, as depicted in Figure 1. The Receive Module provides an interface for the terminal devices to send their IAA data files. Moreover, this module is also responsible to parse and validate incoming files. In our current IAASP im-

```
<?xml version="1.0" encoding="UTF-8"?>
<watchTV country="UK"
         startDate="2012-08-11T08:30:00.0"
         endDate="2012-08-11T17:25:00.0">
  <head>
    <location zip="L75 1 AA" lat="53.40418"
              long-2.99023"/>
    <user birth="10-11-1988" genre="male"> ... </user>
  </head>
  <interaction type="channelChange"
               time="2012-08-11T08:30:00.0">
    <key code="CH_UP"/>
    <channel code="04" name="BBC">
      <program code="160"
               name="Olympics"
               category="Sports" age="10"/>
    </channel>
  </interaction>
  <interaction> ... </interaction>
</watchTV>
```

Figure 4: Sample IAA data file (XML).

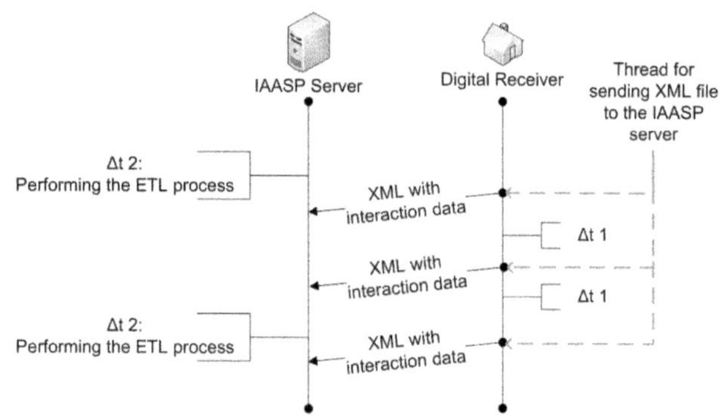

Figure 5: Communication between IAASP and terminal device.

plementation we used a web service as the Receive Module, so that most terminal devices can easily send their data to the service provider.

This approach enables the reception of IAA data captured from any platform, even where the target is not video consumption. In this way, if, for example, a TV show is also transmitted through the Web or if it has some content at a website, every audience and interaction data for this show can be seamlessly measured from each platform, by sending to the IAASP the same IAA data file of the interactions following the proposed standard.

As soon as the file is received and validated, its content is stored at a relational database by the IAASP's Storage Module. The objective of this preliminary storage without any change of the data is mainly to accelerate the process of sending and receiving of IAA data files and thus to maximize the number of concurrent requests supported by the service. In a system like the IAASP, which has a potential of millions of concurrent accesses, scalability is a predominant factor that drove this design decision. In addition to greater speed and scalability in communication between the terminal device and IAASP, prior storage provides other benefits such as the possibility of restoring each file sent to IAASP, preserving the original data.

At each Δt2 time interval, the processing module performs an ETL process, with the objective of inserting external data to the base. These additional data are important for a thorough and complete interaction analysis and may include programming schedules, website maps etc. The transformed data, along with data from external sources are stored at a data warehouse[4] (DW)[5]. Both the data stored at relational database and the DW can be accessed as soon as the IAA data files are received and the ETL process is performed, respectively. Figure 5 ilustrates this process.

5.1 The ETL Process

As shown in Figure 5, every Δt2 time interval the IAASP's Processing Module performs the ETL process, enabling data recently received by IAASP to be included in DW. This process is computationally expensive, which makes it slower and causes the Δt2 period to be carefully chosen. Some factors that influence the decision of this period are:

- The volume of received data;
- The processing capacity of the server/cluster;
- The amount of ETL processes to be performed;
- THe clients' needs when accessing data from DW.

The possibility of performing the ETL process by demand, as soon as the customer needs, is of great value. Noticeably, as time goes by, there will be a growth of relational databases, and thus the ETL process tends to get slower at each run. It may be necessary to increase the Δt2 time between each ETL process for accomplishment. However, if a customer wishes to conduct queries on data that is more recent than the completion time of the last process, and if there is no support for processing ETL on demand, the customer will be required to wait for Δt2, until the next process is accomplished.

Figure 6 presents a simplified schematic of the dimensional model of the DW. The DW is designed to validate the process and illustrates some applications that may become available in the IAASP. This DW has as main activity the aggregation of interaction data held in a terminal device.

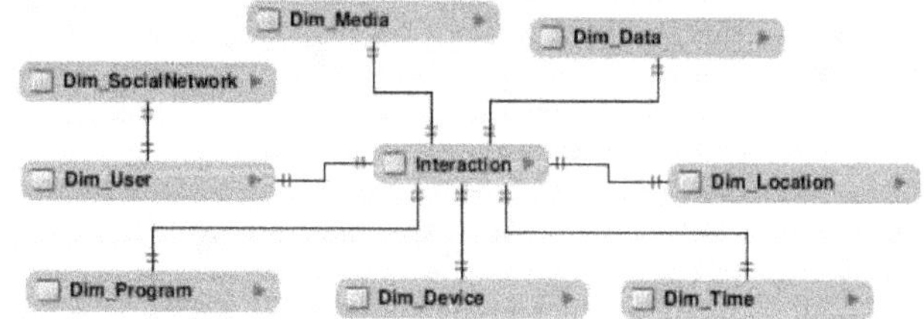

Figure 6: DW dimensional simplified model.

In the ETL process of this DW most data comes from the IAASP relational database, only the dimensions Date (Dim_data) and Location (Dim_Location) are filled with data coming from external sources. The Date dimension contains information such as holidays, weekends and other data that bring peculiarity for each day of the year. Location dimension includes street, neighborhood, city area, city name,

[4]Data warehouse is a central repository created by integrating data from multiple sources, in a consolidated way. The analysis of large volumes of data allows for reacher reports to obtain strategic information.

state and country, from the information contained in the tag Location (XML file header sent by terminal device).

After running the ETL process, data is updated in the DW and then made available to be consulted by IAASP subscribers.

The Access Module makes available the stored data through high level queries. Some, but not all, possible and implemented queries are:

- Amount of online terminal devices in a given period and in particular channel/program (Figure 7);
- Amount of online terminal devices in a given period and in particular channel/program, during the weekend within a time slot;
- Amount of online terminal devices in a given period and in particular channel/program, on a specific holiday;
- Applications running in a given period;(Figure 8)
- Amount of online terminal devices in a certain time slot, in particular channel/program;
- Applications running in a certain slot;

Beside these queries, it is possible to export data from a specific DW through an XML file available for download.

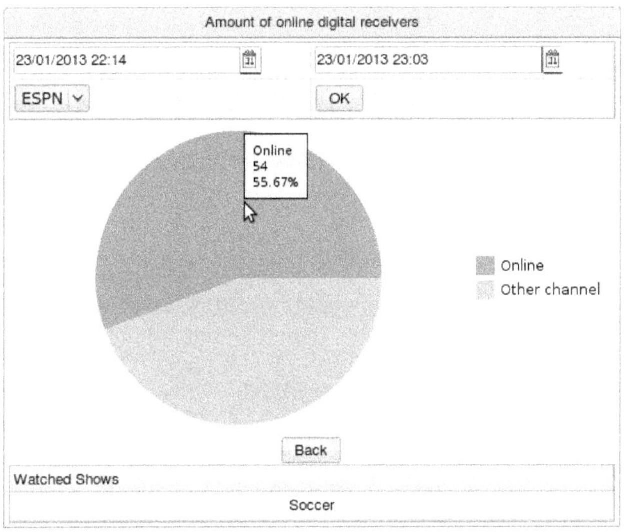

Figure 7: Amount of terminal devices online in one period and in particular channel/program.

6. COMPARISON OF THE PROPOSED DATA CAPTURE METHODS

In this paper we present two methods for data capture that can be adopted for interaction and audience analysis in terminal devices. Certainly, these are not the only possible methods, but these are efficient solutions to improve the audience measurement currently done by research institutes. Table 1 compares the IAA data captured methods.

An important factor is the implementation cost for each method. The Middleware approach is a solution cheaper

Applications running in a given period

23/01/2013 10:58 AM — 23/01/2013 12:00 AM — OK

Applications	Amount of executions
Classification - Libertadores da América	5
Statistics - Libertadores da América	6
Finances Today	2
Prices - US dollar	3

Back

Figure 8: Applications running in a given period.

than the peoplemeter method, because the cost of a terminal device is smaller than the peoplemeter. The Capturer Application method is the cheapest method because it has no hardware cost to its deployment. The only requirement is that the viewer's terminal device is able to run interactive applications.

A shortcoming of the Capturer Application method is its dependence of the TV service provider, which must insert a base-application into the data stream as well as modify interactive applications so interactivity can be measured. Both the peoplemeter and the Extended Middleware are completely independent of the TV service provider.

In terms of sample space size, the Capturer Application method is much better than the other methods, because every viewer with an iTV terminal device is a potential participant of analyses. Peoplemeters are owned by research institutes and their cost makes the sample space reduced and localized. Extended Middleware may achieve more viewers than peoplemeters, but the need for installation of sensitive software in the terminal device and the variety of platforms prevent it to scale as the Capturer Application.

Capturer Application is a portable solution, because it is written in a standardized language, and depending on the chosen language (e.g. NCL and HTML) it may be supported by multiple TV distribution networks.

The type of return channel demanded by our proposals is the ones connected to the Internet, preferably broadband, since the IAASP is implemented as a Web Service. But this was a design decision for the implementation. The architecture doesn't oblige this. Peoplemeters may directly connect to the research institute or through the Internet.

The generated traffic is different among the methods. Online peoplemeters and the Capturer Application generates more requests, each one of a small size, because every interaction is promptly sent to the IAASP. The Extended Middleware approach may include some buffering and aggregate more data, generating fewer requests os a larger size.

Depending on the chosen method, different ways of identifying a viewer among users of a digital receiver are possible. The peoplemeter requires that each registered viewer, whenever a session is started, identify himself with their previously registered number. With the Extended Middleware approach, this form to identify the user is also possible, but more elegant solutions can be developed. Using the computational resources of the digital receivers, is possible to automatically identify the user by his behavior. Using the Capturer Application approach is necessary that the user actively identify himself, but not necessarily by a previously registered number. We propose that instead of a number, the name of the users be stored at the digital receiver configuration file. Whenever the base-application starts it asks if the current viewer is the one that was at the last session. If not, the viewer must chose his name at a list of the registered users or register. If he is the last one, he just wait some seconds and the message disappears and the base-application supposes that he is the last one.

Finally the kind of capturable events are totally different among the methods. As a part of the native terminal device software, Middleware Extension can capture any viewer interaction. The peoplemeter just captures channel switching and the Capturer Application captures zapping and interactivity.

Table 1: Comparison between the capture models ExM: Extended Middleware CApp: Capturer Application

	ExM	CApp	Peoplemeter
Deployment Cost HW/SW	Medium	Low	High
Television Service Provider Dependence	No	Total	No
Sample Space Size	Small	Huge	Very Small
Portability	No	Yes	No
Return Channel Type	Internet	Internet	Internet or Dedicated Channel
Generated Traffic Characteristics	- transfers. $>$ size	+ transfers. $<$ size	+ transfers. $<$ size
Capturable Event	Zapping, interactions,	Zapping, interactions	Zapping
User identification	automatic or manual	manual	manual
Types	volume, EPG...		

7. CONCLUSION AND FUTURE WORK

The specification and implementation of IAASP takes into account functional aspects of software. In the current design, we sought to ensure that requirements identified for the IAASP were filled with the needed robustness for a proof of concept. In this aspect, the IAASP is presented as a viable solution that allows for research institutions to go beyond the traditional audience analysis, by also analyzing viewers' interactions with interactive applications.

Moreover, we achieved a harmonization of the audience measurement service that can now be used not only by TV consumption platforms but also for general media consumption platforms.

We present two methods for IAA data capture, one extending the middleware standards with software installable modules and another one strictly following the standards, using interactive applications. With the Extended Middleware method, costs could be diminished and rich interaction analysis could be easily made, without major changes at the current business model. In its turn, the Capturer Application method is potentially able to multiply the sample space used for the audience analysis with no deployment costs. In both proposals it would not be required the use of expensive devices peoplemeter.

As future work, there are several improvements that can be done at the IAASP tools, as discussed in the text. Performance measures and the study of non-functional requirements, like scalability for the various combinations of approaches to data capture and analysis are ongoing.

Many other researchs can be done with a large scale implementation of the proposed approachs. One is a thorough user behavior modeling. This modeling can be used for simulations and generation of synthetic data e.g. Analyzing the users behavior on multiple platforms is also possible, and with that, infer how the users attention is distributed.

Currently, some ITU-T recommendations are being finalized for the data standardization and audience measurement services operation on IPTV, but nothing has yet been defined in the context of interaction analysis. As the architecture presented in this paper complies with ITU-T [1], we also intend to contribute to normative specifications.

8. REFERENCES

[1] Recommendation ITU-T H.761, Nested Context Language (NCL) and Ginga-NCL for IPTV, 2011.

[2] Recommendation ITU-T H.741.0, IPTV application event handling: Overall aspects of audience measurement for IPTV services, 2012.

[3] F. Alvarez, C. Martin, D. Alliez, P. Roc, P. Steckel, J. Menendez, G. Cisneros, and S. Jones. Audience measurement modeling for convergent broadcasting and iptv networks. *Broadcasting, IEEE Transactions on*, 55(2):502 –515, june 2009.

[4] I. Jennes and J. Pierson. Audience measurement and digitalisation: digital tv and internet. In *Proceddings of the 9th international interactive conference on Interactive television*, EuroITV '11, pages 97–100, New York, NY, USA, 2011. ACM.

[5] R. Kimball and J. Caserta. *The Data Warehouse ETL Toolkit*. Wiley, 1 edition, 2004.

[6] N. Simons. Television audience research in the age of convergence: challenges and difficulties. In *Proceddings of the 9th international interactive conference on Interactive television*, EuroITV '11, pages 101–104, New York, NY, USA, 2011. ACM.

[7] C. A. C. Teixeira, E. L. Melo, R. G. Cattelan, and M. d. G. C. Pimentel. User-media interaction with interactive tv. In *Proceedings of the 2009 ACM symposium on Applied Computing*, SAC '09, pages 1829–1833, New York, NY, USA, 2009. ACM.

[8] L.-C. Yeh, C.-S. Wang, C.-Y. Lin, and J.-S. Chen. An innovative application over communications-asa-service: Network-based multicast iptv audience measurement. In *Network Operations and Management Symposium (APNOMS), 2011 13th Asia-Pacific*, pages 1 –7, sept. 2011.

Vox Populi: Enabling Community-Based Narratives through Collaboration and Content Creation

Janak Bhimani*, Toshihiro Nakakura, Ali Almahr,
Masaki Sato, Kazunori Sugiura, Naohisa Ohta
Keio University, Graduate School of Media Design
Yokohama, Kanagawa, Japan
*janak@kmd.keio.ac.jp

ABSTRACT

The traditional models of content delivery and creation as well as the roles of content producer and content consumer are being challenged in an environment of increasing user generated content. While the amount of user generated content is growing, there is an imbalance with respect to its quality and variety. In this paper, we discuss our research into two real collaborative documentary productions. We present two collaborative models of digital narrative content creation that demonstrate the way in which users have more control over their content and have a more immersive experience with the content. Curation of crowdsourced assets after a major natural disaster is explored in the production of "lenses + landscapes". We discuss crowdsourcing as a means to utilize existing media assets to deliver a narrative experience that consists of many small parts, but whose overall impact as a whole is greater than the sum. Co-located collaborative content creation is examined in the production of "places + perspectives". The implementation of large, remotely located touchscreens in collaborative sessions is illustrated and analyzed. By presenting the findings and details of our two experiments on collaborative storytelling, our goal is to demonstrate different methods for creating and curating collaborative content in and over remote locations to allow for a unique and enhanced user experience for content creators, collaborators and viewers.

Categories and Subject Descriptors

H.5.1 Multimedia Information Systems; H.5.2 User Interfaces; J.5 Arts and Humanities

General Terms

Management, Design, Human Factors

Keywords

Collaboration, User Generated Content, Narrative

EuroITV'13, June 24–26, 2013, Como, Italy.

1. INTRODUCTION

Narratives are intertwined with the human experience. Stories, recipes, traditions, history, and knowledge have been passed down and shared for generations by people of various backgrounds through narratives. Narratives have taken the form of spoken stories through oral tradition, visual stories through primitive drawings, sculpture and later, paintings. Some even consider science and research as another form of narrative, further solidifying narrative's incorporation into our human fabric [11]. With advances in technology, the use of photographs, audio recordings, and motion pictures combined many of the earlier narrative models into the forms available today [13].

In the last decade, video technology, particularly video cameras and video capture technology, has advanced dramatically in terms of media and capabilities. No longer are videotapes and videotape playback machines necessary to create and watch video content. Besides the streamlining of equipment and accessories needed to capture video, the technology has reached a point of affordability and portability where almost anyone can possess some type of video recording device and have it on their person at almost any time. Video cameras, which once were a heavy burden on wallet and strain on one's arm, now cost the same amount as a dinner for two and can fit into a pocket. By 2006, close to half of all households in America owned a camcorder [25]. With camera-enabled mobile phones and smartphones becoming more prevalent, almost anyone anywhere has the tools necessary to record digital video.

The proliferation of the internet through the expansion of broadband infrastructures and high(er) speed mobile internet in the past decade has complemented the advancements in video technology [9]. In the mid-2000s, the rise of internet video sharing sites has made it possible for anyone to share and distribute their digital video content with people anywhere at any time. These recent digital innovations have brought new trends in the forms and shapes that narratives can take. Video sharing sites in the last few years, such as YouTube and Vimeo, have made it easy for almost anyone to share their experiences, likes, dislikes, opinions, and creative work with the whole world. Never has it been easier for people to show their neighbors, next door and around the world, what and how they see the world. This new found convenience of effortlessly sharing videos has caused a shift in the content producer and content consumer paradigm.

In the past, the viewership had to make the best of what was offered in terms of video content. Motion picture content, whether in the form of film or television, was a result of a creative process fully controlled by the producer. Many times, it was the content rights' holders (major film studio, TV network, etc.), not the artists or creative forces behind the narrative, who dictated the way in which the production was carried out. This centralized model of authorship meant that the audience played a passive role and was at the mercy of the producers in terms of content choices available to them.

Until recently, the criteria for judging content as 'interesting' depended on the choices made available by the content producers. Now, however, the audience has the choice and equipment to create content that they find interesting. The consumers are also producing and changing the landscape of the video content market through the use of digital video technology. Content distribution, as well as authorship, is shifting form a hierarchical model to include collaborative models that are based on shared authorship [4]. Instead of only relying on existing or traditional modes of narrative expression such as books, magazines, radio, television and movies for entertainment and information, people are now presented with a wider variety of content options from which to choose from. There are multiple distribution channels to search and watch contents, and a legion of devices by which contents can be consumed.

The transition to more video-based contents has increased the number of content choices for people. YouTube, the most widely recognized video sharing platform, has 72 hours of video added every second [26]. There is no doubt that selection has exponentially increased quantitatively in a short period of time. Despite the increase, the quality of available videos online is not commensurate to the quantity. This is demonstrated by the fact that 30% of videos on YouTube draw 99% of the audience [23].

The disproportion in the relation between content quantity and content quality is reflective of a prevalent issue with user generated content. While the proliferation of video technology and the increase in broadband infrastructure into households accounts for the rise in the number of contents produced by users (i.e. non-traditional contents producers), the content has yet to reach its potential in providing qualitative value for the audience. User generated content is expanding the viewer's horizons as to what is happening around them and the world. However, when content production comes to mind, many viewers are not only looking for more resources from known or available sources, but many also they also want more engaging content. In order for the content to engage the audience, the audience needs to have a genuine connection with the content.

2. OVERVIEW

In order to form a connection between the audience and the content is for the audience themselves to become content creators. In this paper we discuss narrative creation from two different collaborative approaches: narrative from content already available and narrative from content produced by co-located teams through remote collaboration. Both approaches, synchronous and asynchronous, represent different possible scenarios of collaborative content creation and curation that can enhance the user experience for collaborators as well as viewers.

3. GROWING DOCUMENTARY

Modern technology has enabled more people to assist in times of disasters in ways never thought of before. An example of this are the "Voluntweeters" [22] who answered the call to (digital) arms after the 2010 Haiti Earthquake by using Twitter to facilitate and organize recovery and rescue efforts on the ground in the Caribbean from all over the world. In times of disaster, the natural human inclination is to help those affected. However, when disasters occur, especially those of a great magnitude, coming to the aide of people who have suffered is often overshadowed by the scope of the tragedy itself.

After the Great East Japan Earthquake and Tsunami on March 11th, 2011 (3.11), this urge to help in any way shape or form with the recovery and rebuilding efforts was very strong not only for Japanese citizens but for people everywhere. Everyone's eyes were transfixed by the unbelievable images being broadcast on television and the internet. News media, both Japanese and international, were transmitting real-time images of the havoc caused by the earthquake, aftershocks, and tsunami on the Tohoku region of Japan. Additionally, social media allowed average people to share images taken from their digital cameras, mobile phones and smartphones with the rest of the world instantaneously. The world, at once was inundated with a barrage of shocking real unfiltered images of the catastrophe unfolding in front of them and the trail of destruction it left in its path.

In order to create the and facilitate a collaborative environment for visual media production, the Growing Documentary concept was introduced as a result of 3.11. The Growing Documentary is a platform for computer supported cooperative work (CSCW) using user-generated content to produce digital video contents that can be remixed, reworked and built upon as the story and story tellers change and adapt. We discuss the Growing Documentary as a framework to test and observe the production of user-generated content into a narrative representation. Not only did the events of 3.11 have an impact on Japan and the rest of the world because of the unprecedented level of damage caused by both terrestrial and oceanic natural disasters; but also because news of the impact was delivered to people everywhere with unprecedented expedience and quality via a variety of media.

3.1 "lenses + landscapes"

3.1.1 Background

Inspired by how traditional media outlets were utilizing new social communication tools to broadcast their contents and the way in which people in the most devastated areas, who had no electricity or public services, were using their smartphones or feature phones to receive information and communicate their stories to the outside world, graduate students at Keio University were inspired to create the documentary film "lenses +landscapes" [18]. The students set out to document the reality of the earthquake, its aftermath, reconstruction and recovery with 4K motion pictures in order to make an impact by providing a lasting impression on people so that the memory of the disaster would not fade [20]. In 16 weeks, the students, with a novice level of documentary film production, were able to create a high quality visual narrative that reflected the immediacy of the events surrounding the disaster through the curation of user generated content.

3.1.2 Pre-Production Challenges

In the early planning phase of the project, many logistical problems arose. The greater Tokyo area, where the students' university was located, was not very badly affected, in terms of structural damage, injury and human casualties. Access to the hard-hit Tohoku region, however, was greatly hindered. Many basic services that one takes for granted like electricity and transportation infrastructure were very limited, if not completely wiped out, in the heavily affected areas. Once it was realized that it would very difficult to bring, transport and maintain all the equipment and even travel to the local areas in Tohoku for filming and interviewing, a new production plan was devised.

3.1.3 Production

Building upon the notion that crowdsourcing can play a pivotal role in disaster management [18], high-quality digital still images from three amateur and professional photographers were collected in the production of the documentary. Photos were chosen as the message medium of the narrative for "lenses + landscapes" to communicate the impact of the tragedy while and after it was happening. Still images captured singular moments in time. By stitching and blending the images and moments, snapshots of the disaster were transformed into a visual storyboard of people's lives, loss and the rebuilding thereof.

The three photographers -- Max Hodges, Kensuke Mori and Yukoh Nakamura -- all had a personal connection to the affected areas (Figure 1). Max Hodges was an American professional documentary photographer living and working in Tokyo. With virtually no information available, Max hitchhiked his way up to the Tohoku region to capture the raw images of the immediate aftermath of the situation. From the top floor of his former school, Kensuke Mori, then a 19 year old university student, watched and documented his hometown being wiped away by the massive force of the tsunami. In a matter of moments, Kensuke witnessed and captured his life change. Similar to Kensuke, Yukoh Nakamura was also from the Tohoku region. Yukoh's hometown of Morioka in Iwate prefecture was far away from the pacific coastline of Rikuzentakata and other coastal towns which were heavily damaged, if not almost completely annihilated (Figure 2). Yet Nakamura felt the need to go to the devastated areas and document what had happened for the sake of those who could not and for people everywhere to know what took place.

By utilizing the images taken by the three photographers, the same story about the earthquake and tsunami and how it damaged land and lives was given a multifaceted point-of-view. The three men served as storytellers adding feeling and emotion to complement the images. Although the three were from different backgrounds, the images and stories they provided through their photographs added layers to the narrative of the tragic event and drew on similar messages of hope for the recovery and rebuilding of the affected areas and people.

Over 2300 hi-resolution digital still images were procured from Max, Kensuke and Yukoh. Taking principles from crowdsourcing, curation of the *human-sourced* the digital assets was facilitated with a cloud server. As the images were high resolution, a dedicated cloud server was setup to allow the photographers to send their files. Due to the large file size, volume per upload and connection fidelity issues, a proprietary and controlled cloud server proved to be a better choice rather than commercial cloud services or email. The cloud server played an important role for both the photographers and the production team. The three collaborators were able to simultaneously upload multiple files to a secure server. The production team was able to easily sift through the files and choose the images for the documentary while each photographer's data was being transferred to their own folder (Figure 3).

Figure 1. From left: Max Hodges, Yukoh Nakamura, and Kensuke Mori.

Figure 2. A map showing the Tohoku region in Japan denoted in dark green.

3.1.4 Mix-Media/ Hybrid Approach

In order to deliver the most vivid and real imagery to the audience, 4K resolution [1] was chosen as the output resolution of the film. 4K was an ideal format because the digital still images taken by the three photographers were within the range of the pixelation required for 4K resolution (4096×2160 pixels). This allowed for little or no alteration, modification or enhancement of the original assets. The audience was able to see the images in their rawest and purest form with crisp clarity. The reality of the situation could be best understood visually through this high resolution medium. 4K has been and continues to be adopted as a digital standard by both the film and television industries.

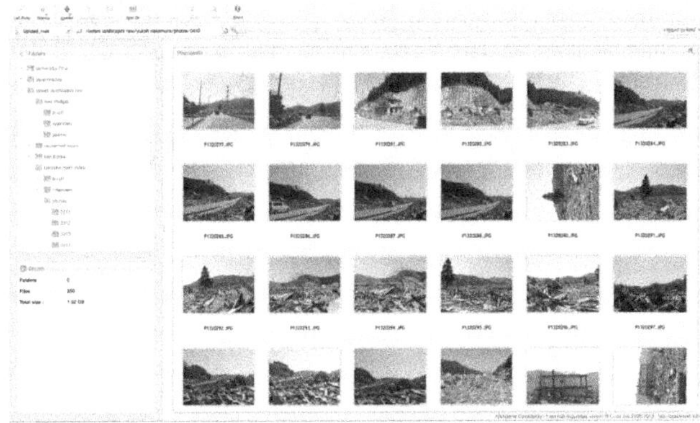

Figure 3. Screenshot from the cloud server web interface.

One of the greatest challenges facing the production of "lenses +landscapes" was the lack of 4K recording equipment. Due to time and budget constraints, a 4K camera could not be acquired for the production of the documentary. As a workaround, a combination of high-resolution digital still images from the photographers combined with 1080p (1920×1080 pixels) HD video interviews and background footage of the photographers was implemented to create a mixed media ultra high-resolution film. In the documentary, still images were juxtaposed with other still images and/or HD video to create a hybrid visual narrative (Figure 4).

As the HD video scenes were of a lesser resolution than 4K, the screen was divided into 4 quadrants: upper left and right, lower left and right. HD video scenes would be in one or more of these quadrants at any one time. This division of the screen enabled

Figure 4. A screen-grab from "lenses + landscapes".

content to be simultaneously presented on the screen at once. By choosing 4K as the output resolution, the multiple scenes on the same screen did not confuse or overwhelm the audience as would be a likely effect with lower quality assets. Through this hybrid approach, the documentary gained a level of narrative depth that would not have been otherwise possible. The stories, like the individuals who shared them with the audience, were dynamic and unique representations of the human condition.

3.1.5 Screening and After-life

"lenses + landscapes" was shown at a special screening during the 2011 Tokyo International Film Festival. Audience members, as well as those who did not see the movie, also had a chance to interact with the film via Scalable Adaptive Graphics Environment (SAGE) [19] OptIPortables. Multiple SAGE "walls" were joined together in San Diego [7]. Through the use of high photonics networks, the film streamed in 4K across the Pacific, and the audience in San Diego was also able to connect with the production team in Japan and interact with them as well as the contents (Figure 5). People had a chance to immerse themselves in the images, videos and sounds of "lenses + landscapes" (Figure 6) while interacting with others in Japan in real-time to re-mix and create their own short versions of the story [2].

Figure 6. SAGE interaction with the media assets during CineGrid 2011.

3.2 "places + perspectives"

3.2.1 Background

After the interest shown in the interactive demonstration of contents from "lenses + landscapes" at the Cinegrid 2011 meeting, graduate students from KMD and undergraduate students from UCSD (the University of California, San Diego) decided to start the next iteration under the Growing Documentary framework. The students started work on the short documentary "places + perspectives" [15]. The short documentary "places + perspectives" was a co-located collaborative filmmaking project.

The majority of participating students were within the 20-30 age demographic and came from diverse backgrounds. Thus, the production was international both in location and people involved. The team sizes changed depending on time and availability on the KMD side. There were at least 4-7 people depending on their time restrictions and the production needs. The UCSD team had more students with around 11 members, and that number also varied as this project was done on a voluntary basis at UCSD.

3.2.2 Production Process

The Japanese and American teams met online through HD streaming sessions made possible by the 10 GigE network between KMD and UCSD (Figure 5). The early discussions revolved around finding a documentary theme, discussing the production process, deciding on the equipment, and defining a style for the piece. The frequency of consecutive meetings was

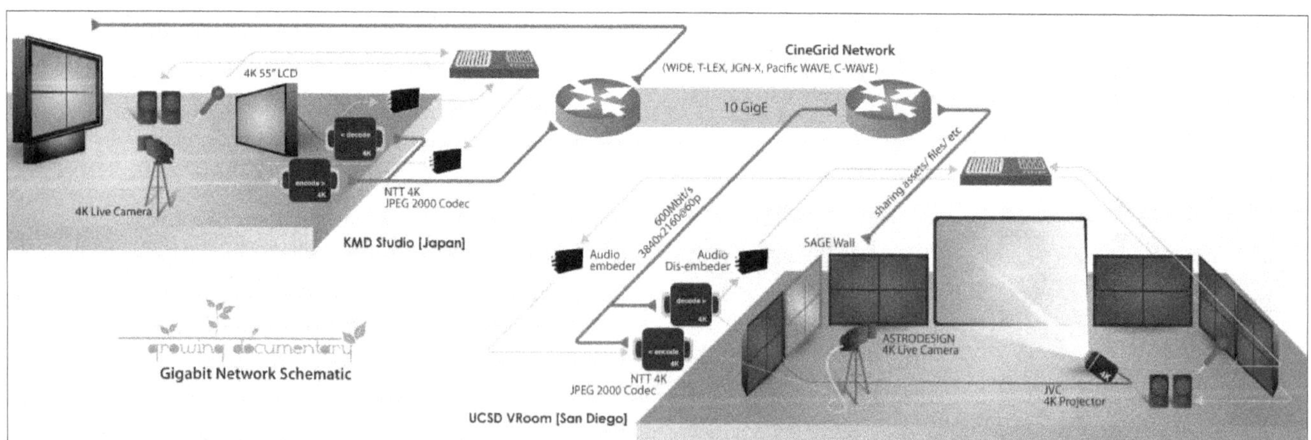

Figure 5. A network schematic showing the live HD/4K streaming and SAGE configurations between KMD and UCSD.

determined as twice per week until March 2012; the goal at this stage was to collect as much content for the 11th Annual ON*VECTOR Photonics Workshop (February 29 - March 1, 2012). After the event, the meetings were reduced to once per week until the summer of 2012.

The original theme that both sides agreed on was "displacement" and how it affects people who choose to live abroad. Both teams also discussed their equipment and chose DSLR (digital single-lens reflex) cameras with 1080/24p HD recording capability as the main video capture devices; a setup is considered to be a hot trend amongst independent documentary makers nowadays. To establish visual unity in the content from both teams, it was necessary to keep the filming process as similar as possible. This was easier to achieve by choosing to film with the Canon 60D and 5D DLSR bodies and 70-200mm zoom lenses, 16-35mm fixed zoom lenses, 24-70mm lenses, 115mm lenses and 100mm lenses for special effects in the b-roll.

Most of the people involved in making this documentary were fresh to the many aspects of documentary making in general, and DSLR shooting in particular. Consequently, this gave a good viewpoint into some of the difficulties that new filmmakers and amateurs face during the entire film production.

There were set production roles such as the director, director of photography and production manager while others such as cameraman, audio engineer and set assistant were split amongst different people depending on their availability. For each interview, the following workflow was conducted:

- The majority of the interviewees were chosen through contacts and social media. The production manager contacts the person of interest via social networks, email, or personal phone call. The prospective interviewee approves and provides a date, time, and place for the interview.
- The production team arrives and sets up the equipment. The interview is recorded with two cameras and an audio recorder.
- After getting the interview footage, b-roll footage of interviewee is taken afterwards.

Table 1. Cloud Server Specifications

	KMD (Yokohama, Japan)	UCSD (San Diego, California, USA)
CPU	Intel Xeon X5365@3.0GHz 4 Core x 2	Intel Xeon X3350@2.66GHz 4 Core x 1
Memory	4GB	
Storage	SATA RAID5 1TBx5	SATA RAID0 2TBx3
OS	Linux kernel 2.6.18	
Software	Apache, AjaXplorer	

3.2.3 Social Media & Sharing Infrastructure

Various methods of communication were used aside from the live streaming sessions. Social media communication tools such as regular email, Google Groups, Google Drive, and Facebook were used to accommodate for the time differences. These communication tools were used due to their familiarity and availability to the participants. Moreover, cloud servers were also employed in the making of this documentary with one on KMD's side and another on UCSD's side (Table 1).

In order to manage the assets properly through metadata and annotations, the PIX System was used [14]. This professional system is used by many professional production studios to allow for better project management and streamlined online collaboration. It could be used to cover the entire production process but the implementation of PIX within the production of "places + perspectives" came at the later stages after the assets were already gathered.

The introduction of both the cloud sharing and PIX System allowed both UCSD and KMD production teams to focus on the narrative and creative process and less about file management and asset tracking. All these tools had one common goal: to communicate the vision of the production clearly for the best possible outcome (Figure 7).

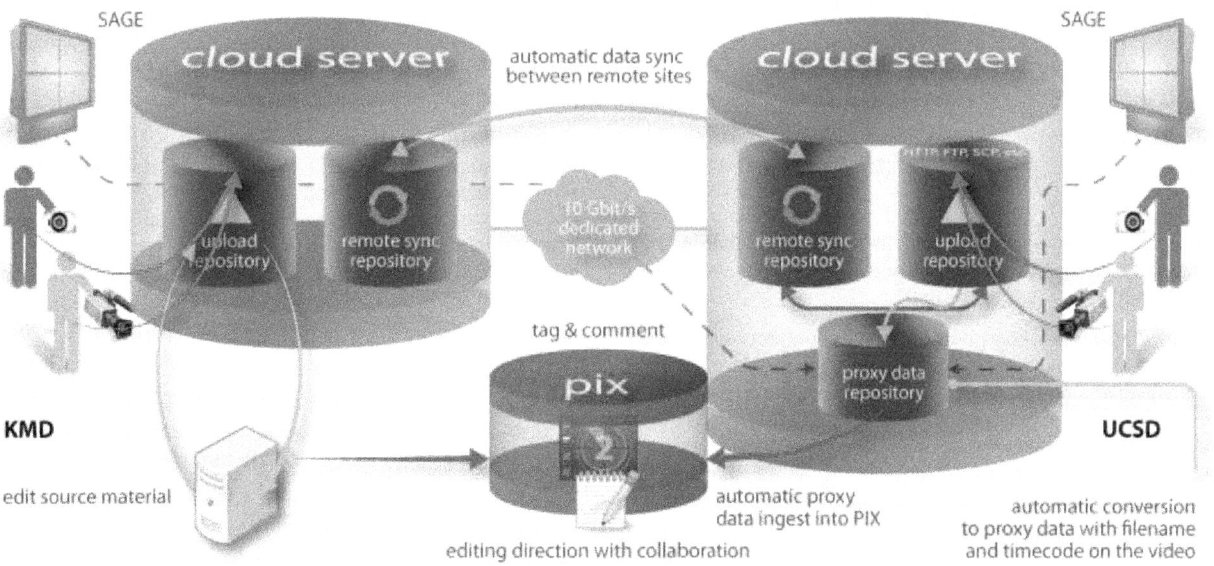

Figure 7. The network infrastructure, tools, and hardware used in "places + perspectives".

Figure 8. A SAGE wall at UCSD showing the video assets collected for "places + perspectives" being used for visual storyboarding

Figure 9. A screenshot from PIX System application. The left panel shows the notes attached to a specific frame of the video displayed on the right.

3.2.4 SAGE Implementation

Due to the nature of documentaries, the storyboarding stage usually starts after collecting the assets to find the connection [21]. This aspect allowed the experimental use of SAGE as a visual storyboarding tool to view the assets and interact with them more openly than in a file browser or Non-Linear Editor (NLE) environments [5]. SAGE was essential for keeping the communication channels open, and allowed each group to easily share video contents (Figure 8).

One distinctive element in the production of "places + perspectives" was the use of SAGE OptiPortals throughout the different stages of production (pre-planning, ideation, and, later, storyboarding). As it was designed to display a large number of media assets, SAGE was a good tool for visual brainstorming and a convenient method for sharing photo and video contents [17]. The SAGE units at KMD and UCSD were connected over the a 10-gigabit network with its high bandwidth pipelines.

During the brainstorming stage, SAGE was essential in sharing sample clips of still image and documentary film shooting styles. By visualizing the brainstorming sessions over the SAGE units, participants had a better shared understanding of the direction of the project. The large real estate allowed for the simultaneous viewing of the different movie clips and the interactivity of the screens allowed the clips to be rearranged in different sequences without much effort. This method granted the users with a view of the story flow that is different than viewing it on a smaller PC screen.

3.2.5 Facilitating Collaboration

The MOV (.mov) file format was chosen for the various contents filmed and collected by both teams. Having MOV as the only file format simplified the editing process, and streamlined the production process. Additionally, MOV is one of the supported file formats on SAGE. This meant that the original file could be shared directly, without the need to convert the contents. Sharing the contents in this manner during the weekly meetings established unity in shots.

One of the realizations throughout the production process was the difficulty of the organizing and editing stages. In order to find common themes amongst the collected videos, the PIX System was used for video annotations and metadata tagging (Figure 9). Yet, it was the editing phase that caused the biggest bottleneck to the production. The majority of the production was done in a collaborative manner with constructive exchange between both parties, however, the collaboration level gradually decreased at the editing stage.

The editing of the project took place in many stages. The post-production of an average documentary film typically involves reviewing the acquired contents and assembling a story out of seemingly unrelated interviews and other clips. As such, each team conducted a separate internal review and editing session before discussion at a meeting. The editing session was greatly aided by the input of notes and tags on the PIX System. The tags on all the videos are not only searchable, but they also have the added function of being able to be exported into a spreadsheet software, such as Microsoft Excel, so that the tags can be visualized.

After narrowing down the videos of interest, a thematic timeline listing the desired theme and topic order was created by UCSD as a guide for the story sequences. From the biggest umbrella tags, the general categories were established. The proxy videos on PIX System were then re-tagged, with each possible scene from all of the interviews being assigned into one of the general categories. The post-production process was identical to conventional film making, with the contents being arranged in storyboards, then subsequently edited in an editing suite.

4. DISCUSSION

4.1 Synchronous/Asynchronous Collaboration

In both "lenses + landscapes" and "places + perspectives", collaborative approaches to building the narrative were implemented in the production process. Through the use of user generated content, both works were examples of a shared authorship model of collaboration. The type of collaboration differed with respect to how and when it occurred during production and in their collaborative approach.

Collaboration in "lenses + landscapes" was a result of implementing elements of crowdsourcing techniques on small, controlled scale. Digital assets procured from various collaborators were managed by the production team. Time and cost factors decreased while quality and efficiency increased as a result of crowdsourcing [8]. The asynchronous collaboration was facilitated by the curation of assets by the production team. Crowdsourcing proved invaluable as time was a major factor in the production of the documentary. There was a sense of urgency in sharing the film as soon as possible in order for people to retain a connection with the events of the earthquake and tsunami. Also, as is the case with breaking news, crowdsourcing allowed for the acquiring of assets immediately after the event.

In order to produce a high quality film, both in visual quality and message, crowdsourcing was the best option. By tapping into the talent of the "crowd", the assets acquired not only made "lenses + landscapes" cohesive as a narrative, but it also brought together

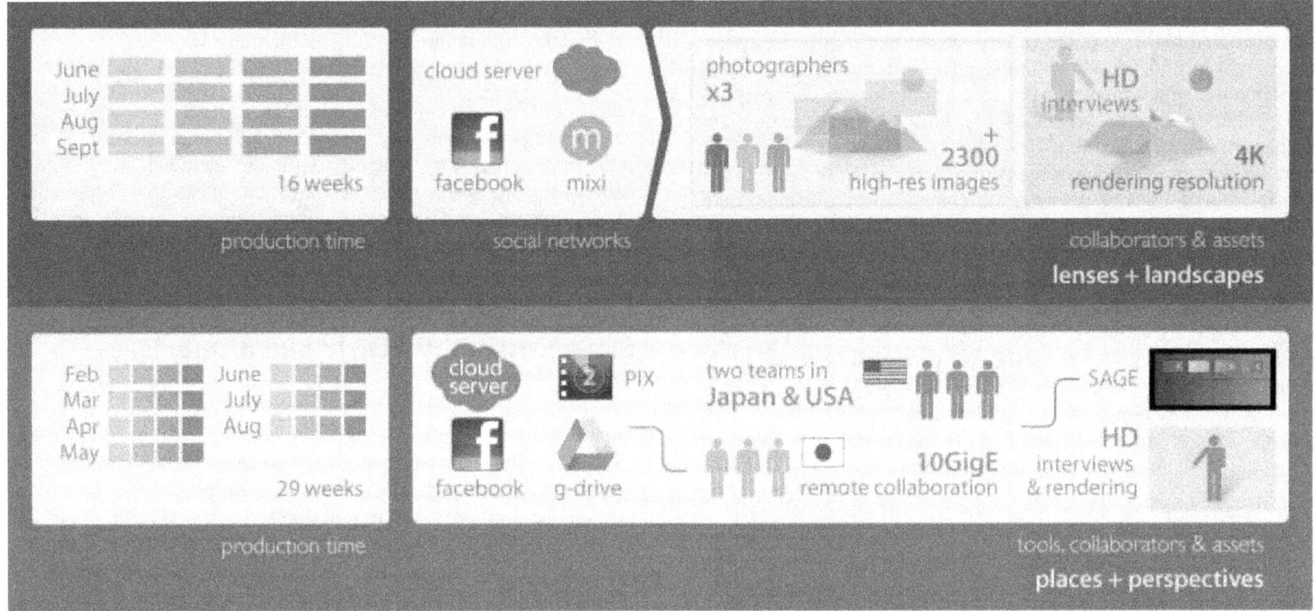

Figure 10. An overview graphic showing the different approaches employed in the production of both documentaries.

the three photographers and their stories and solidified the narrative structure by giving it depth and dimension otherwise not possible within the limited time frame.

"places + perspectives" employed collaboration not only in the acquisition of contents, but also in the production process and people involved in the production. The teams and team members were engaged in active, real-time collaboration from the pre-production stage of the documentary. Although typical production roles were assumed, the creation of the narrative was not based on a hierarchical structure. The collaborators in both Japan and America, while separated by physical distance, were able to connect and share ideas virtually through the use of technology. The virtual presence did not hinder human interaction during the joint collaborative sessions. This close contact allowed the collaborators to think together, make mistakes together and, most importantly, learn together.

4.2 Democratization of Tools

In both of the aforementioned examples of the Growing Documentary, synchronous and asynchronous collaborative filmmaking techniques were utilized to make deliverable and presentable projects. Both iterations of the Growing Documentary, "lenses + landscapes" and "places and perspectives" were made possible by the democratization of tools and equipment. Each of the cameras used by the three photographers in "lenses + landscapes" were readily available consumer/prosumer (professional + consumer) models that could be purchased easily from many shops at competitive price points. Although there is a noticeable difference in the quality, and detail with respect to quality, of parts and sensors in relation to the cost of the equipment, the technology is at such a degree that the digital still images captured from these three wide ranging and different digital cameras can appear together, blended into an ultra high-resolution documentary film.

In "places + perspectives", teams on both sides of the Pacific ocean were able to overcome a viscous time difference and collaboratively produce a multifaceted documentary piece. The democratization of technology played an essential role in maintaining the consistency of the narrative as well as its look. By using the same or similar DSLR cameras, both the UCSD team in California and the KMD team in Japan were able to discuss the settings before and after filming sessions. Not only was this beneficial for the editing process with respect to color correction and lighting, but it also created another sharing aspect to the collaborative process. The teams had the ability to assist their own team members with filming as well as provide advice or even get help from the remote team located half a world away (Figure 10).

4.3 A More Collaborative Workflow

In a traditional film setting, the film production process usually follows this sequence: pre-production, storyboarding, production, post-production, and distribution. Documentaries, however, have the production process preceding the storyboarding process. After the production and collection of assets, they are examined by the filmmakers to oversee the content flow (Figure 11). This method allows for more freedom in terms of narration, yet could also hinder the storytelling when a narrative thread cannot not be identified easily [21].

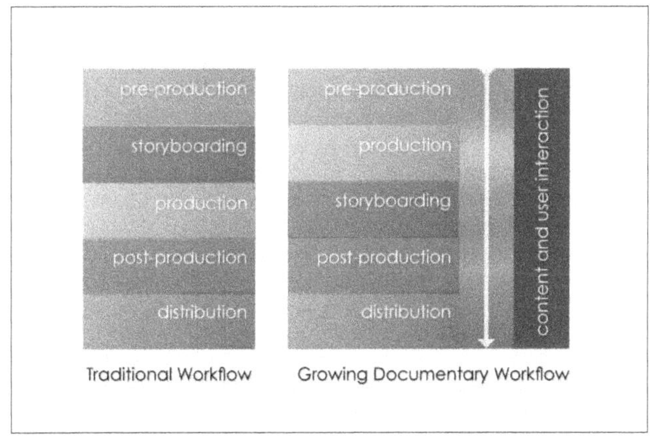

Figure 11. Traditional film workflow compared to the Growing Documentary workflow

Similarities between our research and commercial works is demonstrated in the he documentary "Japan in a Day". "Japan in a Day" was produced using short videos submitted by various participants, which later evolved into a long format documentary [24]. The production processes of "lenses + landscapes" and "Japan in a Day" in parallel show that both films were done with very little intervention from the filmmakers during the production phase. The filmmakers allowed the content to take center stage. Additionally, most of the assets were acquired from user generated content. Their function was only directing and guiding the flow and cohesion of the narrative [18]. Consequently, the filmmakers of Japan in a Day gave up some of their traditional roles to allow for the user's involvement. In "lenses + landscapes" this loosening of control was evinced by the curatorial role of the production team. Depending on the photographers to provide them with still images of the Tohoku area, the production team saw the images that would make up the narrative for the first time when they received them from the collaborators.

The production "places + perspectives" was conducted in two different timezones and locations. This opened up different possibilities for the entire film. First, the production relied on the familiarity of the production team to their surroundings and people within their respective geographic locations. Thus, scouting locations and contacting people happened smoothly and simultaneously in Japan and the USA. This meant more leads and resources within a shorter time frame and lower costs overall. Second, tasks, especially those in the post-production phase, were divided amongst these co-located teams allowing the production to span 24 hours if needed. Third, accountability was achieved with the implementation of the PIX System. The Pix System allowed the tracking of every change to the assets as well as the user responsible for that change.

4.4 Feeling the Story

Both "lenses + landscapes" and "places + perspectives" took advantage of SAGE OptiPortals at some stage of the production process. In "lenses + landscapes" their was a topic on which the storyline was based — 3.11earthquake and aftermath — but there was a lack of assets to support the narrative. After the production was completed, the potential of taking advantage of and utilizing "extra" or "unused" footage along with portions that made the final cut were made evident by introducing SAGE walls. The SAGE units gave the narrative a new life and direction by putting the power of the storytelling in the hands of the audience.

The scalability, interactivity, and computing power of SAGE along with the large real estate allows a simultaneous view of different media assets. By applying these strengths during the production of "places + perspectives", SAGE was utilized as a storyboarding tool when discussing the collected assets from KMD and UCSD. The interactive touchscreen enabled the video clips and images to be rearranged in different sequences more easily and flexibly than in an editing suite. This permitted both sides to work on the direction and narrative flow of the movie. The capability to share multiple laptop/desktop screens on the unit's touch display provided opportunities to use different software applications that were not natively supported by the OptIPortable environment such as non-linear editors.

In both situations, having the touch capability not only altered the way in which the story was created or the contents accessed, but it also transformed the way in which the concepts were consumed. The user experience became much more immersive by introducing the SAGE OptIPortables. The ability of users, creators and collaborators to literally handle content and the story at their fingertips creates an intimacy between the contents as well as other collaborators interacting with the touch interface.

The Growing Documentary is a step in furthering the field of collaborative content creation [2]. The framework allows collaborators to not only interact with one another in the production of a story, but also with the elements of the story. Moreover, the story itself becomes an interactive experience: ideas, images, videos and sounds can be shared and re-contextualized by the storytellers as well as the audience.

5. FUTURE RESEARCH

5.1 Accessibility - Distributed Interfaces

SAGE OptiPortables are ideal research tools for studying and experimenting with the possibilities of non-linear narrative creation and co-located collaborative content creation and sharing. SAGE for all its merits, however, has some issues pertaining to space, power and cost demands that make it prohibitive. Due to its size and amount of power required to run it, a SAGE unit may not be practical for institutions or individuals everywhere. Building on this, we tried to evaluate sharing between different interfaces, which led to the use of UME Board (Universal Media Board).

5.1.1 Integration of SAGE and UME Board

UME Board is a multimedia device that can be used "out of the box". Like SAGE, UME Board allows touch interaction through the use of infrared sensors. Unlike SAGE, which is an open-source and multi-platform environment, UME Board is (currently) preloaded with Microsoft Windows 7 (Table 2). The hardware also contains a free-writing memo feature which complements joint projects conducted remotely by allowing real-time annotation. Due to its size and (relative) affordability, UME Board is a practical and feasible tool for people who would like to work collaboratively in visual production.

Table 2. SAGE & UME Board Comparison

	SAGE	UME Board
Size	Scalable	55" Fixed
Sensors	infrared sensors	
Setup	Requires technical skills	Easy
OS	SAGE Commons	Windows 7
Sharing	Native sharing function	Requires apps/ services

In order to enhance the collaboration level and number of collaborators, we decided to integrate SAGE and UME Board. As both systems were developed to communicate within their own environments, the initial challenge was to make the two systems interoperable. The goal was to capitalize on the strengths of both systems to maximize the collaborative potential. For the initial experiment, a SAGE protocol-based unidirectional connection was established between an UME Board in Berlin, Germany and a SAGE OptIPortable in Yokohama, Japan. The experiment was conducted in real-time with assets transferred from the UME Board to SAGE (Figure 12).

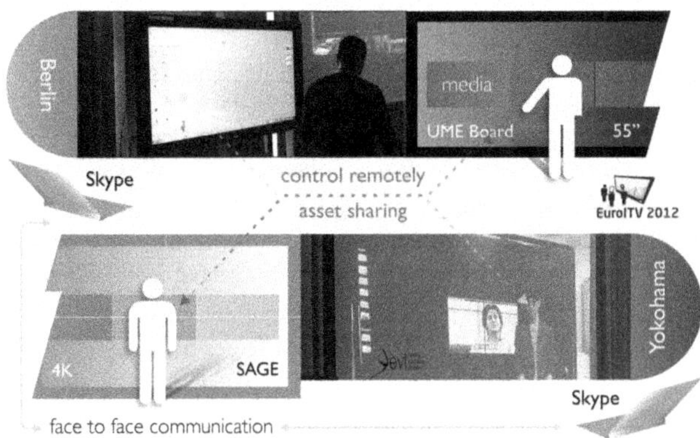

Figure 12. Network configuration and settings used between UME Board (Berlin, Germany) and SAGE (Japan).

5.2 Large Screens as Productivity Tools

Part of the appeal of both SAGE and UME Board is the large estate they provide for users. The ability to view content and assets on these large screens at the same time is quite different than working with a smaller desktop or laptop screen. With the proliferation of HDTVs in the last few years, more people incorporated large screen devices into their lifestyles. Yet, the majority of uses are bound to content consumption and viewing rather than content creation. Even with the recent trend of Internet-connected SmartTVs, the uses of such devices is mostly for video viewing [3]. Furthermore, there has been a strong emphasis on second-screen devices as companions to viewing content on a large screen. Despite this, large screens have not reached their potential in terms of application development and implementation.

Based on what was done with the Growing Documentary and with further research and development of large screen solutions, future HDTVs especially Ultra HDTVs could offer users a new chance to rethink TVs as a more productive tool rather than being solely consumption devices or large screen windows to existing applications. Combined with second screens, the potential for interaction with existing content and collaborative content creation can be realized.

6. CONCLUSION

The work presented in this paper aims to demonstrate how collaborative techniques can be implemented into user generated content creation to produce community based narratives. The communities are not just made up of people who live within a certain physical distance of one another. Rather, communities for both digital natives and digital immigrants [16] consist of people (regardless of geographical location) who share similar ideas, beliefs and passions. Members of these communities want to share their thoughts and feelings within and outside their community.

As a result of ubiquitous presence of social media, along with technological and digital innovations, people can now express themselves through their own content. The collaborative projects discussed in this paper aim to serve as a guide for individuals and content communities [10] to engage in their own narrative creation and add to the quality of user generated content.

Media is a collection of applications, tools and phases. Interaction must take place not only between people, but also hardware and software in order to bring about collaborative content creation. Technology should allow people to use the tools and resources within their reach and means and those with which they are comfortable. To promote shared authorship, future research will focus on the development and implementation of middleware, such as a user interface, that can facilitate collaborative content creation on multiple platforms while adapting to the needs of the users and the specifications of their devices.

There is still much work to be done in the field of co-located collaborative storytelling. While research needs to be conducted on technical and technological challenges for collaborative content production, there must also be, more importantly, a realization that the intellectual challenges of multiple individuals and identities working together cannot be overlooked. The work presented in this paper is meant to serve as a guide to the possibilities of user generated content creation through the employment and re-use of existing assets or over remote locations . With further research and study, we hope to discover and present more findings on the ways in which user generated content can serve communities and be for the people, by the people.

7. ACKNOWLEDGMENTS

This research is supported by the Singapore National Research Foundation under its International Research Center Keio-NUS CUTE Center @ Singapore Funding Initiative and administered by the IDM Program Office. We thank NTT-AT for their continued support of our research with SAGE OptiPortables. We are grateful to Daisuke Shirai and NTT Innovation Laboratories for their support and guidance. We thank Orion Electric Co., Ltd. for allowing us to use and demo the first prototype of UME Board. We also would like to thank the University of California, San Diego for their active role in the Growing Documentary and the electronic visualization laboratory (University of Illinois at Chicago) for their support in implementing tools into SAGE.

8. REFERENCES

[1] "4K Resolution". 2013. PCMAG.com Encylopedia. n.d. Web. (Jan. 26, 2013) http://www.pcmag.com/encyclopedia/

[2] Bhimani, J. Mahdia, A. Almahr, A. Shirai D. and Ohta, N. 2012. Growing Documentary: Creating a Collaborative Computer-Supported Story Telling Environment, *SIGGRAPH 2012*, Los Angeles.

[3] Buffone, J. 2012. Internet Connected TVs Are Used To Watch TV, And That's About All. *The NPD Group Blog* (Dec 26, 2012). https://www.npdgroupblog.com/internet-connected-tvs-are-used-to-watch-tv-and-thats-about-all/

[4] Cesar, P. and Chorianopoulos, K . 2008. Interactivity and user participation in the television lifecycle: creating, sharing, and controlling content. In *Proceedings of the 1st international conference on Designing interactive user experiences for TV and video (UXTV '08)*. ACM, New York, NY, USA, (2008) 125-128.

[5] CineGrid 2011. http://www.cinegrid.org/

[6] Cisco White Papers. 2012. *Cisco Visual Networking Index: Forecast and Methodology*, 2011-2016 White Paper, (May 30, 2012). www.cisco.com/en/US/solutions/collateral/ns341/ns525/ns537/ns705/ns827/white_paper_c11-481360.pdf

[7] Cioccaa, G., Olivob, P., and Schettinia, R. 2012. Browsing museum image collections on a multi-touch table. *Information Systems*, Volume 37, Issue 2 (April 2012), 169–182.

[8] Howe, J. 2006. The Rise of Crowdsourcing. Wired Magazine, 14, 6 (June, 2006). http://www.wired.com/wired/archive/14.06/crowds.html

[9] Jenkins, H. 2006. *Convergence Culture: Where Old and New Media Collide.* New York, NY: New York University Press.

[10] Kaplan, A. and Haenlein, M. 2010. Users of the world, unite! The challenges and opportunities of Social Media. *Business Horizons*, Volume 53, Issue 1 (Jan.–Feb., 2010), 59-68.

[11] Landau, M. 1984. Human Evolution as Narrative. *American Scientist,* 72 (May-June, 1984), 262-268.

[12] "lenses + landscapes". 2011. http://vimeo.com/31093347

[13] Pimenta, S. and Poovaiah, R. 2010. On Defining Visual Narratives. *Design Thoughts* (August 2010), 25-46.

[14] PIX System. http://www.pixsystem.com

[15] "places + perspectives". 2012. http://vimeo.com/47917016

[16] Prensky, M. 2001. Digital Natives, Digital Immigrants. On the Horizon, 9, 5 (Oct. 2001). http://www.marcprensky.com/writing/prensky%20-%20digital%20natives,%20digital%20immigrants%20-%20part1.pdf

[17] Renambot, L., Rao, A., Singh, R., Jeong, B., Krishnaprasad, N., Vishwanath, V., Chandrasekhar, V., Schwarz, N., Spale, A., Zhang, C., Goldman, G., Leigh, J. and Johnson, A. 2004. SAGE: the Scalable Adaptive Graphics Environment. In *Proceedings of WACE 2004 (*4th Workshop on Advanced Collaborative Environments, Nice, France, Sept. 23-24, 2004).

[18] Reyes, A. 2010. Towards Crowdsourcing for Disaster Management. *UbiComp10* (2010) - Workshop.

[19] Scalable Adaptive Graphics Environment. http://www.sagecommons.org/

[20] Shirai D., Kitamura, M., Fuji, M., Takahara, A., Kaneko, K. and Ohta. N. 2011. Multi-point 4K/2K layered video streaming for remote collaboration. *Future Generation Computer Systems,* Volume 27, Issue 7 (July, 2011), 986-990.

[21] Sørensen, K., Bahnsen, M., Holch, H., Hvid, G. and Otte, L. 2002 Guidelines for producing a short documentary, *p.o.v.: A Danish Journal of Film Studies,* No. 13 (March 2002),126-127.

[22] Starbird, K. and Palen, L. 2011. "Voluntweeters": Self-Organizing by Digital Volunteers in Times of Crisis. *Proc of CHI 2011*, 1071-80.

[23] Whitelaw, B. 2011. Almost all YouTube views come from just 30% of films. *The Telegraph.* (April 20, 2011). http://www.telegraph.co.uk/technology/news/8464418/Almost-all-YouTube-views-come-from-just-30-of-films.html

[24] Wigley, S. 2012. Eastern thinking: Philip Martin on Japan in a Day. *British Film Institute*. (Oct. 19, 2012). http://www.bfi.org.uk/news/eastern-thinking-philip-martin-japan-day

[25] Willett, R. 2009. *In the Frame: Mapping Camcorder Cultures in Video Cultures: Media Technology and Everyday Creativity edited by David Buckingham and Rebekah Willet* (New York, NY: Palgrave Macmillan), 1-22.

[26] YouTube Statistics. http://www.youtube.com/t/press_statistics

Transmedia Phenomena: New Practices in Participatory and Experiential Content Production

Mariana Ciancia
Department of Design
Politecnico di Milano
Via Durando 38/A, Milan, Italy
mariana.ciancia@mail.polimi.it

ABSTRACT

Nowadays we live in a society rich of a growing number of messages spread by multichannel and multimodal devices [9, 27, 21, 18]. A scenario that outlines the need of a designer able to manage the coexistence of multiple factors using his tools to overcome the problems of understanding information and access them.

Starting from the complex nature of communication flows, we can define a main objective: to encourage a dialogue among people. For this reason, it is important to study the evolution of the existing communication systems, in order to understand the new collaborating and participative processes, and how media and languages have broken their historical isolation and incompatibility. Thanks to a large amount of theoretical assumptions, and according with thoughts and projects developed during these years [10, 13, 26, 2, 28, 5, 6], the aim of this work is to figure out a new form of literacy that allows us to develop content able to foster a rich user experience through the merge of different languages, media, technologies, and to rethink the participatory process.

First of all, the work will highlight the developments in the communication field that are affecting society, exploring three main areas: audience, technology and user engagement. After this the paper will focus on *The Spiral*, a pan-European Transmedia co-production, as an example of potential activities in order to do reflections on what are challenges and opportunities of the *Transmedia Phenomenon* within the broadcast production.

Categories and Subject Descriptors

I.7.2 [**Document And Text Processing**]: Document Preparation – *Multi/mixed media.*

General Terms

Design

Keywords

Communication Design, Transmedia, Storytelling, Audience Engagement.

1. INTRODUCTION

The communication could be seen as an environment where generative flows converge. Therefore it is important to study the evolution of the existing communication systems, in order to understand how media and languages have broken their historical isolation, and how to design communication processes able to foster transformation and produce brand new and shared visions.

In the contemporary scenario it is possible to highlight the developments in the communication field that are affecting society, exploring three main areas: audience, technology and user engagement.

The first change is related to audiences and to their use of the communication systems. According to Toschi [30], this is a process whose construction is taking place nowadays in what we could define *script's society*. During the last decades we were witness to the developments that have given a new relevance to the storytelling practice. A practice born in the dawn of mankind history and grown up during the centuries, which allowed people to build and share the meaning of their common experiences, to communicate and to structure the surrounding reality. Compared to last decades, the strength of contemporary days is the rising importance of narrative structures with stories able to provide not only a linear path - with beginning, middle and end – but also multiple points-of-entry: therefore the people, looking for an increasing number of information and experiences, can cross the boundaries of a single-line story [22]. Instead from a designer perspective, it is possible to recognize two phases in the project of the contemporary communication systems. The former is the identification and understanding of the audiences in order to offer them the right content that is spread over appropriate devices. The latter concerns the definition of the user engagement «allowing our viewers, now users, become players to engage in active participatory building of the future» [7].

The second change concerns the technological field: a development that lead us to interact with a digital vision of reality, and whose new textuality can be an important tool to envision our future. Nowadays the remarkable acceleration in technological development has provided unimaginable instruments to the communication strategy. Hence, we need to understand how we can make a proper use of these tools in order to develop communication systems able to combine different media (on-line/off-line) to engage audiences. Because «New Technology creates new opportunities» [1, pp. 167-168] «but there is no reason to believe that the increased equality for all subjected to the technology» [1, p. 168]. In fact, our aim will be sustain an «ongoing conversation and engagement with the viewer» [2, p. 54] choosing the proper media for our content.

The third change refers to the participatory processes. In the last decades we have been witnessing of a shifting from a *platform-centric* vision [2, p. 137], to a *user-centered* vision, but now it is necessary to push the development toward a *community-centered* vision to allow the growth of the collective consumption and the community engagement. As a matter of fact, while in the last decades the user interaction was related to technology, nowadays it depends on the time and space that people are willing to devote to content. This makes possible classifying audiences into two categories of experience: users that don't interact with the content (Lean Back Experience) and people who are very active (Lean Forward Experience) [2]. This doesn't mean we must design the highest possible degree of engagement. Thus, shifting the focus from technology to user and community experience, it is possible to find a balance between the two engagement degrees: as a consequence the audiences can enjoy the best content spread by the suitable devices, taking advantage of the opportunities of the contemporary cross mediated world. For these reasons, the designer has to learn how to manage the new communication processes: shifting his position from an executive role to a supporting role, while the generative process increases. This is because people are becoming aware of their lead role in the contemporary communication activities.

1.1 Contemporary Communication Trends

This work highlighted three fields – Audience, Technology, User Engagement - that are living in a situation of interchange within the age of convergence [17] and whose development has given rise to conceptual spaces that affect the *mediascape*.

In the light of these evolutions, it is possible to outline three trends that are characterizing our contemporary society: *Storytelling Space*, *Participative Space* and *Experiential Space*.

Storytelling Space and Participative Space refer to narrative structures that give us more tools for tell and listen to stories. As a consequence there are new generations of interactive narratives able to change the connection between the mass media (*top-down*) and the participatory culture (*bottom-up or grassroots*). Instead, Experiential Space is related to how the people, thanks to developments in technology and media habits, can interact with each other creating collaborative networks: a way to experience narratives spread out by non-linear communication systems through a collective consumption [19].

1.2 Communication Systems

The next step in the process of understanding the current scenario is the identification of the main communication systems and related projects. Developing a matrix that takes into account the channels involved (Single/Multiple) and the narrative structure (Story/Story world) it is possible to identify a first distinction between *Multimedia* and *Multichannel* (see Figure 1).

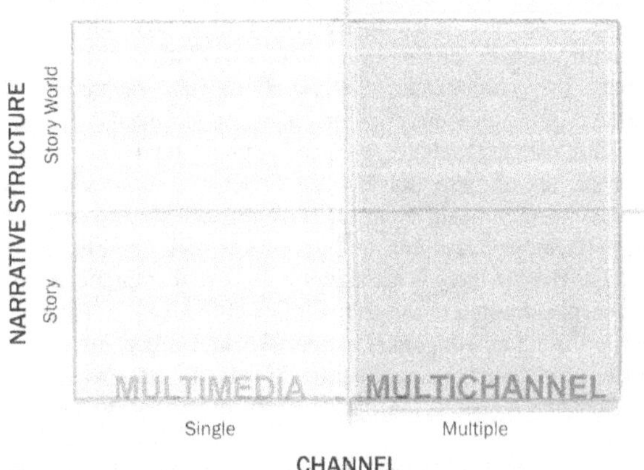

Figure 1. Matrix for the communication systems' identification.

1.2.1 Multimedia

Multimedia is a content that uses a combination of different forms and languages (a mix of text, audio, image, moving image, animation, or interactivity content forms) spread by a single channel: for this reason it is also possible to use the term for all those objects that act as aggregators of different types of content. Multimedia may be broadly divided into linear and non-linear classes related to the user engagement:

- Systems with linear structure, in which the content progresses often without any navigational control for the users (more viewers).
- Systems with non-linear structure (e.g. hypermedia), in which the objects are represented as nodes in a network and the users are allowed to navigate, from one node to another, within the rules of the program.

1.2.2 Multichannel

With the term *multichannel* we refer to the transmission mode and consumption of the content itself that can be adapted and distributed across different types of channel.

Focusing on the multichannel field it is possible to highlight two phenomena: *Crossmedia* and *Transmedia* (see Figure 2).

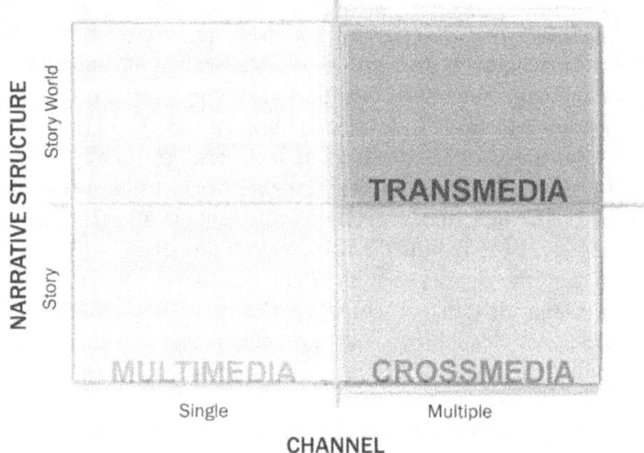

Figure 2. Matrix for the communication systems' identification: focus on multichannel systems.

With the term C*rossmedia* we refer to the adaptation of the same text, or proto-text - which can be a content of any nature - in different media: namely progressive or contemporary adaptations on different channels. This paradigm allows us to develop systems in which a single story is interpreted independently in different media and the content is self-contained (e.g *The Lord of the Rings*[1]). In this work the definition of crossmedia is close to the *adaptation* concept developed by Geoffrey Long «Retelling a story in a different media type is adaptation» [20, p. 22] and that Christy Dena defined «A practice that can have many functions and has been practiced to varying degrees of success for centuries» [6, p. 163]. Usually the reason for the repurposing is to enhance or focus the user's experience according to the devices' features and functions.

Instead *Transmedia* is a phenomenon able to support the construction of a *human landscape*, relying on the storytelling

[1] *The Lord of the Rings* is a fantasy novel written by J. R. R. Tolkien (1954-1955) that has been adapted for radio, film, stage and game (video games and MMoRPG).

ability to foster multiple perspectives and allowing people to become aware of their lead role in the contemporary mediascape.

But what really means the term *transmedia*? Henry Jenkins [17, p. 293] described it like «Stories that unfold across multiple media platforms, with each medium making distinctive contributions to our understanding of the world». Robert Pratten [28, pp. 1-2] said that transmedia «is telling a story across multiple media [...] with a degree of audience participation, interaction or collaboration» and that «the engagement with each successive media heightens the audience' understanding, enjoyment and affection for the story». Christy Dena [6, p. 331] defined it as «a working implementation of unity in diversity». Or again, Jeff Gomez advised us to become orchestra directors because «What transmedia does is it brings a few of those instruments together and attempts to compose music that allows for talented people to play them in some kind of concert» [8].

What emerges from the state of the art is the definition of *Transmedia* like a cultural paradigm [15] that allows audiences not only to access content in a different way, but also to participate in the meaning-making process, with a subsequent changing in the relationship between the mainstream media (top-down) and participatory culture (bottom-up or grassroots).

2. TRANSMEDIA APPROACH

We stand in front of a practice that is capable of answering all the needs that are developed in the convergence era [17], and that is able to transform the contemporary weaknesses into strength points, thanks to the subsequent characteristics:

- *Involvement of different media* | The projects involve more than one medium, splitting the message across different channels, and allow users to piece the story fragments together in order to experience the whole story universe.
- *Transmedia storytelling* | The projects are characterized by a complex "mythology" [16], story worlds that are built by story architects able to design the flow of content across different channels.
- *Audience Engagement* | This generation of transmedia works «takes advantage of [that] technology and uses it as part of the narrative canvas.» [23, p. 12]. So we need to make a clever use of technology and distribution, and be aware of the contemporary media habits, in order to «start a dialog with the[ir] audience in such a way as to activate them, validate their participation and get them to advocate on behalf of the story [...]» [24, p. 11].

Not to mention the permanence of two basic needs: the importance of the audience media habits and the economic sustainability of cultural and entertainment industry [11]. That's why the moments where «a creative project isn't influenced in some way by economics» [6, p. 38] are sporadic, and the «economic concerns are part of a complex design ecology» [6, p. 38].

In conclusion, whatever is the content we are working on, we don't forget that telling stories across multiple channels is an opportunity, not the rule, and that we have to respect the audience and the story we're telling [25, p. 15].

But what happens to the television content? It is possible to identify new practices in network serial production?

2.1 Transmedia Television: The Spiral

According to Clarke [3], that coined the term *tentpole TV* to describe networks able to experiment alternative forms of organization and ancillary texts, this work describes *The Spiral* [29]: one of the latest television projects within the transmedia field, in order to demonstrate how this phenomenon is affecting the network serial production, becoming a real practice.

The Spiral is a pan-European Transmedia project coproduced by 7 broadcasters: VRT (Belgium), SVT (Sweden), TV3 (Denmark), YLE (Finland), Arte (Franco-German), NRK (Norway), VARA (Netherlands).

The thriller story starts with Arturo, a notorious street artist, who plans with six young persons, part of an art community in Copenhagen (*The Warehouse*), to steal six paintings throughout Europe in order to create the most valuable work of art.

The *Spiral* is a participative project that combines traditional storytelling (a five parts TV-series) with audience engagement solutions (LARPing, on-line scavenger hunt, on-line art community) in order to allow the audience to join the hunt of six artworks that have been hidden in www.thespiral.eu.

A transmedia project that was born and bred in a scenario characterized by changes in media consumption and audience expectations, that debuted on August 21st, 2012 and was broadcasted simultaneous across Europe in September 2012.

The story world was accessible through a variety of channels and the action, according to Phillips [26], was conveyed over multiple media through *Story Archaeology* (the ability of the audience to piece together information spread across different media), *Live* and *Delayed Coverage*: for example, the final event in front of the European Parliament was filmed and inserted in the final episode that was broadcasted two days later.

A project that tried to leave people to decide their own participation path in the story world, a space "where everyone can participate the way they like and the way they feel" [12]. In this case the audience engagement reflected the 90-9-1 principle, and was conducted through 4 phases. The first one was related to the building of the online artistic community (*The Warehouse*) with the help of 40 practiced *LARPers* (1% of heavy contributors). The second and the third phases were characterized by the opening of the online game on the platform www.thespiral.eu: with the possibility of testing games and challenges in preview for *The Warehouse* community, till the real opening of the hunt on the 21st August, with the physical removal of the six paintings from the involved museums (9% of intermittent contributors). Finally, during the last phase was broadcasted the TV series that helped the story world understanding of the online players, and that allowed to the great amount of the audience the passive enjoy of the plot (90% of lurkers).

Therefore, *The Spiral* acknowledged the importance of a dialogue between mainstream media (*top-down*) and participatory culture (*bottom-up*), building the community's growth through different stages, and focusing on the audience members that even more «self-sort themselves into group more or less willing to pay for ancillary texts or "invest" in repeat viewings» [3].

3. CONCLUSION

Trying to answer to the contemporary challenges of the mediascape, we understand that it is necessary to find new collaborative ways for creation and consumption.

One of the possibilities that emerges from the multichannel media habits is the *Transmedia Thinking* (*and Practice*): a phenomenon that is able to expand and share the production of meaning between mainstream channels and audience members.

A first aim of this work is to suggest that it is possible to build transmedia projects not only for the big Hollywood productions, but also at a European level in order to change the way we create, develop and distribute stories. In fact, the briefly illustrated case study and many works developed all over the world, demonstrate how it is possible to involve the audiences at different levels: people that, according to their participation desires, could be engaged not only in the meaning-making process of the story world. As a matter of fact, we are witnessing of the development of more and more collaborative processes: e.g. co-design processes and crowdsourcing (micro or small tasks that don't ask any investment or the sharing of the design focus) that nowadays are adopted by the design teams. In this case we could see an application of both practices: in fact the project utilized a collaborative approach between the seven European broadcasters and a crowdsourcing process, as evidenced by the second and third phase in the community development.

Not to mention the possibility of involve the audience in the financing of the projects, as demonstrated by *Iron Sky* [14] or *The Cosmonaut* [4]: two productions that include crowdfunding strategies.

In conclusion, it is possible to demonstrate that we can answer to all the changes that are affecting society through the design of content able to deliver a rich user experience. The result will be the development of a widespread creativity thanks to phenomena (e.g. Transmedia) based on storytelling, collaboration and networks creation.

4. REFERENCES

[1] Aarseth, E. 1997. *Cybertext: Perspective on Ergodic Literature.* Johns Hopkins University Press, Maryland.

[2] Bernardo, N. 2011. *The producer's guide to Transmedia. How to develop, fund, produce and distribute compelling stories across multiple platforms.* BeActive Books, Lisbon-London-Dublin-Sao Paulo.

[3] Clarke, M. J. 2013. *Transmedia Television. New trends in network serial production.* Bloomsbury. New York – London

[4] *The Cosmonaut* 2013. Nicolás Alcalá. Spain.

[5] Davidson, D. 2010. *Cross-Media Communications: an Introduction to the Art of Creating Integrated Media Experiences.* ETC Press, Italy.

[6] Dena, C. 2009. *Transmedia Practice: Theorizing the Practice of Expressing a Fictional World Across Distinct Media and Environments.* Doctoral Thesis. University of Sydney.

[7] Dinehart, S. 2010. DAREtoENGAGE. *TEDxTransmedia*, 30 September 2010 – Geneva, Switzerland. Retrieved January 22, 2013, from TEDxTalks YouTube Channel http://www.youtube.com/watch?v=gz9fZJatIQw

[8] Dinehart, S. 2009. Creators of Transmedia Stories[tm] 3: Jeff Gomez. *The Narrative Design Explorer. A publication dedicated to exploring interactive storytelling.* Retrieved January 22, 2013, from http://narrativedesign.org/2009/09/creators-of-transmedia-stories-3-jeff-gomez/

[9] Galbiati, M. and Piredda, F. 2010. *Design per la Web TV. Teorie e tecniche per la televisione digitale.* Franco Angeli, Milano.

[10] Giovagnoli, M. 2013. *Transmedia Storytelling e comunicazione.* Apogeo, Milano.

[11] Grasso, A. and Scaglioni, M. (edited by) 2010. Televisione Convergente. La tv oltre il piccolo schermo. LINK idee per la televisione, Milano.

[12] Hurst, B. S. 2012. Interview with Peter De Maegd. *Interactive TV Today.* Retrieved January 29, 2013, from http://vimeo.com/48494252

[13] Ibrus, I. and Scolari, C. A. (edited by) 2012. *Crossmedia Innovations. Texts, Markets, Institutions.* Peter Lang, Frankfurt am Main-Berlin-Bern-Bruxelles-New York-Oxford-Wien.

[14] *Iron Sky* 2012. Timo Vuorensola. Finland-Germany-Australia.

[15] Jenkins, H. 2011. Transmedia 202: Further Reflections. *Confessions of an Aca-Fan. The official weblog of Henry Jenkins.* Retrieved January 27, 2013, from http://henryjenkins.org/2011/08/defining_transmedia_further_re.html

[16] Jenkins, H. 2009. The Aesthetics of Transmedia: In Response to David Bordwell (Part One). *Confessions of an Aca-Fan. The official weblog of Henry Jenkins.* Retrieved April 20, 2013, from http://henryjenkins.org/2009/09/the_aesthetics_of_transmedia_i.html

[17] Jenkins, H. 2006. *Convergence Culture.* New York University Press, New York.

[18] Kress, G. and van Leeuwen, T. 2001. *Multimodal discourse: the modes and media of contemporary communication.* Hodder Education, London.

[19] Lévy, P. 1996. *L'intelligenza collettiva. Per un'antropologia del cyberspazio.* Feltrinelli, Milano.

[20] Long, G. 2007. *Transmedia Storytelling: Business, Aesthetics and Production at the Jim Henson Company.* Degree Thesis, Massachusetts Institute of Technology.

[21] Manovich, L. 2001. *The Language of New Media.* The MIT Press, Cambridge.

[22] Murray, J. 1997. *Hamlet on the Holodeck. The future of narrative in cyberspace.* The Free Press, New York.

[23] Perrone, D. 2012a. Alan Lee. La vita multi-screen nella visione del fondatore del Transmedia dream team mondiale. In *Subvertising. La sovversione della pubblicità* (year VI, number 49). Retrieved January 27, 2013 from http://subvertising.it/

[24] Perrone, D. 2012b. What Jeff has not said. In *Subvertising. La sovversione della pubblicità* (year VI, number 50). Retrieved January 27, 2013 from http://subvertising.it/

[25] Perrone, D. 2012c. Transmedia wonder woman: Andrea Phillips. In *Subvertising. La sovversione della pubblicità* (year VI, number 51). Retrieved February 4,2013 from http://subvertising.it/

[26] Phillips, A. 2012. *A creator's guide to transmedia storytelling: how to Captivate and Engage Audiences Across Multiple Platforms*, McGraw Hill.

[27] Piredda, F. 2008. *Design della comunicazione audiovisiva. Un approccio strategico per la "televisione debole".* Franco Angeli, Milano.

[28] Pratten, R. 2011. *Getting started in Transmedia Storytelling. A practical guide for beginners.* CreateSpace, USA.

[29] *The Spiral* 2012. Hans Herbots. Belgium-Netherlands-Sweden.

[30] Toschi, L. 2011. *La comunicazione generativa.* Apogeo, Milano.

TV Program Detection in Tweets

Paolo Cremonesi
Politecnico di Milano, DEI
P.zza Leonardo da Vinci, 32
Milan, Italy
paolo.cremonesi@polimi.it

Roberto Pagano
Politecnico di Milano, DEI
P.zza Leonardo da Vinci, 32
Milan, Italy
pagano@elet.polimi.it

Stefano Pasquali
Politecnico di Milano, DEI
P.zza Leonardo da Vinci, 32
Milan, Italy
stefano.pasquali@mail.polimi.it

Roberto Turrin
Moviri S.p.A., R&D
Via Schiaffino, 11
Milan, Italy
roberto.turrin@moviri.com

ABSTRACT

Posting comments on social networks using second screen devices (e.g., tablets) while watching TV is becoming very common. The simplicity of microblogs makes Twitter among the preferred social services used by the TV audience to share messages about TV shows and movies. Thus, users' comments about TV shows are considered a valuable indicator of the TV audience preferences. However, eliciting preferences from a tweet requires to understand if the tweet refers to a specific TV program, a task particularly challenging due to the nature of tweets - e.g., the limited length and the massive use of slangs and abbreviations.

In this paper, we present a solution to identify whether a tweet posted by a user refers to one among a set of known TV programs. In such process, referred to as *item detection*, we assume the system is given a set of items (e.g., the TV shows or movies) together with some features (e.g., the title of the TV show). We tested the solution on a dataset composed by approximately 32000 tweets, where the optimal configuration reached a precision of about 92% with a recall equals to about 65%.

Categories and Subject Descriptors

I.2.7 [**Artificial Intelligence**]: Natural Language Processing—*Language parsing and understanding, Text analysis*; I.5.2 [**Computing Methodologies**]: Pattern Recognition—*Classifier design and evaluation*

General Terms

Algorithms

EuroITV'13, June 24–26, 2013, Como, Italy.

Keywords

Item detection, Twitter, SVM, ROC, implicit elicitation, recommender systems

1. INTRODUCTION

Sharing TV viewing experience through a second screen is getting very common [10, 16, 17]. An increasing number of users watching television desire to be part of a more social experience, either reading other viewers' messages or sharing their own opinions, e.g., commenting an episode of the TV show currently on air. Simplicity makes microblogs among the preferred social services to post short messages while watching TV[8], so that Twitter[1] - the most known and used microblog - has been recently mentioned as a possible measure of television audience [11, 19] and as a tool for preference elicitation in recommender systems [9].

In order to elicit user preferences from a microblog comment, we need to recognize if the user is writing about specific TV shows or movies. However, automatically detecting the subject of tweets - i.e. comments posted on Twitter - is a challenging task. In fact, the 140-character limit of tweets strongly affects the way people write on Twitter. Minimal contextualization, use of slang, grammatical exceptions, acronyms, abbreviations, and tiny URLs are characteristics common in tweets, making them very noisy and difficult to be analyzed. As an example, the tweet

> *Anyone going to @cstn this year? @theperfectfoil will be presenting BLJ in the Imaginarium tent on Thurs evening, July 5th*

refers to a talk that would be given by Steve Taylor ('The Perfect Foil') about the movie *Blue Like Jazz* ('BLJ') during the Cornerstone music festival in Illinois ('cstn') on Thursday ('Thurs').

In this work we propose a solution to analyze the content of tweets and to identify whether they refer to one (or more) known items - e.g., TV programs or movies. Given a set of known items (e.g., a Video on Demand catalog) and the tweet of a user, the proposed solution tries to identify if the user is commenting about some of the items. In the following we will refer to this process as *item detection*.

Item detection is a variant of topic detection, a text classification task. Topic detection tries to assign a text to one

[1] `http://twitter.com`

or more known categories; similarly, item detection tries to assign a tweet to one or more known items. However, item detection differs from standard topic detection mainly because of:

(i) **The number of classes**. Classes correspond to categories - e.g., the genres of movies - in topic detection and to items - e.g., the movies of a Video on Demand provider - in item detection. Thus, while the number of categories in topic detection is small, the number of items in typical TV applications can exceed some thousands elements.

(ii) **The variability of classes**. In typical item detection problems, items are constantly added to/removed from the catalog. This prevents the use of traditional classifiers, used in topic detection tasks, that are designed to work when the set of possible classes is defined a priori and static. In fact, such classification algorithms must be initially trained with a set of classified samples (e.g., tweets) representing all possible classes. Once trained, the classifiers can be used to predict the classes of unclassified tweets. Unfortunately, traditional classifiers require to be re-trained when some classes (i.e., items) are added or removed, being not suitable for applications where classes frequently change.

In order to overcome these two problems we opted to use a battery of one-class supervised classifiers [7]. A one-class classifier is a classification algorithm that is trained using only the objects of a given class c; once trained, it tries to distinguish class-c objects from all other possible objects. In our application domain, each one-class classifier is trained using only tweets referring to a single item. We implemented a battery of classifiers composed by a one-class classifier for each item in the catalog. When a new item (i.e., class) is added to the catalog, we simply need to train an additional one-class classifier, while no updates are required for the remaining one-class classifiers previously trained. Given the tweet of a user, we detect the items it refers to by interrogating all the one-class classifiers forming the battery; hence, the tweet is assigned to the items related to the classifiers that positively classify the tweet.

We tested the classification performance of our solution on a dataset composed by approximately 32000 tweets on 40 movies, reaching a precision of 92% and a recall of 65%.

Finally, we implemented a set of additional experiments in order to analyze the impact on the performance of (i) the number of tweets per item, (ii) the number of items in the dataset, and (iii) the presence of out-of-scope tweets, i.e., tweets not related to any item in the catalog or even not related to movies at all. The results demonstrate a good robustness of the item detection system with respect to all the considered variables.

The rest of the paper is organized as follows. In Section 2 we discuss related and supporting work. In Section 3 we formalize the problem and design the architecture of the item detection problem, together with a description of the three stages of the developed system. Section 4 describes methodology, metrics, and dataset used. In Section 5 we discuss more in detail about the third stage of our system, analyzing the main parameters that it requires. The main findings of our research are discussed in Section 6, where we also explore the hypothesis that a tweet refers to only one item. Finally, Section 7 draws some conclusions and presents some directions for future work.

2. RELATED WORK

Item detection shares several characteristics with topic detection, a traditional classification task that assigns a text document to one or more topics on the basis of its content. The analysis of the topic of a text is usually performed by means of statistical and mathematical models, assuming that there is a strong correlation between the terms (and their frequencies) used within the text and the arguments under discussion.

Blei et al. [3] describe *Latent Dirichlet Allocation* (LDA), an unsupervised approach based on hierarchical Bayesian models and the Expectation Maximization algorithm (EM) to classify text documents into predefined topics. Following this work, several other approaches based on LDA (e.g., [1, 2]) have been proposed. Makkonen et al. [18] describe *Topic Detection and Tracking* (TDT), a technique that analyzes a stream of news articles, taking into consideration its temporal nature, to extract topics or specific events. Griffiths et al. [13] use short-range syntactic correlations and long-range semantic dependencies in a Monte Carlo Markov Chain Model to classify documents into topics.

Support Vector Machines (SVM) have shown to yield good generalization performance on several classification problems, including text categorization [14, 15, 26, 27].

In the case of short texts, topic detection results more complex. In a very short text like a microblog post, in which the user message must be compacted within few sentences, finding statistically-relevant correlations among words is uncommon. With reference to Twitter, a tweet typically contains retweets, mentions, and hashtags. Retweets - abbreviated as 'RT' - are used to cite a tweet of an other user; mentions - defined by the prefix '@' - are used to address a tweet to a specific Twitter account; hashtags are free labels used to mark tweets.

Ramage et al. [22] try to identify the topics of a set of tweets using Labeled-LDA, a supervised learning model which extends LDA. Labeled-LDA has been applied to Twitter comments, where hashtags, emoticons, and generic social signals (e.g., the presence of questions or mentions) have been used as labels. Cataldi et al. [4] detect the emerging topics within a flow of short messages posted by the community of users. They take into account not only the term frequency but also the social network relationships (based on the well-known Page Rank algorithm). Finally, they generate a topic graph that connects emerging terms with semantically-related words, in order to form sets of emerging topics.

Part-of-speech (PoS) tagging [5] has been applied to microblogs by Gimpel et al. in [12], leading to an accuracy of 90% on about 1800 tweets manually annotated.

Wakamiya et al. [25] analyze Twitter messages to evaluate TV audience. Tweets are mapped onto TV shows - assuming that each tweet can be at most mapped onto a TV program - on the basis of the data available from the EPG (Electronic Programming Guide). The matching criteria takes into consideration word-based similarity in the Japanese language between tweet terms and the TV program title. Our work extends the work of Wakamiya et al. in the following directions:

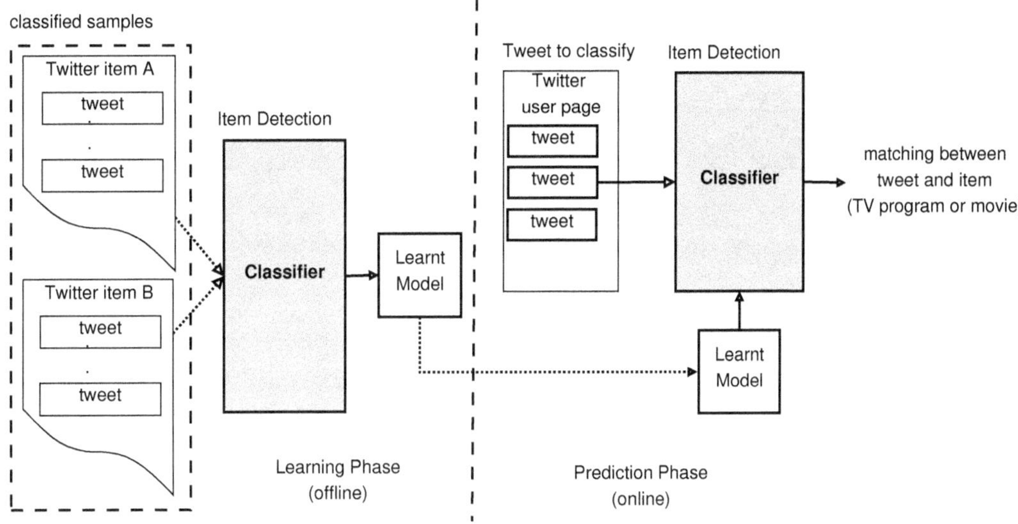

Figure 1: Item detection framework: learning and classification phases

- language: our classifier has been designed and tested to work with English tweets
- scalability: the structure of the system described by Wakamiya et al. does not allow to run the classifier in real-time applications; our classifier is intrinsically parallel and scales linearly with the number of available cores
- multiple classes: our classifier can detect if a tweet refers to more than one item
- our classifier is able to directly classify tweets that do not contain any word from item title or description

3. ITEM DETECTION

We define the *item detection* problem as a classification problem whose purpose is to detect whether a tweet refers to one of the items of a given catalog.

Classification systems are machine learning algorithms capable of assigning an item to a class. In our settings, the classes are the items in catalog, i.e. movies or TV shows. Typically, the classification task can be decomposed into two parts: *learning* and *prediction*. During the *learning* phase, as detailed on the left-hand side of Figure 1, the classifier takes a set of classified samples (i.e., tweets whose class is known) as input and computes a classification model - e.g., a set of parameters - optimized to accurately classify the input samples. Learning is usually performed off-line by a batch process. Once the classification model has been computed, the classifier can predict the class of any unclassified tweet, as shown on the right-hand side of Figure 1. Prediction can be usually performed on-line by a real-time process.

Detecting the item discussed in a tweet can be difficult, even for human. As an example, we reported in Table 1 three different tweets. Tweet 1 is very easy to classify and it clearly refers to the movie 'Harry Potter'. However, there can be ambiguous tweets, as tweet 2. In this case the tweet is comparing two movies: a correct classifier should assign it to two different classes. Some tweets, as tweet 3, require additional knowledge, not included in the message. Tweet 3 refers to the movie 'Blue Like Jazz' being Steve Taylor the movie's director.

We can note from the previous examples that a tweet can refer to one or more known items and can express either a positive, negative, or neutral opinion (the polarity). In this work we will neglect the tweet polarity and we will focus on the item detection task.

We designed the item detection system as a battery of one-class classifiers. A one-class classifier is a particular classifier that is trained with samples (tweets) belonging to a single class. The classifier, once trained, will be able to recognize whether an unknown tweet belongs or not to that class. Thus, the output of a one-class classifier can be either positive (the tweet is 'classified') or negative (the tweet is 'unclassified'). The choice of using a battery of one-class classifiers offers different advantages over a single multi-class classifier [24], such as:

1. **Scalability**. This solution easily scales with the number of items (i.e., TV programs or movies). In fact, since each classifier is independent from the others, the system can be straightforwardly parallelized in separated processes running on multiple cores.
2. **Incremental training**. When new items are added to the catalog, we simply need to train a new one-class classifier, with no impact on the rest of the system. Similarly, the removal of an item simply implies excluding the associated classifier.
3. **Multiple item classification**. Each classifier returns its own decision value, allowing a tweet to be assigned to multiple classes. For example, tweet 2 (Table 1) should be classified by two one-class classifiers, associated to the movies 'Twilight Breaking Dawn' and 'The Amazing Spiderman'.

Let $\mathcal{I}$ be the set of items available in the catalog and $|\mathcal{I}|$ its cardinality. Given a tweet t, the one-class classifier associated to item $i \in \mathcal{I}$ estimates the score $c_i \in \mathbb{R}$, which represents the degree of membership of such tweet to item i (i.e., how much the tweet refers to the item). The whole

ID	Tweet content	Item	Note
1	*I must be the one to kill #HarryPotter.*	Harry Potter	trivial
2	*#BreakingDawn Part 2 trailer ahead of The Amazing Spiderman! cheering: from the (female) crowd!*	Twilight: Breaking Dawn The Amazing Spiderman	ambiguous
3	*Special Q&A with Director Steve Taylor tomorrow night following the 7:40 show at Regal Cinema 8!*	Blue Like Jazz	difficult

Table 1: Sample tweets.

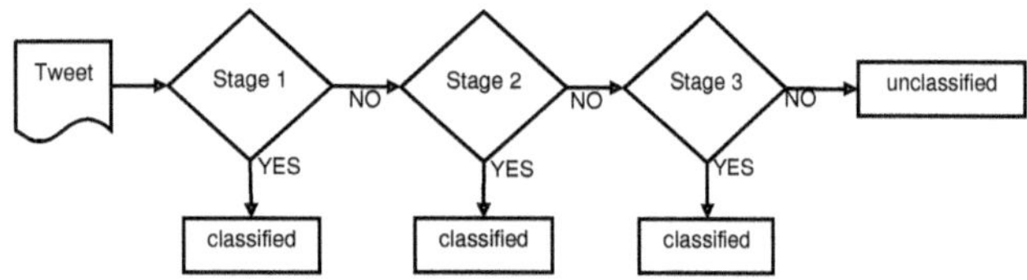

Figure 2: Architecture of a one-class classifier.

system is composed by a battery of $|\mathcal{I}|$ one-class classifiers. Its classification function $c(t)$ can be written as:

$$c(t) : t \to \mathbb{R} \times |\mathcal{I}| \quad (1)$$

Given the classification function (1), we can detect which items are commented in a certain tweet t.

We have decided to implement each one-class classifier as a pipeline composed by three stages, as detailed in Figure 2. An unknown tweet is initially processed by stage 1, that can either assign the tweet to its associated class or it can discard the tweet. Only tweets not classified at stage 1 are processed by stage 2. Similarly, only tweets not classified by stage 2 are processed by the last stage. Tweets not even classified by stage 3 remain not classified by this specific classifier. We used two keyword matching algorithms in stages 1 and 2: these algorithms are based on regular expressions. As for stage 3, we implemented a more advanced approach based on the SVM algorithm.

In the next sections we will describe the three stages in detail. Without loss of generality, we will assume the catalog of items is formed by movies, each one described by the movie title.

3.1 Stage 1: Exact Match

This stage deploys a simple keyword matching on the basis of the movie title, denoted as *exact match*. Given the item-i classifier, this algorithm classifies the tweet as belonging to class i if the text contains exactly the title of movie i.

3.2 Stage 2: Free match

Stage 2 of the one-class classifiers is composed of a keyword matching algorithm, denoted as *free match*. Differently from the keyword matching implemented in stage 1, stage 2 (i) searches for single words of the title within the tweet and (ii) searches for words of the title contained in other words in the tweet.

For instance, this stage is able to find a matching between the tweet

> *the future is back to the present in the past*

and the movie 'Back to the Future'. Still, this approach allows us to match the movie title also with hashtags and mentions. For instance, the tweet

> *Happy Birthday Elizabeth Reaser! #BreakingDawn*

is not recognized by stage 1, but it is associated by stage 2 to the movie 'Breaking Dawn' because it contains the hashtag '#BreakingDawn'. Similarly, the tweet

> *@abduction Just got home and it was awesome!!!!*

is matched by stage 2 because it contains the mention '@abduction', but not by stage 1. Moreover stage 2 is able to detect some cases of misspelling, as in the tweet

> *@OnlyHumour Sara had a life in the city, loopers are obviously not a secret in the club scene.*

in which the movie title 'Looper' is found in the word 'loopers'.

Stage 2 computes a matching score as the percentage of words in the item title that have been found in the tweet. The tweet is classified by stage 2 if the matching score is above a fixed threshold. With a low threshold (approaching 0%) we accept also tweets only partially matching with the terms in the movie title; contrarily, with a high threshold (approaching 100%) we accept only tweets strongly matching with the terms in movie titles.

3.3 Stage 3: SVM-based

Stage 1 and 2 search for the title of the related movie within the tweet. Thus, tweets reaching stage 3 are those not containing the title of the movie. For this reason, the classification task tackled by stage 3 is very difficult and requires a classification solution more advanced than the trivial keyword matching implemented in the previous stages.

We opted to implement in stage 3 a classification algorithm based on SVMs. In particular, we used the one-class SVMs derived by Cortes and Vapnik in [7]. In the following we will refer to this approach simply as SVM.

Differently from the previous two stages, the SVM algorithm requires also a learning phase to tune its parameters for a specific class. Learning is performed using a 5-fold cross validation [7] on a set of classified tweets. The output of the SVM algorithm is a real number, referred to as *decision value*. Only tweets with a decision value over a fixed threshold are classified. Training the SVM requires representing the tweet content: the following section describes how to represent tweet content as vectors of features.

3.3.1 Feature Extraction

According to most text processing works [21], we represented tweets as vectors of features. In such feature vector space, similar items - i.e., items that share common features - are closer than dissimilar items.

Before proceeding with feature extraction some text preprocessing must be performed because tweets are free text and, partially due to the 140-character limit, they often include slangs, abbreviations, and acronyms. Furthermore, flexed forms of verbs and conjugations of adjectives and nouns should be normalized in order to be attributed to the same feature.

We implemented common text processing tasks such as tokenization, stop word removal, expansion of contracted forms, punctuation cleaning, and term stemming [21]. To better understand the next sections, we detail here two of the techniques we have implemented: tokenization and stop word removal. Tokenization splits text into tokens, i.e., individual entities such as terms. In addition, each language has words used much more than the others (e.g., articles). Such words - denoted as stop words - should be removed since they do not add any information useful to classify items, but they rather add noise.

Once the text has been preprocessed we can build a vector of features. Each possible value of a feature corresponds to a dimension of the vector. Each dimension has a numeric weight which is related to its importance in representing the tweet content. We extracted the following four types of features:

Unigrams correspond to the single terms. Each distinct unigram is represented by a dimension in the feature space. Unigrams formed by stop words are discarded.

Bigrams are pairs of adjacent terms. Each unique bigram corresponds to a dimension in the feature space. Note that only bigrams formed by two stop words have been discarded. For instance, this allows to extract bigrams such as 'The Matrix' ('the' is a stop word) while discarding 'I will' (both 'I' and 'will' are stop words).

Hashtags represent the tags given by a user to a tweet. However, due to the lack of constraints and common criteria, users can freely use hashtags, either reusing existing tags or defining new ones. Any different hashtag is represented with a dimension in the feature space.

Title is not linked to specific terms, but it refers to the degree of matching between the tweet content and the item title. Each possible title - related to an item - is represented by a dimension in the feature space.

In our test dataset composed by 32000 tweets - described in Section 4.1 - we have approximately 5400 unigrams, 14000 bigrams, 700 hashtags, and 40 titles for a total amount of 20000 features. Note that we discarded features occurring just in a single tweet.

3.3.2 Feature Normalization

Features extracted from a tweet must be normalized [6] in order to differentiate their importance in representing the tweet content. This task can be achieved in different ways. We evaluated the use of (i) binary weighting and (ii) the TF-IDF schema.

In the case of *binary weighting*, a feature can be either 1 (if present) or 0 (if absent), without differentiating among terms with different significance.

Classic text processing [21] uses the TF-IDF weighting schema to discriminate the different features [23]. According to this model, the TF (term frequency) component takes into account the repetitions of a term inside a document, while the IDF term (inverse document frequency) considers how common a term is in the whole collection of documents.

Property	Value
Items	40
Tweets	31694
Avg tweets per item	792.4
Avg terms per tweet	15.3
Avg stop words per tweet	5.2
Tweets with URLs	46.2%
Tweets with hashtags	40.4%
Tweets with exactly 1 hashtag	30.0%
Tweets with exactly 2 hashtags	7.6%
Tweets with at least 3 hashtags	2.8%

Table 2: Dataset statistics

In our normalization schema we ignored the TF term, assuming the absence of repeated terms because of the limited length of tweets. Given a tweet t, the IDF term for its feature w is given by

$$\mathrm{idf}(w, T) = \log \frac{|\mathcal{T}|}{|\{t \in \mathcal{T} : w \in t\}|} \tag{2}$$

where $|\mathcal{T}|$ is the cardinality of the set of sample tweets and the denominator is the number of documents in which feature w occurs. After having calculated this value, groups of features of the same type (unigrams, bigrams, etc, ...) are normalized by dividing for the maximum value of the group, so that the value for each feature is in the range [0,1].

We excluded the features of type title from TF-IDF normalization.

4. EXPERIMENT SETTINGS

In this section we describe the testing methodology and the metrics used to tune and validate our solution, together with the dataset of tweets collected from Twitter.

4.1 Dataset

For the purpose of automatically testing the performance of our classifier, tweets are required to be classified. However, recent Twitter policies have limited the availability of public datasets of tweets. Thus, we collected a dataset of classified tweets of movies[2]. In order to automatically classify the tweets retrieved, we assumed that, given the official Twitter page of a movie[3] (or a similar page, such as the movie fan page), all tweets posted on such page refer to that movie. On the basis of this assumption, confirmed by a manual analysis on some Twitter pages, we identified 40 Twitter pages concerning 40 recent movies. Hence, using the Twitter RESTful Web Services we collected 800 tweets for each item, among the most recent, for a total of 32000 tweets. The dataset has been collected in September 2012 and is composed by English tweets. Table 2 reports its main statistical properties.

[2]The dataset is available for download at `http://home.dei.polimi.it/cremones/recsys/Microblog_Item_Detection.zip`

[3]E.g., the official Twitter page of the movie Harry Potter is related to the account '@HarryPotterFilm' hosted at `http://twitter.com/HarryPotterFilm`

4.2 Testing Methodology

The performance of the classification system have been measured using a standard hold-out dataset partitioning.

For each one-class classifier, we randomly selected 200 tweets to form the *test set*. The class (i.e., the item) of tweets in the test set is assumed to be unknown to the classifier and is used to verify if the predicted class corresponds to the actual class. Part of the remaining tweets has been used to form the *training set*, used by stage 3 as sample tweets for the learning phase.

Taken a generic class-c classifier, its performance has been measured on the test set computing common classification accuracy metrics: recall, precision, and f-measure [21]. *Recall* is defined as the percentage of class-c tweets that have been classified as class c. *Precision* is defined as the percentage of tweets classified as class c that are actually class-c tweets. Finally, *F-measure* is defined as the harmonic mean between precision and recall. In our study, in order to give more importance to precision with respect to recall, we have used a special implementation of F-measure, referred to as $F_{0.5}$-measure:

$$F_{0.5} = \frac{(1+0.5^2)(P \cdot R)}{0.5^2 \cdot P + R} \quad (3)$$

where precision (denoted by P) weights twice as much as recall (denoted by R).

The global performance of the battery of one-class classifiers has been computed as the macro-average of the single metrics, i.e., (arithmetically) averaging the performance indicators of each single one-class classifier. An ideal classifier should have recall, precision, and f-measure equal to 1. We graphically represented the performance indicators using precision versus recall plots.

5. STAGE 3: PARAMETER VALIDATION

In this section we focus on stage 3, the only component which requires a learning phase. Furthermore, stage 3 operates on the non-trivial tweets that the first two stages were not able to classify.

Different combinations of features can lead to different performance of the SVM classifier; in Section 5.1 we experiment several variants of feature vectors in order to find the optimal settings. Finally, in Section 5.2 we will study the impact of different factors on the classification performance.

5.1 Feature analysis

In this section we look for the optimal combination of features used to classify tweets with the SVM algorithm.

Initially, we considered one type of features at a time. Hence, we trained the SVM using such set of features and compared the classification performance both in the case of binary and TF-IDF schema normalizations. The TD-IDF normalization has been used only for unigrams (uni), bigrams (bi), and hashtags (HT).

As for the features of type 'title', we tested two weighting schemas, percentage (%) and binary weighting (bin). Percentage weighting is the matching score computed by stage 2, while binary weighting assigns 1 if the matching score is at least 75%, 0 otherwise. The choice of this threshold derives from the results of the experiments performed on stage 1 and 2. Table 3 reports the recall and precision values.

In these experiments we selected the configurations leading to the highest precision, regardless of the recall. Unigrams, bigrams, and hashtags have been proven to reach the best performance using the TF-IDF schema, which outperforms the binary weighting - in terms of precision - of about 10-13%. As for the title, we selected the binary weighting, being the one with the best precision and recall. It is worth noting that the precision obtained using only bigrams is slightly higher than the one computed by using only unigrams, while the recall is lower; in fact, a bigram is more significant than an unigram, but much less frequent.

Feature	Weight	P	R
uni	idf	0.636	0.122
	bin	0.505	0.110
bi	idf	0.671	0.037
	bin	0.576	0.050
HT	idf	**0.807**	0.176
	bin	0.704	0.166
title	%	0.228	0.086
	bin	0.501	**0.779**

Table 3: Feature comparison. Metrics are reported in correspondence of the maximum precision point. The table reports the binary (bin) and the TD-IDF normalization (idf) for the features: 'unigrams' (uni), 'bigrams' (bi), and 'hashtags' (HT). As for the feature 'title' the table reports the binary (bin) and the percentage (%) weighting schema.

Feature	P	R	$F_{0.5}$
HT	0.777	0.169	0.452
HT+T	**0.789**	0.300	0.595
HT+T+Bi	0.695	0.403	**0.607**
HT+T+Bi+Uni	0.464	**0.628**	0.490

Table 4: Comparison of the best feature combinations. Metrics are reported in correspondence of the maximum precision point.

Finally, note that all recalls are very low. For instance, the use of hashtags (with idf weighting schema) led to the highest recall (as well as precision), which, however, is only 18%. For such reason, in these further experiments we took into account also the other features, focusing on the combination leading to the highest $F_{0.5}$-measure, so to improve recall while maintaining high values of precision.

Therefore, we considered pairs of features, by combining hashtags with either unigrams, bigrams, or title. The combination hashtags+title provided the best $F_{0.5}$-measure, with a precision equals to 80% and a recall equals to 60%. Once established that the best two-feature combination is given by hashtags and title, we explored the addition of a third feature, either unigrams or bigrams. The experiments showed that the best three-feature combination is hashtags+title+bigrams, with an $F_{0.5}$-measure equals to 0.607, thanks to an increment of precision (88%) that compensates a significant decrease of recall (down to 12%).

We finally compared the best single feature, the best two-feature combination, and the best three-feature combination with the use of all four kinds of features. The results are reported in Table 4. The best combination of features - in terms of $F_{0.5}$-measure - is given by hashtags+title+bigrams.

# tweets per item	P	R
100	**0.730**	0.216
200	0.741	0.314
400	0.695	0.403
600	0.690	**0.475**

Table 5: Sensitivity analysis of the SVM algorithm (stage 3): impact of the number of tweets per item. Metrics are reported in correspondence of the maximum precision point.

# items	P	R
5	**0.956**	**0.541**
10	0.769	0.488
20	0.733	0.432
40	0.695	0.403

Table 6: Sensitivity analysis of the SVM algorithm (stage 3): impact of the number of items. Metrics are reported in correspondence of the maximum precision point.

Therefore, in the following we will use this combination of features for stage 3.

5.2 Sensitivity analysis

Several factors can impact the performance of the SVM algorithm. In the next sections we answer to the following three questions: (i) "how many tweets are required to obtain a good classification quality?", (ii) "does performance degrade by adding more items?", and (iii) "how are out-of-scope tweets classified?".

Number of tweets per item.

In all previous experiments we used a fixed training set. However, the *number of tweets per item* can affect the SVM algorithm performance. Indeed, the larger the sample of classified tweets, the higher the confidence of the classifier.

We fixed the test set as composed by 200 tweets per item and we used part of the remaining tweets to train stage 3. We varied the number of training tweets per item in the range $[100, 600]$. The results are reported in Table 5 and show that performance increases with the number of tweets up to 400. Increasing the number of tweets per item from 400 to 600 only slightly affects the performance.

Number of items.

This set of experiments evaluates how the number of items in the catalog can affect the quality of stage 3.

In fact, the higher the number of items the harder the classification task. We tested the SVM algorithm by varying the catalog size in the range $[5, 40]$. Table 6 confirms that the performance decreases with the number of available items. However, we can observe that the decrease between 20 and 40 items is minimal, proving a good robustness of the classifier.

Out-of-scope tweets.

In the rest of the paper we have tested the classifiers in an uncontaminated environment, where we predicted the classes only of tweets associated to the items in the catalog. However, if we take into consideration the tweets posted by a typical Twitter user, only a small part of his/her comments is likely to refer to one of the movies in the catalog.

# out-of-scope tweets	P	R
0	**0.695**	**0.403**
2000	0.674	0.400
4000	0.652	0.395
8000	0.638	0.398

Table 7: Sensitivity analysis of the SVM algorithm (stage 3): impact of out-of-scope tweets. Metrics are reported in correspondence of the maximum precision point.

Thus, in this section we verify how out-of-scope tweets (i.e., tweets not referring to any known item) are classified. We collected a set of further 8000 tweets retrieved from 10 different general-purpose, popular Twitter pages (e.g., NHL, IBM, etc.). These tweets have been added to the test set, which now contains 16000 tweets, 8000 regarding the 40 known items and 8000 regarding *out-of-scope* subjects.

Basically, the experiments showed that the presence of out-of-scope tweets does not impact the performance of the SVM algorithm, whose recall, precision, and fallout resulted almost identical to the previous tests, as reported in Table 7.

6. RESULTS

In this section we present and discuss the main results of the proposed item detection system. For this set of experiments we trained stage 3 with a training set composed by about 400 tweets per item using the three types of features derived in Section 5.1: hashtags, title, and bigrams. Figure 3 reports the precision vs recall comparing the quality of the stage, the quality of stage 1 and stage 2, with the quality of the overall pipeline composed by the three stages.

Stage 1 (solid diamond) obtains a high precision (96%), against a recall of only 11%. This result confirms that tweets containing the exact title of a movie effectively refer to that movie. On the other hand, the low recall suggests that only a few tweets contain the exact movie title. In fact, the title might not have been used at all, or it might have been contracted, misspelled, or mangled.

The dashed line shows the performance of stages 1+2. The line has been obtained by varying the acceptance threshold of stage 2. The optimal value (circle point) is in correspondence of a threshold equals to 75%: recall is 58% and precision 96%. Stage 2 allows to increase recall while maintaining the same precision. The power of stage 2 is given by its capability of searching for the title also within other terms (e.g., hashtags, mentions, abbreviations)

The solid line in Figure 3 shows the performance of the whole pipeline composed by the three stages. Stage 2 threshold has been set to 75%, while the threshold on the decision value of stage 3 has been varied. We note that stage 3 does not significantly affect precision, but it improves recall.

6.1 Aggregation policies

The proposed item detection system allows to assign a tweet to multiple items. This happens when multiple one-class classifiers classify the tweet.

However, the dataset we have collected contains only tweets related to one single item (see Section 4.1). In order to check the correctness of the results obtained in the previous sec-

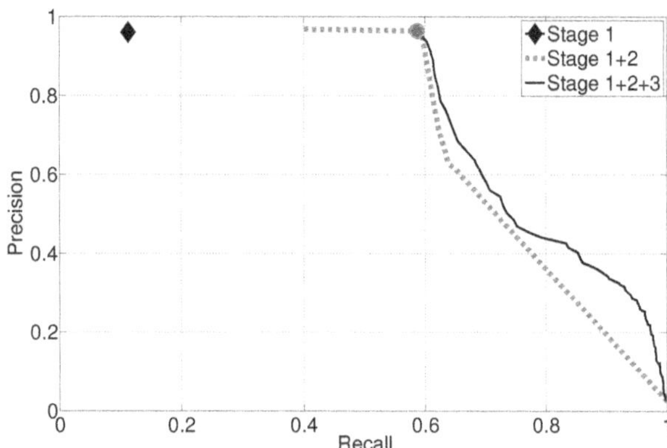

Figure 3: System performance of the battery of one-class classifier in term of precision vs recall

Aggregation policy	P	R	$F_{0.5}$
No aggregation	**0.950**	0.595	**0.849**
Max precision	0.922	0.598	**0.832**
Max decision value	0.917	**0.654**	**0.849**

Table 8: Comparison between the two aggregation policies, with respect to the baseline approach where no aggregation is applied. Metrics are reported in correspondence of the maximum precision point.

tions, we run a set of experiments where we forced the system to return one class at most. In the following, we verify that the presented performance does not depend on the fact that the item detection system allows a multiple-item assignment, while the tweets we classified referred to a single item. We tested two possible aggregation strategies: (i) max decision value and (ii) max precision policies.

Given a tweet and a subset of items it has been classified into, the *max decision value policy* assigns the tweet to the class with the highest decision value resulting from stage 3.

Alternatively, the *max precision policy* uses a subset of the input sample tweets to estimate the precision of each single one-class classifier. Therefore, in the case of ambiguous assignment, it prefers the classifier with the higher estimated precision. Thus, for each one-class classifier, we used the 400 tweets of the training set to train the SVM algorithm, and an additional set of 200 tweets (whose class is known) - referred to as *validation set* - to optimize the SVM parameters. Parameters are set so to maximize the precision computed on the validation set. Afterwards, we classify the tweets in the test set (composed, again, by 200 tweets whose class is assumed to be unknown) using the parameters computed on the validation set. In the case multiple classifiers classify a tweet, the tweet is assigned to the class corresponding to the classifier with the highest precision on the validation set.

Table 8 reports the outcome of these experiments and confirms that the performance reported in the previous tests were not affected by the multiple-assignment hypothesis. In fact, neither maximum precision nor maximum decision value policy bring a significant change - in term of $F_{0.5}$-measure - with respect to not applying any aggregation strategy: in particular, both aggregation strategies bring to a decrease of precision, compensated by a slight increase of recall.

7. CONCLUSIONS

In this work we defined and evaluated a three-stage item detection system. The system is based on a battery of one-class classifiers for the matching between tweets and TV programs.

Empirical tests on 32000 tweets have proven the ability of the system to accurately classify tweets. The scalability of the system allows for an extensive usage in VoD and Linear TV platforms, both as mechanism for preference elicitation and recommendation and as audience meter.

Interesting for future work would be the inclusion of temporal information (e.g., matching the time message was posted with the TV program scheduling in Electronic Programming Guide) and the analysis of tweets as conversations (e.g., a sequence of tweets with questions/answers). This additional information could be used, for example, to disambiguate two movies with the same title but released in two different years. Furthermore, item detection associated with opinion mining (e.g., [20]) will allow to infer the polarity (e.g., positive or negative) of user preferences.

8. REFERENCES

[1] D. Andrzejewski and X. Zhu. Latent dirichlet allocation with topic-in-set knowledge. In *Proceedings of the NAACL HLT 2009 Workshop on Semi-Supervised Learning for Natural Language Processing*, SemiSupLearn '09, pages 43–48, Stroudsburg, PA, USA, 2009. Association for Computational Linguistics.

[2] D. Andrzejewski, X. Zhu, and M. Craven. Incorporating domain knowledge into topic modeling via dirichlet forest priors. In *Proceedings of the 26th Annual International Conference on Machine Learning*, ICML '09, pages 25–32, New York, NY, USA, 2009. ACM.

[3] D. M. Blei, A. Y. Ng, and M. I. Jordan. Latent dirichlet allocation. *J. Mach. Learn. Res.*, 3:993–1022, Mar. 2003.

[4] M. Cataldi, L. Di Caro, and C. Schifanella. Emerging topic detection on twitter based on temporal and social terms evaluation. In *Proceedings of the Tenth International Workshop on Multimedia Data Mining*, MDMKDD '10, pages 4:1–4:10, New York, NY, USA, 2010. ACM.

[5] E. Charniak, C. Hendrickson, N. Jacobson, and M. Perkowitz. Equations for part-of-speech tagging. In *In Proceedings of the Eleventh National Conference on Artificial Intelligence*, pages 784–789, 1993.

[6] J. Chen, R. Nairn, L. Nelson, M. Bernstein, and E. Chi. Short and tweet: experiments on recommending content from information streams. In *Proceedings of the SIGCHI Conference on Human Factors in Computing Systems*, CHI '10, pages 1185–1194, New York, NY, USA, 2010. ACM.

[7] C. Cortes and V. Vapnik. Support-vector networks. *Machine Learning*, 20(3):273–297, Sept. 1995.

[8] C. Courtois and E. D'heer. Second screen applications and tablet users: constellation, awareness, experience, and interest. In *Proceedings of the 10th European conference on Interactive tv and video*, EuroiTV '12, pages 153–156, New York, NY, USA, 2012. ACM.

[9] P. Cremonesi, F. Garzottto, and R. Turrin. User effort vs. accuracy in rating-based elicitation. In *Proceedings of the sixth ACM conference on Recommender systems*, RecSys '12, pages 27–34, New York, NY, USA, 2012. ACM.

[10] E. D'heer, C. Courtois, and S. Paulussen. Everyday life in (front of) the screen: the consumption of multiple screen technologies in the living room context. In *Proceedings of the 10th European conference on Interactive tv and video*, EuroiTV '12, pages 195–198, New York, NY, USA, 2012. ACM.

[11] M. Doughty, D. Rowland, and S. Lawson. Who is on your sofa?: Tv audience communities and second screening social networks. In *Proceedings of the 10th European conference on Interactive tv and video*, EuroiTV '12, pages 79–86, New York, NY, USA, 2012. ACM.

[12] K. Gimpel, N. Schneider, B. O'Connor, D. Das, D. Mills, J. Eisenstein, M. Heilman, D. Yogatama, J. Flanigan, and N. A. Smith. Part-of-speech tagging for twitter: annotation, features, and experiments. In *Proceedings of the 49th Annual Meeting of the Association for Computational Linguistics: Human Language Technologies: short papers - Volume 2*, HLT '11, pages 42–47, Stroudsburg, PA, USA, 2011. Association for Computational Linguistics.

[13] T. L. Griffiths, M. Steyvers, D. M. Blei, and J. B. Tenenbaum. Integrating topics and syntax. In *In Advances in Neural Information Processing Systems 17*, pages 537–544. MIT Press, 2005.

[14] B. Han and T. Baldwin. Lexical normalisation of short text messages: makn sens a #twitter. In *Proceedings of the 49th Annual Meeting of the Association for Computational Linguistics: Human Language Technologies - Volume 1*, HLT '11, pages 368–378, Stroudsburg, PA, USA, 2011. Association for Computational Linguistics.

[15] T. Joachims. Text categorization with suport vector machines: Learning with many relevant features. In *Proceedings of the 10th European Conference on Machine Learning*, ECML '98, pages 137–142, London, UK, UK, 1998. Springer-Verlag.

[16] M. Lochrie and P. Coulton. Mobile phones as second screen for tv, enabling inter-audience interaction. In *Proceedings of the 8th International Conference on Advances in Computer Entertainment Technology*, ACE '11, pages 73:1–73:2, New York, NY, USA, 2011. ACM.

[17] M. Lochrie and P. Coulton. Sharing the viewing experience through second screens. In *Proceedings of the 10th European conference on Interactive tv and video*, EuroiTV '12, pages 199–202, New York, NY, USA, 2012. ACM.

[18] J. Makkonen, H. Ahonen-Myka, and M. Salmenkivi. Simple semantics in topic detection and tracking. *Inf. Retr.*, 7(3-4):347–368, Sept. 2004.

[19] Nielsen. Nielsen and twitter establish social tv rating. http://www.nielsen.com/us/en/insights/press-room/2012/nielsen-and-twitter-establish-social-tv-rating.html, 2012.

[20] B. Pang and L. Lee. Opinion mining and sentiment analysis. *Found. Trends Inf. Retr.*, 2(1-2):1–135, Jan. 2008.

[21] M. F. Porter. Readings in information retrieval, 1997.

[22] D. Ramage, S. Dumais, and D. Liebling. Characterizing microblogs with topic models. In *Proceedings of the Fourth International AAAI Conference on Weblogs and Social Media*. AAAI, 2010.

[23] G. Salton and C. Buckley. Term-weighting approaches in automatic text retrieval. *Inf. Process. Manage.*, 24(5):513–523, Aug. 1988.

[24] D. Tax. *One-class classification - Concept-learning in the absence of counter-examples*. PhD thesis, TU Delft, 2001.

[25] S. Wakamiya, R. Lee, and K. Sumiya. Towards better tv viewing rates: exploiting crowd's media life logs over twitter for tv rating. In *Proceedings of the 5th International Conference on Ubiquitous Information Management and Communication*, ICUIMC '11, pages 39:1–39:10, New York, NY, USA, 2011. ACM.

[26] M.-C. Yang, J.-T. Lee, and H.-C. Rim. Using link analysis to discover interesting messages spread across twitter. In *Workshop Proceedings of TextGraphs-7: Graph-based Methods for Natural Language Processing*, pages 15–19, Jeju, Republic of Korea, July 2012. Association for Computational Linguistics.

[27] J. T. yau Kwok. Automated text categorization using support vector machine. In *In Proceedings of the International Conference on Neural Information Processing (ICONIP*, pages 347–351, 1998.

WeSlide: Gestural Text Entry for Elderly Users of Interactive Television

Nathan Godard
Université de Lorraine, LCOMS
Ile du Saulcy, 57045 Metz
France
nathan.godard@univ-lorraine.fr

Isabelle Pecci
Université de Lorraine, LCOMS
Ile du Saulcy, 57045 Metz
France
+33 387315442
isabelle.pecci@univ-lorraine.fr

Poika Isokoski
School of Information Sciences
FIN-33014, University of Tampere
Finland
+358-40-706 3377
poika@cs.uta.fi

ABSTRACT
Interactive television provides useful services for older people. These include social networking tools, video on demand, and broadcast TV. Many of the Internet-mediated services require text entry. The usual multi-tap text entry supplied with TV remote control is not suitable to many older people. In this paper, we evaluate WeSlide, a gestural text entry technique that uses the Wiimote as the input device. We conducted a study to compare WeSlide with the multi-tap technique. WeSlide was faster and less error prone and users strongly preferred it over multi-tap.

Categories and Subject Descriptors
H.5.2 [**Information interfaces and Presentation**]: User Interfaces – *Graphical user interfaces, Input devices and strategies and Interaction styles.*

General Terms
Design, Experimentation, Human Factors.

Keywords
Text entry. Gestural interaction. Interactive television. Wiimote. Elderly users.

1. INTRODUCTION
Many older people want to live at home as long as possible, but they may need assistance to do so [8]. Interactive television (iTV) is a way to supply them with many services: video on demand, voting, social networking, bank services etc. Unfortunately, it is difficult for older people to access these services [2]. Some of the interaction techniques are not easy to use. Our research is focused on text entry in this context. The usual text entry techniques with iTV are virtual keyboards [18] and the multi-tap technique [3]. The TV remote control is used as the input device for both techniques. Its directional buttons make it possible to highlight a key on the virtual keyboard and another button is pressed to activate it. This is a very simple way to enter text, but the selection of the buttons can be slow because the long distance between them leads to many key presses per entered character. The multi-tap keyboard reduces this problem (see Figure 1). In multi-tap a button matches multiple characters. The number of consecutive presses on the key determines the character to enter: 3-4 letters are shown on each key. Often, more characters are accessible on each key depending on language setting. For example in French the key "2abc"is used to access characters "a b c 2 à â ç". Users of mobile phones are familiar with this technique and some of them even like it [10]. However, not all elderly users are familiar with text messaging on mobile phones. Therefore, the multi-tap means a learning process for them. Working memory tends to decline with age and it can be difficult to remember the set of characters linked to a key, the present state of the system, and even the number of key presses already executed.

Marshall et al. [10] claimed that the best solution for iTV text entry was a physical keyboard, but Lee et al. [9] argued that the desktop metaphor must be avoided. In the iTV context the user may be interacting with the device in different postures, a desktop keyboard is difficult to handle in many postures. Consequently small keyboards embedded in the TV remote control [1] have been proposed. However, the user needs to look both at the device and at the TV screen. This divided attention has an impact on the quality of the interaction [6] and it is especially difficult for elderly users with reduced cognitive capacity [14]. Moreover, Epelde et al. showed that a TV remote control is unsuitable for older users: buttons and the text on them are too small and there are too many buttons [4].

Figure 1. The TV Remote control (left) and the Wiimote (right).

Several researches have argued that older people can easily use the Wiimote [8, 13]. They easily understand the pointing, but some buttons have been found to be too small and the "B" button behind the remote is sometimes activated involuntarily [13]. However, even if pointing is easy to understand, reduced physical abilities may limit accuracy [2, 13].

2. WESLIDE
The design of the WeSlide technique has been described in a demonstration abstract in a French conference [16]. For convenience we give a brief description here in English.

2.1 The main interaction technique

We chose the Wiimote as the input device based on [8, 13]. By reducing the use of buttons and requiring only approximate pointing, we envisioned that the Wiimote might be a good device for text entry in the iTV context for older people. It satisfies many requirements of the iTV context: it is a small one-handed device [6], usually used for fun, easy to use (not many buttons), usable in many postures, reducing efforts of visual accommodation required when switching between remote device and screen [14], and convenient for use in a dark room.

A main goal in the design was character selection without accurate pointing because older user may have reduced physical abilities [14]. Two existing text entry techniques MessagEase [12] and Claviature [11] inspired us. They utilized an interaction technique consisting of pointing and sliding to select a character. They associate several characters with each key. The user first selects a key by landing the finger/stylus on the key that contains the desired character. Then, to select the character among those on the key the finger/stylus is dragged to a direction indicated by the position of the characters on the key before lifting it.

MessagEase and Claviature were designed for pen-based computers and touch-screens. For Wiimote use the interaction needed to be adapted. Because the hand of an older user may be shaking, the eight directions used in MessagEase demand too much precision. The final selection is done by crossing one of the sides of the key. We call this design "WeSlide" to highlight sliding, the main action to select a character (it is also a translation of WeGliss, the name used in [16]).

2.2 The keyboard

The keyboard used in our experiment is shown in Figure 2. Each key had four characters. To select a character, for example 'G' in Figure 2, the user pointed at the key containing the character and pressed "A" button on the Wiimote. Then moved the pointer crossing the side closest to 'G', and finally released the button to enter the character.

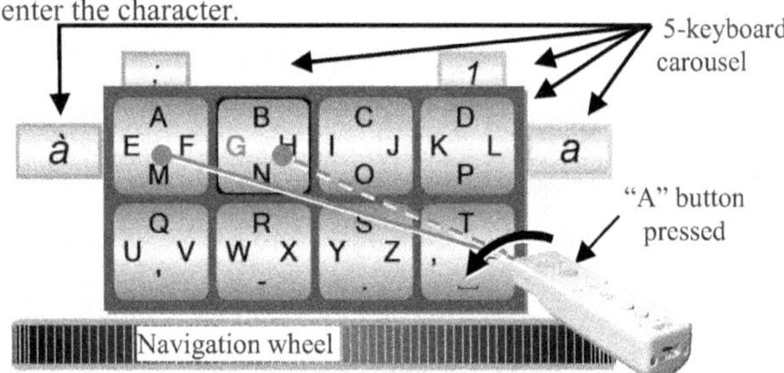

Figure 2. WeSlide: selection of 'G'

Instead of the QWERTY layout an alphabetic layout was chosen for WeSlide. This is against the recommendation of Norman et al. [15], but it is due to the design: there are positions for 4 letters on the first line, 8 on the second, 4 on the third, etc. This does not match the QWERTY layout. In addition to the alphabet, many more characters are needed for a fully functional text entry method. We designed alternative character sets in different keyboards arranged in a "carousel" system as shown in Figure 2. We believed that introducing explicit interaction with visual feedback to change the key layout would be less confusing than introducing new technique to select special characters on the keyboard. The user can navigate through the carousel with a virtual "scroll wheel" (see bottom of Figure 2). The wheel is used by pointing it, pressing "A" on the Wiimote, and moving the Wiimote to the left or to the right. WeSlide had five keyboards: lower-case letters, upper-case letters, numbers, accented letters, and punctuation symbols. All keyboards included the most common punctuation symbols and the return characters. Each keyboard was associated with a different background color to help user in differentiating them.

Feedback on the state of the system was implemented as follows: when the user selected a key, the outline of the key was highlighted in black. When the pointer left the key while pressing the "A" button, the color of the character that would be selected after releasing "A" button was shown in red (see Figure 2).

3. METHOD

We conducted a study to: (1) gather subjective feedback of the WeSlide keyboard versus the standard multi-tap text entry of the TV remote control (see Figure 1) and (2) to assess the performance of the WeSlide technique.

We use multi-tap as a baseline because it is often mentioned in iTV text entry literature. The experiment of Iatrino and Modeo [5] compared multi-tap, multi-tap with visual feedback and a virtual keyboard controlled by the arrows keys of an interactive keypad. Thirty-six subjects participated: eighteen were inexperienced in text entry with multi tap and aged 43-80 and eighteen had experience in text messaging (aged 18-50). In this study, the best interface was the multi-tap with non visual feedback. Moreover, we did not compare our method to a remote control with mini-qwerty keyboard because the user needs to divide visual attention between the remote and the TV screen more than with multi-tap and small buttons are unsuitable for older users [4]. Finally, we did not compare to a static alphabet matrix pointed to with the Wiimote because in our pilot testing comparing this keyboard (with young users) WeSlide was better [16]. Moreover, direct pointing to an ordered matrix requires quite high accuracy. Older user could have had difficulties in this pointing task.

3.1 Participants

Six participants, 4 females and 2 males, took part in the user study. Their ages ranged from 63 to 83 with a mean of 71 years. All the participants were right-handed. One participant suffered from a chronic wrist pain and another from numbness in fingers in intensive tasks. None of the participants had prior experience with virtual keyboards or the Wiimote. They were all users of television and its remote control, but not to enter text.

3.2 Procedure

The experiment was divided in 10 sessions with a break of about one day between the sessions. Before the first session, participants had a practice trial to get used to the WeSlide keyboard. They were instructed on how to point a key, how to select a letter by sliding the pointer, and on how to use the navigation wheel to change keyboard. For the multi-tap text entry method they tried out the selection of characters on the TV remote control. Each participant was given 10 minutes for initial practice but none of them spent all this time.

Each session lasted for 20 minutes: 10 minutes of text entry with WeSlide and 10 minutes with multi-tap. The order of the text entry method was inverted between sessions. The task was text transcription. The phrases to enter were presented on screen and the task was to enter the same phrase in a text box on screen. The sentences were presented in random order and came from iTV context (title or information of TV programs).

To complete a sentence the participant needed to enter the return character. If the length of the entered phrase was the same as length of the presented phrase (± one character), the task was reset with a new presented phrase. Participants were instructed to enter the phrases as quickly and as accurately as possible. Error correction was available only by erasing the text back to the point of error. For WeSlide, the cross button of the Wiimote was used to delete the last character and for the multi-tap keyboard, it was the "clear" button on the TV remote control (see Figure 1).

After the last session, all participants answered a survey using a Likert-type scale. The questions were about: (1) the device i.e. the Wiimote or TV remote control (tiredness, ease of use); (2) the design of WeSlide (colors, size and layout of characters, carousel); (3) the selection of a character with WeSlide or multi-tap (feedback, ease of use) and (4) the comparison between the two keyboards (feeling of speed, error rate, overall preference).

3.3 Apparatus

The participants were in a quiet room simulating the iTV context. They were sitting down on a comfortable chair positioned 2.86 meters away from the display surface. The display was projected by a video projector and adjusted to a size of 30 inches in diagonal. The resolution of the image was 1280x720 pixels. The bottom of the display showed the WeSlide Input area shown in Figure 2. The top of the display showed the presented and transcribed phrases.

4. RESULTS AND DISCUSSION

4.1 Text Entry Rate

We measured text entry rate in words per minute (WPM). The results are shown in Figure 3. We measured approximately the same entry rate for multi-tap that Iatrino and Modeo [5] reported for their eighteen older novices (2 and 3 WPM).

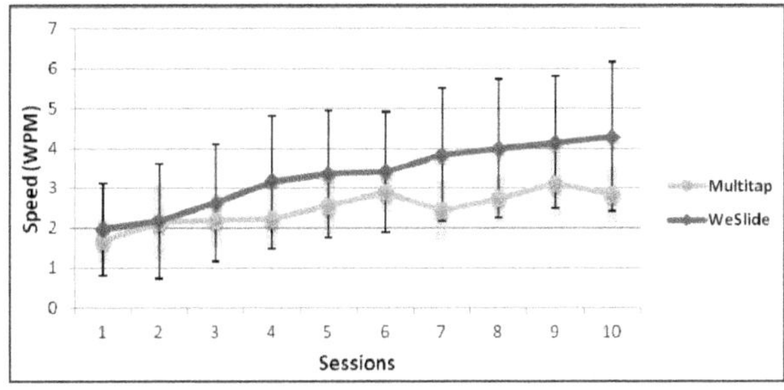

Figure 3. Speed (WPM) of the two keyboards per session

Our ANOVA results were evaluated under Greenhouse-Geisser corrected degrees of freedom. For simplicity, however, we report the results using conventional unadjusted degrees of freedom.

As Figure 3 suggests, the WeSlide technique was faster. An repeated measures ANOVA shows a statistically significant effect of the text entry method ($F_{1,5} = 7.2$, $p = 0.04$). The effect of the session was also statistically significant ($F_{9,45} = 24.7$, $p < 0.0001$) as was the interaction of these factors ($F_{9,45} = 4.1$, $p = 0.001$). Based on Figure 3 we can interpret the interaction as an increasing difference between the methods towards the end of the experiment. During the experiment we observed two issues that slowed down the participants initially with WeSlide. The first issue was the tendency to point at the character instead of the key containing it. The pointing was slow because it required more accuracy than necessary. In addition, some participants were slow to learn the carousel.

4.2 Error Rate

The Total Error Rate (TER) metric defined by Soukoreff et al. [17] is shown in Figure 4. TER is the sum of the Corrected Error Rate (CER) and the Uncorrected Error Rate (UER). The effect of the text entry technique was statistically significant ($F_{1,5} = 9.9$, $p = 0.026$) as well as the effect of session ($F_{9,45} = 4.1$, $p = 0.001$). Based on Figure 4 we can interpret these as lower error rate for WeSlide and an overall reduction of the error rate for both text entry techniques. The interaction of the text entry technique and session factors was not statistically significant.

At the end of the 10 sessions, WeSlide averaged 52% faster with mean text entry rate of 4.3 wpm. The text entry rate with the multi-tap method in the 10th session was 2.8 wpm.

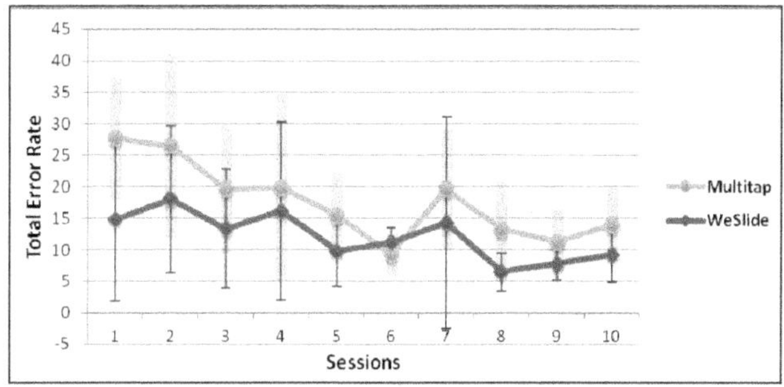

Figure 4: Total Error Rate of the keyboards per session

Figure 5 shows the mean error rates recorded in the 10th session for the two keyboards. The difference in UER is small (3.6% vs. 4.7%) but the difference in CER larger (5.6% vs. 9.2%). In other words, the difference in TER is mainly because of CER.

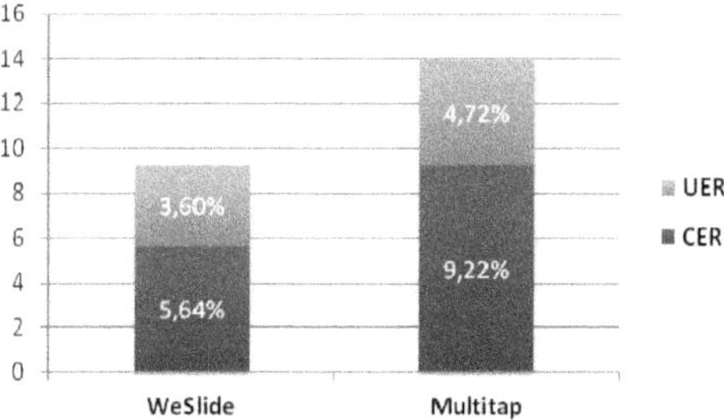

Figure 5: Error rates for the two keyboards in session 10.

Overall, the interpretation of these results is in favor of WeSlide. It was faster and less error prone. A striking observation is also the low overall text entry rate in this experiment. A cross-study comparison to James' and Reischer's study [7], we see that the elderly participants in our experiment reached about one third of the text entry rate of young adults. The young adults in the study by James and Reischer used a mobile phone where the distance to switch attention from the display to the keypad is shorter, but still it is probably fair to say that our participants were very slow.

4.3 Survey results

4.3.1 WeSlide

None of the participants, including the two with physical disabilities, found the use of the Wiimote tiring. The interaction with it was rated as easy. Most participants were confused by the use of the cross button to delete characters. A backspace button on-screen would have been preferred. Only one participant noticed that the character layout was alphabetical. Only a few participants noticed the color coding and the carousel arrangement. Thus, we cannot conclude on the usefulness of these features. Most participants just rotated the carousel until the desired keyboard appeared without understanding the order or structure in it. All participants found the virtual wheel easy to use.

4.3.2 Multi-tap text entry

Only one participant reported that the characters were easy to find. Four participants found it difficult to understand the multi-tap concept. Moreover all participants except one were confused by the need to divide visual attention between the TV remote and the display during text entry.

4.3.3 WeSlide versus Multi-tap text entry

Five participants had a strong preference for WeSlide and one for the multi-tap text entry (although he was faster with WeSlide). This one participant used the multi-tap system on his mobile phone. Four participants believed that they made fewer errors with WeSlide. The others had no opinion.

Our overall interpretation is that there was a preference for WeSlide, but the implementation could have been better.

5. CONCLUSION

We have evaluated WeSlide, a virtual keyboard for text entry, in the iTV context for older people. A preliminary study with 6 older users showed that WeSlide may have better performance than the multi-tap text entry that is often used on a TV remote control. The majority of the participants preferred WeSlide.

6. REFERENCES

[1] Berglund, A. and Johansson, P. Using speech and dialogue for interactive television navigation. *Universal Access in the Information Society*, 3(3) (2004), 224-238.

[2] Coelho, J., Duarte, C., Biswas, P. and Langdon, P. Developing accessible TV applications. In *Proc. ASSETS '11*, ACM Press (2011), 131-138.

[3] Cooper, W. The interactive television user experience so far. In *Proc. of the 1st international conference on Designing interactive user experiences for TV and video*, ACM Press (2008), 133-142.

[4] Epelde, G., Valencia, X., Carrasco, E., Posada, J., Abascal, J., Diaz-Orueta, U., Zinnikus, I., Husodo-Schulz, C. Providing universally accessible interactive services through TV sets: implementation and validation with elderly users. *Multimedia Tools and Applications*, Springer (2011), 1-32.

[5] Iatrino, A. and Modeo, S. Text Editing in Digital Terrestrial Television: A Comparison of Three Interfaces. *In Interactive Digital Television: Technologies and Applications*, ed. George Lekakos, Konstantinos Chorianopoulos and Georgios Doukidis, 224-241 (2007).

[6] Ingmarsson, M., Dinka, D., and Zhai, S. TNT: a numeric keypad based text input method. In *Proc. CHI 2004*, ACM Press (2004), 639-646.

[7] James, C. L. and Reischel, K. M. Text input for mobile devices: comparing model prediction to actual performance. In *Proc.CHI '01*, ACM Press (2001), 365-371.

[8] Lawrence, E., Sax, C., Navarro, K. F. and Qiao, M. Interactive Games to Improve Quality of Life for the Elderly: Towards Integration into a WSN Monitoring System. In *Proc. ETELEMED '10*. IEEE Computer Society (2010), 106-112.

[9] Lee E. And Schmidt, M. Revolutionizing TV with User-Centered Design and Research. *User Experience*, 2(4) (2005), 10-13.

[10] Marshall, D., Foster, J. C. And Jack M. A. User performance and attitude towards schemes for alphanumeric data entry using restricted input devices. *Behaviour & Information technology*, vol.20, no.3 (2001), 167-188.

[11] Microth Claviature website. http://www.microth.com/claviature/

[12] Nesbat, S. B. A system for fast, full-text entry for small electronic devices. In *Proc. ICMI '03*. ACM press (2003), 4-11.

[13] Neufeldt, C. Wii play with elderly people. International reports on socio-informatics, Enhancing Interaction Spaces by Social Media for the Elderly : *a Workshop Report*, 6(3) (2009), 50-59.

[14] Nichols, T. A., Rogers, W. A., Fisk, A. D. Design for Ageing. *In: Salvendy, G. (Edt.). Handbook of human factors and ergonomics.* Hoboken, N.J: John Wiley (2006), 1418-1458

[15] Norman D.A. and Fisher A. Why Alphabetic Keyboards Are Not Easy to Use: Keyboard Layout Doesn't Much Matter. *Human Factors : The Journal of the Human Factors and Ergonomics Society*, 24(5) (1982), 509-519.

[16] Oberst, A., Laas, G.and Pecci, I. WeGliss, clavier pour la télévision interactive. In Proc. IHM 2010, ACM Press (2010), 221-224.

[17] Soukoreff, R. W., & MacKenzie, I. S. Metrics for text entry research: An evaluation of MSD and KSPC, and a new unified error metric. In *Proc. CHI 2003,* ACM Press (2003), 113-120.

[18] Teixeira, C. A., Melo, E. L., Cattelan, R. G., Pimentel, M. D. User-media interaction with interactive TV. In *the Proc. of the 2009 ACM Symposium on Applied Computing*, ACM Press (2009), 1829-1833.

SoundsLike: Movies Soundtrack Browsing and Labeling Based on Relevance Feedback and Gamification

Jorge M. A. Gomes
LaSIGE, Faculty of Sciences
University of Lisbon
1749-016 Lisboa, Portugal
+351217500087
jgomes@lasige.di.fc.ul.pt

Teresa Chambel
LaSIGE, Faculty of Sciences
University of Lisbon
1749-016 Lisboa, Portugal
+351217500087
tc@di.fc.ul.pt

Thibault Langlois
LaSIGE, Faculty of Sciences
University of Lisbon
1749-016 Lisboa, Portugal
+351217500087
tl@di.fc.ul.pt

ABSTRACT

Movies and games are amongst the biggest sources of entertainment, in individual and social contexts. Increasingly, movies and videos are becoming accessible as enormous collections over the Internet, in social media and interactive TV, demanding for new and more powerful ways to search, browse and view them, that benefit from video content-based analysis and classification techniques. Game elements, in turn, can help in this often challenging process, e.g. in the audio, to obtain user feedback to improve the efficacy of classification, while maintaining or improving the entertaining quality of the user experience. In this paper, we present and discuss SoundsLike, a gamification approach to engage users in movies soundtrack labeling, based on relevance feedback and integrated in MovieClouds, an interactive web application designed to access, explore and visualize movies based on the information conveyed in the different tracks or perspectives of its content, especially audio and subtitles where most of the semantics is conveyed, and with a special focus on the emotional dimensions expressed in the movies or felt by the viewers.

Categories and Subject Descriptors

H.3.3 [**Information Storage and Retrieval**]: Information Search and Retrieval – *retrieval models;* H.5.1 [**Information Interfaces and Presentation (I.7)**]: Multimedia Information Systems – *video*; H.5.2 [**Information Interfaces and Presentation (I.7)**]: User Interfaces – *GUI, interaction styles, screen design, user centered design;* H5.5 [**Information interfaces and Presentation (I.7)**]: Sound and Music Computing – *signal analysis, synthesis, and processing;* K8.0 [**Personal Computing**]: *General – games.*

Keywords

Gamification, Relevance Feedback, Audio, Music, Video, Tagging, Bootstrapping, Movies, Entertainment, Engagement.

1. INTRODUCTION

Nowadays, movies, video and audio have a strong presence in human life, being a massive source of entertainment. The evolution of technology has enabled the fast expansion of media and social networks over the internet, giving rise to huge and increasing collections of videos and movies accessible over the internet and through video on demand services on iTV. These multimedia collections are tremendously vast and demand for new and more powerful search mechanisms that may benefit from video and audio content based analysis and classification techniques. Some researchers [2,6] pointed the importance for the development of methods to extract interesting and meaningful features in video to effectively summarize and index them at the level of subtitles, audio and video image. Once this information is collected, we can try to use it for a better organization and access of the individual and collective video spaces.

EuroITV'13, June 24–26, 2013, Como, Italy.

The approach described in this paper was designed for the VIRUS - Video information Retrieval Using Subtitles - project [4] to help towards the resolution of a cold-start problem in data acqui-sition for content classification. This project aims to provide users the access to a database of movies and TV series, through a rich graphical interface. MovieClouds [4], an interactive web application was developed, adopting a tagcloud paradigm for search, overview, and exploratory browsing of movies in different tracks or perspectives of their content (subtitles, emotions in subtitles, audio events, audio mood, and felt emotions). The backend is based on the analysis of: video image, and especially audio and subtitles, where most of the semantics is expressed. In the present paper, we focus on the analysis of the audio track for the inherent challenge and potential benefit if addressed from a game perspective to involve the users.

In this context, our objective is to provide an overview or summary of the audio and access the video moments that contain audio events (e.g. gun shots, telephone ringing, animal noises, shouting, etc.) and moods to users. To this end, we build statistical models for such events that rely on labeled data. Unfortunately these kinds of databases are rare. The building of our own dataset is a huge task requiring many hours of listening and manual classification - often coined the "Cold Start" problem. If we have models that perform relatively well, we could use the models to collect data similar to the audio events we want to detect, and human operators would be asked to label (or simply verify the classification assigned automatically from) a reduced amount of data. This idea of bringing the human into the processing loop is similar to the interaction used in Relevance Feedback (RF). This is a technique used in Information Retrieval (IR) that takes into account users' feedback to a given query while computing answers of subsequent queries, e.g. [13]. Here, we use the RF approach in a different way, since the system provides the queries and the users provide the answers. Our proposal consists in adapting this kind of approach with a gamification perspective, to help solve the cold start we have to deal with, integrated in the context of movie watching.

Gamification can be defined as the use of game design elements in non-game contexts [3]. Game elements can be designed to augment and complement the entertaining qualities of movies, motivating and supporting users to contribute to the content classification, combining utility and usability aspects [3,7]. Main properties to aim for include: Persuasion and Motivation, to induce and facilitate mass-collaboration, or crowd-sourcing, in the audio labeling; Engagement, possibly leading to increased time on the task; Joy, Fun and improved user experience; Reward and Reputation inspired in incentive design. Here attention must be paid to the cultural contexts and values [7], involving balanced intrinsic and extrinsic motivations and rewards. Some examples for the use of gamification as a game with a purpose can be found in the Related Work section.

In the following sections we present and discuss design options of our approach. Section 2 makes a review of most relevant related work, section 3 introduces SoundsLike interaction design, while section 4 highlights the gamification process and design options. The paper ends in section 5 with brief conclusions about SoundsLike, MovieClouds and user feedback, and perspectives for future work.

2. RELATED WORK

Regarding the audio stream analysis, most related work is in Computational Auditory Scene Analysis, where the research goal is illustrated by the "cocktail party problem" describing a hypothetical room with several people talking and other audio events (musical background, phone ringing etc.) [14], to identify different sound sources. The difficulty depends mainly on recording conditions. If the number of microphones is superior or equal to the number of audio sources, Independent Component Analysis techniques may be used to decompose the mixed signals; otherwise, the problem becomes extremely difficult. Our case (video's audio stream analysis) corresponds to the worst case scenario, because the recording conditions are unknown and the audio signal is likely to have been engineered. There have been several proposals of techniques for detecting specific sound events in surveillance systems, differing by the set of features extracted and by the kind of classifier used [11], applied to recognize specific sounds (e.g. crying, knocking, speech, music, impact, alarms, gun shots) or auditory environments. More recently, Chu [1] described a new feature extraction method to classify audio environments (not speech or music). The Music Information Retrieval community has recently seen an increasing nb of propositions toward the resolution of the Autotagging task. Tags usually designate musical genre, instruments used, or mood. The techniques used consist in extracting a set of acoustic features and use Support Vector Machines for the classification [10].

Some previous works use gamification in the context of Information Retrieval (IR) tasks: The ESP game [12] was created as an alternative to the tedious and extremely costly method of manual labeling of images, where players play to get amusement while contributing to reliable image labeling. Pairs of users assign tags to images without knowing the other player's choices. They move to the next image once both agree on the same label (receiving points) or decide to pass, and receive bonus points for completing a set of images without skipping or leaving. [5] presents an image tagging game, where scores are based on the automatic annotation algorithm, users' trust levels and RF obtained from previous plays. In Tagatune game [8] two remote users assign labels to a music, observing the labels defined by the other. At the end, they must decide if they are listening to the same music. All these approaches label entire images or musics. We have the additional challenge of labeling audio events along time, along the sound track. Also related is www.Last.fm, a music recommendation service based on users' current listening habits. It has a tagging system powered by the users that take an important role in the last.fm recommendation system. Users can label artists, albums and tracks with textual tags, and in each of these categories page, the most common tags are dis-played and sorted by frequency. It is a practical example of social audio classification but without any specific kind of gamification or direct rewards, other than users being able of organizing textually their own listening habits and receiving matching recommendations.

3. SOUNDS LIKE INTERACTION DESIGN

SoundsLike gaming elements integrate MovieClouds, to induce and support users' in soundtrack labelling. This section focuses in the interaction design, with the emphasis on the Movie View, where SoundsLike is played (Fig.2). Figures 1-2 display a use scenario in MovieClouds with the movie Back to the Future: In Fig.2a), this movie is playing, in the Movie View (top left), with timelines for the five tracks: subtitles, emotions in subtitles, audio events, audio mood (based on music), and felt emotions, showed below the movie, and the overview tag cloud of the audio events presented on the right. Users may select one of the five tag clouds (in the menu above the tag cloud) that correspond to each track, and interact with them: clicking a tag will give it a color and will mark, with colored dots, the temporal locations of its occurrences in the corresponding track of the timeline. Users may also compare in the different timelines where the selected tags occur along time (e.g. what is happening in the other tracks when the audio has a gunshot event?). Tags on the cloud and dots on the timeline are synchronized with the movie that is playing, highlighting (with increased brightness) the dots and the tags as they appear along the movie (e.g. 'shooting' in audio events).

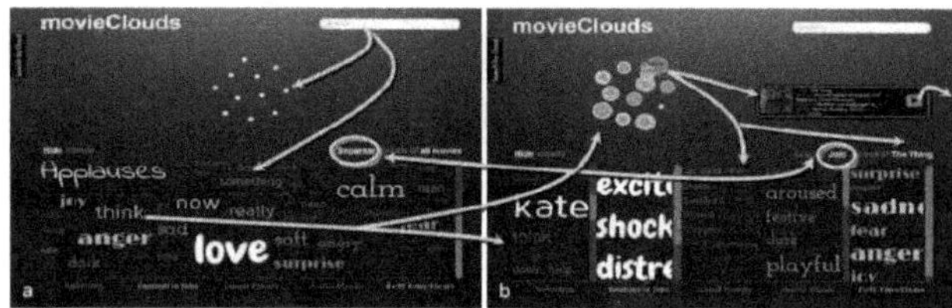

Figure 1. MovieClouds Movie View navigation, leading to the Movie View in Fig.2: a) unique tagcloud for 5 content tracks; b) 5 separated tagclouds for each track. Some tags selected and highlighted in the movies where they appear (top) [4].

Tag clouds were adopted for their power, flexibility, engagement and fun [4], in the Movie View (Fig.2) and the Movies Space View (Fig.1), where movies can be searched, overviewed and browsed, before one is selected to be explored in more detail, and watched, along time, in the Movie View (Fig.1b-2a). If the user is registered, a small HUD is always displayed in the top right corner with the name, accumulated points and rank level (based on points) to the left of the SoundsLike star logo, reminding the user to access it.

3.1 Facing the Challenge

After pressing the SoundsLike logo (Fig.2a-b), a challenge appears: an audio excerpt is highlighted in three audio timelines with different zoom levels below the video, and represented in the center of a non-oriented graph displaying similar excerpts, to the right, with a field for selection of suggested tags below.

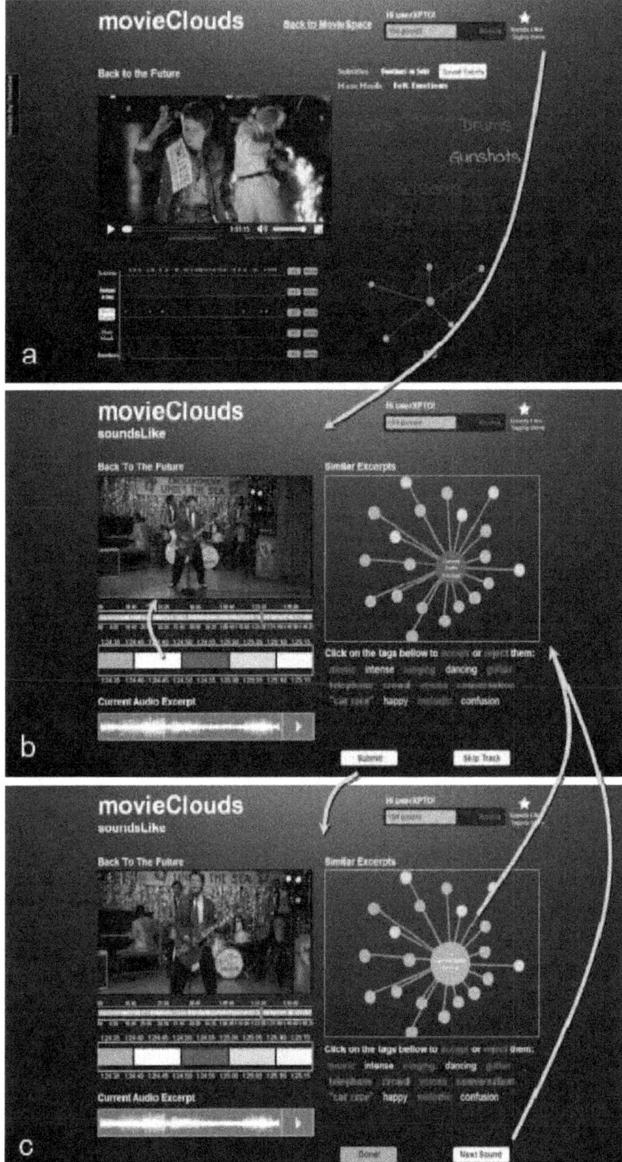

Figure 2. SoundsLike interaction: a) audio events track in MovieClouds Movie View; b) playing with SoundsLike; c) winning from labeling and playing again.

3.2 Audio Timelines

Audio excerpts are segments from the audio event (or mood) track. The audio excerpts are represented as small rectangles in the movie soundtrack timeline (top), and two zoomed timelines are added bellow: one with the current excerpt and surrounding neigh-bors, the other with a close-up of the audio signal of the excerpt, to enable users to navigate and listen to the entire audio excerpt and possibly its context in the movie (by clicking the timelines), to identify and label the current excerpt. In all the timelines, the current excerpt is highlighted in blue, while green (and yellow) refer to excerpts classified before (and skipped) by the user.

3.3 Similarity Graph

This view displays a non-oriented graph, based on a physical particles system, composed of the current audio, highlighted in the middle, and the similarity relations to most similar audio excerpts in the movie (the shortest the more similar). The users can hover on a node to hear the audio excerpts, and see the tags they chose for the nodes they already labeled (in green). The nodes use the same color mapping as the timelines.

A glimpse behind the graph: Due to the great variety of sounds found in movies' audio streams, we define an audio similarity measure between audio excerpts. The audio signal extracted from movies is divided into 20 sec windows and classical audio features are extracted using the yaafe library (yaafe.sourceforge.net) for feature extraction (zero crossing rate (ZCR), Mel Frequency Cepstrum Coefficients (MFCC), Spectral Flux) that result into sequences of vectors with 17 dimensions. The second step consists in computing a discrete representation of the signal, using the k-means clustering algorithm on the set of feature vectors to identify k2 centroids, that constitute our short-term features audio dictionary, used to transform the 20s audio samples into sequences of terms (or symbols), by computing, for each audio frame, the nearest element in the dictionary, using a Euclidean distance. In order to retrieve audio excerpts similar to the one chosen, the query histogram is computed, and the distances between this vector and every audio piece in the database are computed. The top ranked audio excerpts are then presented in the graph [4].

3.4 Labeling and Navigation

The user's task is to choose one or more textual labels, or tags, to describe the current audio excerpt (including the 'none' possibility). Choosing a tag means to highlight it with a click, marking with green to accept it or with red to discard it. The choices are saved only when the submit button is pressed (Fig.2b-c). At any time, the user can skip an excerpt and choose another one, or simply leave the game. The score will remain registered. When choices are submitted, the current excerpt changes color to green in the similarity graph and the timelines, displaying the earned score at the central node of the graph that is also enlarged (Fig.2c). Now, the user may use the Next Sound button to randomly navigate to another segment in the same movie, or may choose a specific excerpt from the similarity graph or from the timeline (Fig.2c-b). Excerpts already labeled (green) cannot be labeled again.

4. THE GAME INSIDE SOUNDS LIKE

This section presents main design options to induce and support users contributing to the movies soundtracks labeling, based on relevance feedback. Users have the objective of labeling correctly the highest possible number of audio segments to get a higher possibility of earning points, also from other users' classifications.

4.1 Assessing User's Skills

Since the labeling of some audio categories is not an obvious task, the users' level of expertise has to be evaluated, to assign a confidence level to their labeling. When users play for the first time, they will be presented with audio excerpts for which the confidence level given by the model is very high. These excerpts will be chosen frequently in the beginning for evaluating skills and trust of beginners, but they will fade with time. The picked control excerpts are seamlessly integrated in the game flow without any visible difference from the other excerpts.

4.2 Involving the User

Once the level of the user is assessed, the system starts to use user's skills. Our models give us some precious information regarding: labels and confidence levels, which ones are probable or unlikely to be assigned, indicating how "difficult" the query is; and an estimate of the similarity between two audio excerpts (independently of labels). Using this info, we can present audio excerpts and possible labels to the user, which represents different level of difficulty, challenge and consequent rewards. During the interaction with the user, it alternates simple and difficult queries to better evaluate the user. When the "correct" labels are unknown, a consensus rule is used between users, reinforcing the confidence of labelings made by a majority of users. Periodically, the new label associations are used to estimate parameters for a new generation of models of the audio backend that will this way benefit from the users' input.

4.3 Rewards

Gamification typically involves some kind of reward to the users. Although the needs for achievement or even cash incentives are often considered, these may be felt as controlling and not aligned with the player's culture [7]. In movies' fans, the sense of belonging and contributing to the classification of movies to allow enriching their access and analysis, along with testing and expressing user classification capacities could be enough reward. But still, to possibly increase sense of challenge, feedback and reward, in SoundsLike, each user is assigned points for different achievements: 1) when their labels correspond to a consensus by partial or entire textual matching, taking in account their confidence level; 2) when the consensual answer correspond to a difficult query; and 3) when previous classifications correspond to new answers. This way, we can differentiate the performance of users and create some competition. The points earned with the game can be just symbols of their expertise and amount of help provided, possibly displayed in rankings, but could also be exchanged for services depending on the applied business model, e.g. gifts, free access to movies, some additional info or special features, or cultural content from the website.

5. CONCLUSIONS AND PERSPECTIVES

We present a gamification approach to involve users in movies soundtrack labeling, while they are watching and browsing movies, and where we are exploring the design of game elements, and visualization techniques such as timelines and similarities graphs for the effective access and navigation in the movies, balancing and complementing utility and entertainment users goals and experiences. Users may help classify the audio, and in turn benefit from more rich and precise info access. Also, gathering considerable amounts of annotated audio will enrich the mining for information based on movies' content, which we expect to be useful for general public and the cinematographic industry. As futures steps, with improved content classification, audio and video features could be used by search engines and recommendations systems to provide accurate recommendations for users based on their preferences, visualization history and current emotional state. A practical example would be the creation of a dynamic preview system where previews would adapt to the user profile by changing the video sequences and soundtrack. The preview could be more successful if it included the scenes to which the user might be attracted the most and the production cost reduced by eliminating the need for producing different trailers aimed at attracting different types of audiences.

In previous evaluations [4] users appreciated the concept of MovieClouds, and the new possibilities and perspectives to search, overview, watch and browse movies from a global perspective down to the timelines where specific events can be found and compared among the different tracks. Overall, the results were very encouraging in terms of perceived utility, satisfaction and ease of use, and users particularly liked the tag cloud paradigm, the representation of movies as particles, the movie view and timelines, the flexible search by tag cloud, and the look and feel, where the similarity graph for audio events was already experimented and appreciated. Future work in this sequence also includes: refining SoundsLike based on a more thorough user evaluation of the more recent gamification design and the labeling effectiveness, and extending the approach to similarities across movies and to other media in the movies, such as music and video visual dimensions, towards richer content classification and even more engaging and joyful user experiences.

6. ACKNOWLEDGMENTS

This work is partially supported by FCT – Fundação para a Ciência e Tecnologia, through LASIGE Multi-annual Funding.

7. REFERENCES

[1] Chu, S., Narayanan, S. and Jay Kuo, C.-C. 2008. Environmental Sound Recognition using MP-based Features. In IEEE Int. Conf. on Acoustics, Speech, and Signal Process.

[2] Daniel, G. and Chen, M. 2003. Video Visualization In Proc. of the 14th IEEE Visualization 2003 (Vis'03). IEEE Visualization. IEEE Computer Society, Washington, DC, 54

[3] Deterding, S., Dixon, D., Nacke, L.E., O'Hara, K., Sicart, M. 2011. Gamification: Using Game Design Elements in Non-Gaming Contexts. In Proc. of CHI EA'11. Workshop. ACM. 2425-2428.

[4] Gil, N., Silva, N., Dias, E., Martins, P., Langlois, T. and Chambel, T., 2012. Going Through the Clouds: Search Overviews and Browsing of Movies. In Proc. of Academic MindTrek'2012. Tampere, Finland, ACM, pp.158–165.

[5] Gonçalves, D., Jesus, R., and Correia, N. 2008. A Gesture Based Game for Image Tagging. In CHI '08 EA,ACM.

[6] Hauptmann, A. G. 2005. Lessons for the Future from a Decade of Informedia Video Analysis Research. Int. Conf. on Image and Video Retrieval, Singapure, July 20-22. LNCS, vol 3568, pp.1-10, Aug.

[7] Khaled, R. 2011. It's Not Just Whether You Win or Lose: Thoughts on Gamification and Culture. In Proc. of Gamification Workshop at ACM CHI'11.

[8] Law, E., and Ahn, L. V. 2009. Input-agreement: A new mechanism for collecting data using human computation games. In Proc.of CHI'2009, ACM.

[9] Li, T., and Ogihara, M. 2003. Detecting emotion in music. In Proc. of the Intl. Conf. on Music Information Retrieval, Baltimore MD, October.

[10] Marques G., Domingues M., Langlois T., Gouyon F. "Three Current Issues in Music Autotagging." Int. Society for Music Information Retrieval (ISMIR) Conference, Miami, 2011.

[11] Pradeep K. Atrey, Namunu C. Maddage and Mohan S. Kankanhalli. 2006. Audio Based Event Detection For Multimedia Surveillance, ICASSP.

[12] von Ahn, L., and Dabbish, L. 2004. Labeling images with a computer game. In Proc. of CHI '04, ACM, 319-326.

[13] Wan, C., and Liu, M. 2006. Content-based audio retrieval with relevance feedback. Pattern Recognition Letters,27(2), 85-92.

[14] Wang, D., and Brown, G.J. Eds. 2006. Computational Auditory Scene Analysis, Principles, Algorithms and Applications, John Wiley & Sons Publishing.

MPEG-DASH Enabling Adaptive Streaming with Personalized Commercial Breaks and Second Screen Scenarios

Stefan Kaiser
Fraunhofer Institute FOKUS
Kaiserin-Augusta-Allee 31
Berlin, Germany

Stefan Pham
Fraunhofer Institute FOKUS
Kaiserin-Augusta-Allee 31
Berlin, Germany

Stefan Arbanowski
Fraunhofer Institute FOKUS
Kaiserin-Augusta-Allee 31
Berlin, Germany

first.lastname@fokus.fraunhofer.de

ABSTRACT

The growing demand for video streaming over the Web has increased the importance of the recently published MPEG standard Dynamic Adaptive Streaming over HTTP (MPEG-DASH). With MPEG-DASH, video delivery will be harmonized across the Internet by enabling consumption and transport of adaptive bitrate content.

Future opportunities that come with this technology are discussed in this paper as well as the challenges that have to be mastered. The discussion is done on the basis of proof of concept implementations. The topics deal with various aspects of MPEG-DASH deployment ranging from the fundamental content creation over applied adaptation to smart use cases like personalized commercial breaks or session mobility-enabled adaptive video streaming.

Categories and Subject Descriptors

H.5.1 [**Multimedia Information Systems**]: Video—*algorithms, design, experimentation, standardization*

Keywords

MPEG-DASH, IPTV, HTTP Adaptive Streaming, Second Screen, Personalized Commercial Breaks

1. INTRODUCTION

Audio and video streaming over Hypertext Transfer Protocol (HTTP) has received much attention in recent months due to its ability to be accessed by any consumer device that is connected to the Web because of increasing bandwidths and affordable provider contracts. In addition, advances in streaming like dynamic adaptation on changed connection quality or serving media resolution depending on the current demands of the user's device are applied to HTTP streaming as well. The recently published ISO standard Dynamic Adaptive Streaming over HTTP (MPEG-DASH) [6] defines media presentation and content formats and is growing in popularity as well as becoming generally accepted by most vendors of pre-existing technologies. It has various possible use cases in the field of audio and video content distribution, including live, on-demand and premium content. This area is as popular as never before with increasing customer numbers for services such as Spotify [3] or Netflix [2] (represents 29.7% of North American Peak Downstream Traffic [4]). In combination with ubiquitous computing, from computer via tablet and smartphone through to TV, over any network, the need for MPEG-DASH support in browsers for these devices is growing as well. Streaming media has become a commodity. Moreover open standards such as HbbTV have adopted MPEG-DASH in recent specification releases (HbbTV Version 1.5 [7]).

MPEG-DASH consists of a XML-based manifest file named Media Presentation Description (MPD) that refers to different representations of one media. Multiple representations of a specific media component are bundled to an adaptation set. The representations are provided in a segmented format and can be switched within an adaptation set adaptively. This format evolved from early efforts done by 3GPP with their proposal of Adaptive HTTP Streaming. At this time proprietary solutions like Microsoft Smooth Streaming, Apple HTTP Live Streaming and Adobe HTTP Dynamic Streaming existed, but only worked in the respective company's own component ecosystem. This led to the existence of multiple different formats that were not interoperable. To overcome this issue the MPEG-DASH specification with minimal innovation and mostly constraints on other existing industry solutions. Further, the standard specifies common encryption, which means that the same short segments of video can be decrypted and decoded by devices using different DRMs. This feature is of great importance for future paid video business models.

2. STATE OF THE ART

The majority of currently available DASH clients consist of native implementations, such as the Osmo player [1], or media player plugins, e.g. VLC [5]. Regarding DASH support in Web browsers contributions of the Web Hypertext Application Technology Working Group (WHATWG) and the W3C HTML Working Group have led to the specification of the Media Source Extension [9] (MSE) and Encrypted Media Extensions [10] (EME) to enable seamless, DRM-interoperable HTTP adaptive streaming using the HTML5 video object. It describes a JavaScript API to dynamically

EuroITV'13, June 24–26, 2013, Como, Italy.

push media segments into an HTML5 video or audio object or respectively, to exchange license keys. Using MSE and EME content providers can build their own MPEG-DASH players for each scenario that they want to support.

The Fraunhofer FOKUS solutions Famium and MPEG-DASH transcoder build a platform for the consumption and delivery of adaptive bitrate content. Famium is a Webkit-based Web browser with integrated MPEG-DASH support. The Famium extensions are based on W3C proposals of MSE and EME. For the usage in delivery networks, the MPEG-DASH transcoder supports MPEG-DASH-compliant content creation, including common encryption for ISO Base Media File Format (ISO BMFF). With an end-to-end solution for browser-based DRM for MPEG-DASH, Fraunhofer FOKUS addresses the need for a common DRM-interoperable encryption and HTTP adaptive streaming.

These components and the related proof of concept implementations are the basis for the discussion of the following opportunities and challenges.

3. OPPORTUNITIES

I. IP-based Broadcast TV

MPEG-DASH enables the delivery of video content over IP. These streams, no matter if on-demand or live video, can be received by multiple devices supporting MPEG-DASH, e.g. laptops, mobile devices and connected TV sets. The latter present a huge opportunity for broadcasters and content providers to transmit their already existing contents to end customers over IP.

During summer 2012 European broadcasters already experimented with a unicast HTTP-based live streaming, but without using MPEG-DASH, as this approach was not supported by HbbTV (1.0) in the corresponding time period. For instance six different Olympic Games' live streams were shown on the red button presence of the German public broadcaster ARD, instead of using dedicated broadcast channels as was done for previous Olympic Games. The user was able to switch between different sport events and see the live streams without leaving the current channel.

It is obvious that this approach is focusing on minimal broadcast and maximal broadband bandwidth. Broadcast channel slots and their corresponding bandwidth are sold by the DVB-T, DVB-C and DVB-S service providers, and are relatively expensive. IP delivery has the potential to be more cost-effective than broadcast, but investment in 'adequate' content delivery infrastructures (e.g. CDNs or multicast), which can scale to meet the demands of replacing broadcast, have to be considered.

The non-adaptive unicast streaming approach leads to several issues and consequently a bad user experience. For example, the video quality had to be decreased significantly to allow a fluent video playback. Moreover freezing and stuttering due to buffering during streaming have not been an exception. These issues may have been caused by an unexpectedly high number of viewers, but are also due to the nature of non-adaptive unicast streaming. Adaptive streaming offers a solution to this problem. It allows devices to seamlessly adapt to media representations (e.g. in terms of resolution, codec or bitrate) that fit the user's device and network connection best. Besides delivering the content in the highest quality available for a given network, faster start up times for media playback can be expected.

As part of our work and with the help of the MPEG-DASH transcoder we provide live DASH content to various organizations to ensure interoperability. Once devices have widely implemented new TV standards (e.g. HbbTV 1.5/2.0) adaptive live streaming using MPEG-DASH can become reality.

The adoption of the new HbbTV 1.5 Standard [7] for IP-based Broadcast TV will improve the user experience by benefiting from MPEG-DASH features, such as adaptive streaming and content protection on connected TV sets. The broadcasted streams can be protected with common encryption for interoperable DRM, thus validating the authorization of the current viewer. Moreover, broadcasters can ensure the delivery of specific content for specific regions, devices or user groups.

II. Personalized commercials using MPEG-DASH

IP-based Broadcast TV can leverage the bidirectional connection to personalize advertising, e.g. based on users' TV watching behavior or ratings. Consequently consumers are only shown commercials that are of interest to them. Thus, a child will not see the regular horror movie trailer but a video showing a toy commercial delivered over IP.

The exchange of ads in an exclusively IP-based environment can use all benefits of MPEG-DASH: a seamless start without video buffering, adaptive streaming and content protection. Therefore advertisement insertion is one of the main topics currently discussed in various standardization organizations dealing with MPEG-DASH. One approach that we have realized for using personalized commercials is to replace the relevant parts in the description file. A MPD-File, which contains the metadata of adaptive video streams to be played back, can be linked to other MPD-Files presenting an ad stream for a pre-, mid- and post-roll [6]. This second MPD-File is generated on demand on a recommendation server and contains the corresponding video information for a specific spot. The decision what content and therefore which description file shall be retrieved server side is based on collected information for the current user.

Taking the law compliance as given, broadcasters must collect usage data to predict the users' behavior and their likings to find the best fitting commercial. One method to get the needed data is the collection of automatically generated feedback such as the watching behavior. Another way is the collection of explicit feedback, as end customers can directly rate actual ads. Collaborative filtering algorithms defined in order to predict ratings and to recommend the potentially best fitting commercials can improve the commercial break experience.

III. Session mobility and Second screen

Second or more screens can leverage the experience of media consumption. While the main content of a live sport event may be presented on the main TV in the living room, the tablets of the people in the room can be used to show additional information synchronously. This can be other camera angles like on board cameras of single car drivers, or always having an overview of the whole racing track of a Formula 1 race. Additional statistics may also be presented, as well as targeted advertisement. Further, session mobility can be easily realized using MPEG-DASH. The media representations in MPEG-DASH use a common timeline, which enables simple timestamp synchronization between devices. For example, this allows a user to watch a movie on the TV, and when leaving the home, the TV session can be transferred to a mobile device to continue watching the movie. Further, session mobility and second screen features are in discussion for HbbTV 2.0. Our implementation of these features serves as proof of concept.

Another opportunity for using MPEG-DASH is the possibility to support additional content metadata. The Media Presentation Description (MPD) therefore supports the declaration of different roles for an Adaptation Set that contains a media component. A role of an adaptation set has an attribute schemeIdUri and a value. For the schemeIdUri "urn:mpeg:dash:role:2011" the following roles are defined: caption, subtitle, main, alternate, supplementary, commentary and dub. Each role shall be used for media content to describe it further, so a MPEG-DASH client can decide which media content to use. Other schemeIdUri and therefore other Role values can be used as well.

Supporting multiple screens with different content is possible using the role description and a single MPD file. This can be for example a car racing video. The MPD file contains one adaptation set with a role value "main". The media files in this adaptation set show the race from an on top view. Additionally other Adaptation Sets in the same MPD file are marked with the role value "alternate" or "supplementary". Media content within these adaptation sets may show videos from the cockpit perspective of different cars. A MPEG-DASH enabled Website and player can now use this extra information to play different content on different devices. Synchronized playback of these videos is one challenge that is addressed in this paper (see section III of the challenges).

There are different possibilities for MPEG-DASH players to choose the role and therefore the media to play. The players can show users the available roles to choose so every user can decide on his own what to see. A second approach is that there are players who know that they are supposed to only play media that is marked with certain Role values. Another way is that the players themselves do not get the whole MPD file.

Second or multiple screen scenarios can take advantage of MPEG-DASH features such as describing cross-platform media formats in one presentation description file (MPD). For example, we have implemented a node.js-based synchronization Web server with WebSocket support. Every device in a household connects to this server and is therefore joining a session. The server now knows the existence of different devices. After selecting a MPD file on one device, this screen shows the media that is described by the main role. All connected devices or devices that join the same server hosting the application after that get custom versions of the original MPD file. This means that some media marked as "alternate" in the original MPD file may now be marked as "main" and the original main Adaptation Set is removed from the MPD file for the other devices. This way the players do not need to decide anything but playing "their" main media.

A second screen scenario can also be realized using special MPEG-DASH players. The decision which part of the MPD should be played can be determined by the client. One approach is that the user manually selects the camera angle of the car race. Alternatively, each player on each device knows what to play. This can depend on other players discovered in the local network using, for example, UPnP or Bonjour. The decision can also depend on device capabilities and properties like the screen resolution or the ability to listen to touch events.

4. CHALLENGES

I. Algorithms for Adaptive Streaming

The first and most crucial challenge of using MPEG-DASH is the adaptivity algorithm of the client. The deployment of MPEG-DASH alone is not sufficient for getting all the benefits of this solution. Since this part is not defined in the standard, it is as distinctive feature of a MPEG-DASH client implementation. In general, the available bandwidth of a client defines the maximum possible bitrate for a representation to be used. An approach that we have implemented for reaching the best quality representation is to start with the lowest bitrate available and gradually increase the quality of the representation. This allows the player to converge to the maximum bitrate and considers the stability of the network condition.

Network condition can vary everywhere the Web is accessed. The most affected devices obviously are mobile phones and tablets connected via a cellular network. The coverage of a mobile network's base station is limited as well as its amount of clients that can be served. This quickly leads to a bottleneck and bandwidth losses for each user. The bandwidth changes in this kind of network can differ in type and intensity and vary from small spikes to long-term changes. Often used with mobile phones and tablets but also with laptops are wireless networks based on the IEEE 802.11 standard. Characteristic for this kind of network are small spikes [8]. In a local area network (LAN) the same Internet connection to the Internet service provider (ISP) is shared with other users. This can lead to unexpected bandwidth drops when other network members cause high traffic loads. This applies to a wireless LAN as well.

The length of the media segments, or more detailed, the distance of key frames, has an impact on the reaction time of the algorithm. Long key frame distances make the client react more slowly than short distances do.

Efficient switching is very complex and difficult to develop. Its behavior depends on the kind of change that is happening, whether it increases or decreases bandwidth, whether it is a small spike, or a long-term change. It also depends on the buffer settings that can be facilitated. And in the end it has to incorporate human aspects of usability. Big steps in quality changes are not expected by the user, therefore the quality should be smoothly shifted to the current possible one. Frequent changes are annoying and thus not recommended to happen. Bandwidth can cause these types of behavior, e.g. by a break-in of bandwidth from highest to the minimum quality level for several minutes on a mobile phone, or oscillating on the bandwidth threshold of a specific quality level. On short spikes that last only for a few seconds, the buffer can possibly compensate the occurred gap, depending on its settings. A proper switching algorithm needs to consider all of these properties to enable a smooth and pleasant viewing experience.

II. Content creation

The counterpart to consuming MPEG-DASH media is its creation. The MPEG-DASH specification is format and codec agnostic. Nevertheless, the standard currently defines rules for the two container formats: ISO Base Media File Format (ISO BMFF) and MPEG-2 Transport Stream

(MPEG-2 TS). These rules are applied by different profiles for each container format and specify restrictions on the manifest file as well as the content.

As the profiles just limit the appearance of the container format, it is possible to convert current content to the MPEG-DASH format by just repacking it along the profile rules. However, adaptive streaming holds some challenges that need to be considered.

First of all, the content has to be encoded into different quality levels, i.e. bitrates. These bitrates should be chosen in an appropriate way to meet the desired target characteristics. The choice of appropriate bitrates is not the only issue that comes with content creation for adaptive streaming. During the creation process, the content will be split into segments and fragments. One segment can contain one or multiple fragments. As mentioned in the previous chapter about algorithms for adaptive streaming, the segment length plays an important role in the behavior of a MPEG-DASH client. Shorter segment lengths mean a more reactive adaptive behavior. At the beginning of a segment, there should be one key frame, i.e. one full picture. Using short segment lengths, the key frame rate might be higher than necessary. This leads to higher storage usage as well as higher bandwidth usage. Altogether, this raises the task of searching for the right tradeoff between effective resource usage and adaptivity reaction time.

MPEG-DASH content is based on existing adaptive streaming technologies like Microsoft's Smooth Streaming, Apple's HTTP Live Streaming (HLS) or Adobe's HTTP Dynamic Streaming. Conversion of content compliant to one of these technologies is less effort than a completely new transcoding/repacking process. The format used in Smooth Streaming is closely related to ISO BMFF using the live profile. Adobe uses the MP4 container format being a specialization of ISO BMFF and therefore also closely related to a MPEG-DASH format. Apple instead uses MPEG-2 TS in its HLS specification, which is related to the MPEG-2 TS profiles defined in the MPEG-DASH standard.

III. Synchronization of multiple devices

Using MPEG-DASH for multiple devices raises the challenge of synchronization, as this is important for the aforementioned second screen and session mobility scenarios. The shown content in the scenarios may need to have the same state. Furthermore some scenarios may need to play media in time synchronization, so actions in the media happen at the same playback time.

Communication and synchronization of different devices is not part of the MPEG-DASH standard. This means there exists no single recommended way of realizing this issue. The program part controlling this mechanism can be part of the MPEG-DASH player or application logic in a Website that communicates with a synchronization server, for example.

A synchronization scenario which cannot rely on extra information, can be implemented by only using current standard Web technologies. Using WebSockets a Website can send and receive data via the TCP protocol. The client devices open the Web application and will be connected to a server. The synchronization server is part of the service provider's delivery network. All information is sent to the server instead. The server then evaluates incoming data such as if a user wants to pause the video and sends this information to all connected clients, so they will pause their media as well.

A centralized controlling mechanism, as we have implemented, has the advantage of one "real" state that is only pushed to other devices. The synchronization server shares a common timeline for synchronization between the clients. The MPEG-DASH players do not need to have special implementations. Only standard functions like play, pause and seek are needed to synchronize the players on all devices.

5. CONCLUSION

Today MPEG-DASH client as well as server implementations are still in development. There are not many native MPEG-DASH clients available and implementation work for built-in browser support or corresponding APIs is still in progress. The paper gives an outlook on future scenarios and advantages that come with them by using MPEG-DASH technology. With our implementation of MPEG-DASH components and scenarios, valuable input to current standardization activities is given in the form of proof of concept implementations.

We described future, as well as already possible scenarios, which can benefit from Internet-delivered television. The ability to have session and context information as well as the possibility to deliver different content parts to different users or devices will greatly improve the user experience. With MPEG-DASH support growing and implementations improving, the opportunities previously mentioned in this paper can be realized. Further, the mentioned challenges need to be handled to improve user experience and to create new business models.

6. REFERENCES

[1] GPAC Osmo4 Player website. *http://gpac.wp.mines-telecom.fr/player*.

[2] Netflix Inc. website. *http://www.netflix.com*.

[3] Spotify Ltd. website. *http://www.spotify.com*.

[4] Global Internet Phenomena Report. *Sandvine*, 2011.

[5] *A VLC Media Player Plugin enabling Dynamic Adaptive Streaming over HTTP*, Scottsdale, Arizona, November 2011.

[6] Dynamic adaptive streaming over HTTP (DASH) - Part 1: Media presentation description and segment formats. *ISO/IEC 23009-1:2012*, 2012.

[7] HbbTV Specification. *http://www.hbbtv.org/pages/about_hbbtv/HbbTV-specification-1-5.pdf*, March 2012. Version 1.5.

[8] S. Akhshabi, A. Begen, and C. Dovrolis. An Experimental Evaluation of Rate-Adaptation Algorithms in Adaptive Streaming over HTTP. *In Proceedings of the second annual ACM conference on Multimedia systems, MMSys*, 11:157–168, 2011.

[9] A. Colwell, A. Bateman, and M. Watson. Media Source Extension. *https://dvcs.w3.org/hg/html-media/raw-file/tip/media-source/meia-source.html*, 2012.

[10] D. Dorwin, A. Bateman, and M. Watson. Encrypted Media Extensions. *https://dvcs.w3.org/hg/html-media/raw-file/tip/encrypted-media/encrypted-media.html*, 2012.

A Virtual Camera Controlling Method Using Multi-Touch Gestures for Capturing Free-viewpoint Video

Junya Kashiwakuma
University of Tsukuba
Tennoudai 1-1-1
Tsukuba, Ibaraki, Japan
+81 298 53 6470
kashiwakuma@image.iit.tsukuba.ac.jp

Itaru Kitahara
University of Tsukuba
Tennoudai 1-1-1
Tsukuba, Ibaraki, Japan
+81 298 53 6470
kitahara@iit.tsukuba.ac.jp

Yoshinari Kameda
University of Tsukuba
Tennoudai 1-1-1
Tsukuba, Ibaraki, Japan
+81 298 53 6470
kameda@iit.tsukuba.ac.jp

Yuichi Ohta
University of Tsukuba
Tennoudai 1-1-1
Tsukuba, Ibaraki, Japan
+81 298 53 6470
ohta@iit.tsukuba.ac.jp

ABSTRACT

This paper introduces a 3D free-viewpoint video-browsing interface that applies multi-touch manipulation to virtual camera control. It is difficult for general users to browse 3D free-viewpoint video because they are not accustomed to controlling a virtual camera from a free-viewpoint perspective. We focus on the multi-touch interface installed in tablet PCs as an input device, which has become a popular interface thanks to its easy and intuitive manipulation. We conduct subjective evaluations to define suitable gestures for virtual camera control. The results reveal which multi-touch gestures users tend to prefer for controlling a virtual camera.

Categories and Subject Descriptors

H.5.2 [Information interfaces and presentation]: User interfaces—Interaction styles, H.5.1 [Multimedia Information Systems] Artificial, augmented, and virtual realities.

General Terms

Performance, Design, Experimentation, Human Factors.

Keywords

3D free-viewpoint video, user study, 3D navigation, multi-touch gesture

1. INTRODUCTION

As computer vision and image media technologies are well developed, 3D free-viewpoint video generation is garnering much research attention [3][10][11][13][16]. The free-viewpoint video technique makes it possible to observe the target scene from various viewpoints including an immersive view and a bird's-eye view. In a large-scale space like a soccer stadium, the advantage becomes remarkable [10][13]. There are well-developed methods for generating a free-view using computer vision and computer graphics. On the other hand, the interfaces for freely controlling the movements of a virtual camera in a 3D environment still have room for improvement. We consider that developing an easy-to-use browsing interface is necessary for popularizing free-viewpoint video. Further, such an interface has to fully demonstrate the merits of free-viewpoint video. Watanabe et al. [22] developed a 3D free-viewpoint video browsing system that controls a virtual camera by using a 3D position sensor and an overview monitor. The user can watch free-viewpoint videos to smoothly set the viewpoint and the gazing point at same time by using both hands. The overview monitor allows the user to understand the context of the scene that is being captured.

Thus, it is possible to accurately track moving objects such as a ball or players in a scene. However, this requires a lot of equipment, such as a 3D position sensor and large displays. Such a setup might not be available in the outdoor environment or in a living room. Inamoto et al. developed an interface using a head-mounted display (HMD) [10]. Users can enjoy a free-viewpoint video of a soccer game by observing an augmented reality (AR) scene in a dioramic soccer stadium. However, the movable area of the virtual camera is constrained by human body motion.

As shown in Figure 1, we propose an interface that utilizes widely used equipment (e.g., a tablet PC) as an input device, which aims to control a virtual camera as the user chooses. These days, watching videos on a smart phone or a tablet PC has become popular; our approach is an extension of these recent video-browsing trends. In order to develop an intuitive interface, we have to solve an important problem caused by the difference in degrees of freedom between the virtual camera motion and the input positional information given by the tablet PC (multi-touch gestures). In other words, we need to control a virtual camera, which has seven motion parameters (i.e., 3D positions, 3D rotations and a zoom), by using a multi-touch interface, which can input 2D positional information and a click.

As the first step to solving the problem, in this paper we investigate intuitive and efficient multi-touch gestures for controlling a virtual camera that captures free-viewpoint videos. We collect subjective evaluations to define suitable gestures for virtual camera control.

Figure 1. Virtual camera control with multi-touch gestures

2. RELATED WORK

In the computer graphics (CG) and virtual reality (VR) research fields, some research has been conducted about virtual camera controlling methods using a two degrees of freedom input-device such as a mouse [4][5][7][12][17][18][19][23][24].

Cohe et al. [2] proposed a method named "tBox" which has a cube-widget around a CG object and users can control its rotation, translation and scale to touch the sides or apexes of the cube.

Hachet et al. [6] proposed an interface, "Navidget", for navigating in a 3D environment using a 2D input device such as a PDA or a mouse. It provided easy 3D navigation by improving point of interest (POI), which is a 3D navigation method proposed by Makinlay et al. [14]. The widgets for the Navidget allowed users to easily control a virtual camera with only one-stroke gestures. This method also provides visual effects and a preview window to compare the displayed view and the virtual camera view after the position/orientation is changed.

Hancock et al. [8] and Martinet et al. [15] proposed a method for controlling CG objects using multi-touch gestures. This method switches the various control methods for CG objects (e.g., 2D and 3D navigation) by referring to the state of specific touch points.

These four approaches work well for controlling a virtual camera in 3D CG space; however, they deal with only static CG objects. In contrast, our target application is for capturing free-viewpoint videos of dynamic scenes such as soccer games. tBox is suitable for realizing CG controlling interface because it is based on formative study to find the most suitable multi-touch gestures. So we propose an intuitive and efficient interface by following the approach. Navidget is a well-developed interface, which has intelligible functions for easy 3D navigation, such as visual widgets and a preview window; however, it is difficult to observe/check the additional windows while capturing free-viewpoint videos in a dynamic scene. We need to realize virtual camera control that allows the user to keep watch on the captured video window. The approach of Martinet et al. [15] for realizing various types of 3D virtual camera control using a 2D input device that refers to the state of touch points is helpful for our development. We apply their proposed method, which focused on controlling CG objects, to controlling a virtual camera to capture free-viewpoint videos.

Human touch gestures and design of virtual camera control in CG space has also been actively researched. Ware et al. [21] conducted experimental evaluations on methods of operating a virtual camera in 3D space. They compared three types of operations, "scene-in-hand", "eyeball-in-hand", and "flying-vehicle", using a 3D position sensor, which has six degrees of freedom. We had to consider how to apply their results to a multi-touch input device, which cannot input as much information at the same time.

In order to realize an intuitive and efficient multi-touch-based interface, Hinrich et al. [9] conducted a study of touch gestures. In the observation, they set up multi-touch tabletop interfaces for browsing visual information in an aquarium, and analyzed the observed users' gestures (e.g., translation, rotation, pushing the buttons, etc.) Purposely, they did not provide a manual for controlling the interface so that they could observe natural human multi-touch gestures. Cohé et al. [1] also displayed 3D objects by using a video projector on a table and asking subjects to control an object without giving any instruction on multi-touch gestures. They analyzed the subjects' multi-touch gestures for controlling rotation, scaling and translation of the 3D object. Although the display device is quite different (i.e., a large projected display and a small touch pad), their approaches are helpful to us. We observed subjects' multi-touch gestures in our free-viewpoint video browsing system without providing instructions for use, and classified the results of the observations in order to identify suitable multi-touch inputs for controlling a virtual camera in each scene. In classifying our experiments, we referred to Cohé's approach that analyzes especially the three kinds of the subjects' gestures, "Form" (the state of the touching finger), "Trajectory" (the locus of motions of fingers) and "Initial Point Location" (start point of fingers). We also analyzed the subjects' touch position, the numbers of fingers they used and the direction of movement.

3. FREE-VIEWPOINT VIDEO BROWSING SYSTEM USING MULTI-TOUCH GESTURES

3.1 Generating free-viewpoint videos

As illustrated in Figure 2, our free-viewpoint video browsing system is based on the 3D live free-viewpoint video system proposed by Koyama et al. [13]. It executes all processes, from capturing multiple images to rendering free-viewpoint video in real-time by using an effective 3D modeling technique, a "player billboard", which represents a target object (e.g., a soccer player) with a single polygon and its texture.

In a scene recognition block, a target scene is captured by using two cameras set in higher places such as rooftops, and the 3D positions of a soccer ball/players are calculated using stereo vision. In a multiple video capturing block, multiple videos to extract the texture information of the players are simultaneously captured. To completely obtain all the texture information without large appearance gaps between the multiple cameras, the system should have a layout of more than eight cameras. The 3D modeling server generates a 3D model of the target objects using the estimated position and the texture information, and transmits the 3D model, which is necessary for rendering an observer's view, to the 3D free-viewpoint browser. In the 3D free-viewpoint browser block, users input a desired viewpoint to observe the soccer action by using a multi-touch interface, which is discussed in this paper. Then the system calculates the camera parameters of the virtual camera and sends them to the 3D modeling server. When the browser block receives a 3D model corresponding to the request, a 3D free-viewpoint image is rendered and displayed using the model.

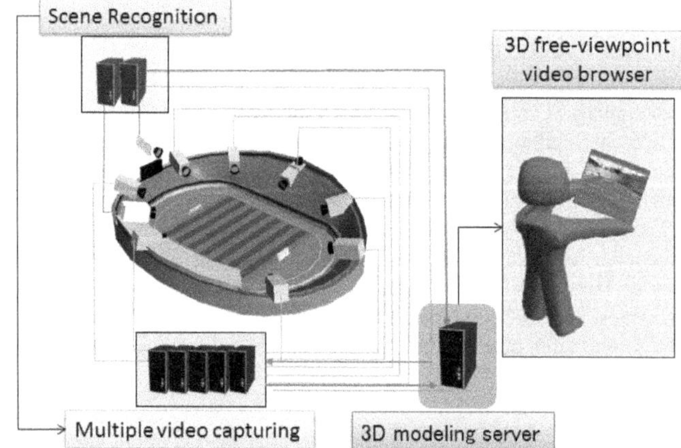

Figure 2. Our proposed free-viewpoint browsing system

3.2 Browsing Interface For Free-viewpoint with Multi-touch Gestures

In this section, we introduce some issues that have to be considered in our free-viewpoint video browsing system, and our approach to solving them.

Our proposed system aims to allow users to comfortably observe free-viewpoint video without having to prepare display equipment. So, as shown in Figure 3, we developed a free-viewpoint video browsing system using a tablet PC as a common information terminal device. Since visual widgets such as GUI controllers sometimes disturb video browsing, and the display size of a tablet PC is much smaller than a PC monitor, we decided not to use any visual widgets, but only multi-touch gestures for control. By using multi-touch gestures, users can input more pointing information at the same time than when using only single-touch gestures. To utilize this feature, we aim to compensate for the difference in degrees-of-freedom between the virtual camera motion and the input positional information given by the tablet PC (multi-touch gestures). A "pinch-in/out gesture", which is commonly used in smartphone and tablet PC interfaces, is a good example.

However, it is difficult to completely compensate for the gap between the virtual camera motion and the input positional information. So we needed to carefully consider the correspondence table between the multi-touch inputs and the virtual camera motion. In the free-viewpoint browsing system, the users control a virtual camera with multi-touch gestures while watching the displayed view. If the user is not satisfied with the view, he/she controls the virtual camera in order to capture a preferred view. In this process, the user has to image motions of the virtual camera to get the preferred view. In order to realize the interface for various types of users, the change of the capturing appearance should be intuitively estimated by multi-touch interface inputs and should be accepted by general consensus. However, there has not been any research aimed at defining multi-touch gestures for controlling a virtual camera for capturing free-viewpoint systems. So in this paper we conduct some subjective evaluation to define them.

Figure 3. Free-viewpoint video browsing system using a tablet PC

4. SUBJECTIVE EVALUATION TO DEFINE SUITABLE MULTI-TOUCH GESTURES FOR CAPTURING FREE-VIEWPOINT VIDEOS

In this section, we introduce our subjective evaluations for defining intuitive and efficient multi-touch gestures for virtual camera control to capture free-viewpoint videos.

4.1 Procedure and Environments of Our Subjective Evaluation

4.1.1 Warming up and preliminary explanations

In order to familiarize the subjects with observing free-viewpoint videos and controlling the virtual camera using the multi-touch interface, 12 types of short (2 or 3 seconds) free-viewpoint videos were displayed on a tablet PC monitor (the display order was random), and for practice, the subjects were asked to input a multi-touch gesture that controlled the virtual camera that was capturing the video sequence. The free-viewpoint video was continuously playing until the subjects finished the gesture input exercise. Before this warm up, the subjects were informed that control of the virtual camera is realized by controlling the viewpoint (position), the gaze direction (orientation) and the zoom, and they were shown example movies so they could easily understand. The subjects were also informed that they could input gestures using two fingers, that the experiment did not have any correct answers so they could freely gesture as they wanted, and that there was no time limit for the experiments.

4.1.2 Video sequences for subjective evaluations

In our evaluations, the subjects observed short video sequences (5 seconds) generated at a fixed viewpoint, and then were asked to perform a specific camera movement using multi-touch gestures. Each video sequence contained 3D objects such as a ball and soccer players as cues for camera control.

As shown in Table 1, we prepared 12 video sequences. They were divided into 2 groups. Of these, 6 of them are the view from a player standing on the soccer field (field view video); the others are the view looking down the soccer field (bird's-eye view video). Each group was classified into three sub-groups, "gaze direction (orientation)", "camera position" and "zoom", by the controlled camera parameters. The gaze direction (orientation) and camera position were divided into another subclass depending on the direction of movements.

As shown in Figure 4, the field view videos reproduced a scene that a soccer player was watching, and these are expected to give immersive presence to users than ordinary telecasted video sequences. When free-viewpoint video telecasting is put to practical use, users may tend to observe the field view video so it is important to investigate virtual camera control methods using multi-touch gestures in the field view video.

Figure 4. Examples of field view videos

One of the typical camera shots in ordinary telecasted video sequences is the view overlooking a large area at a glance, as shown in Figure 5. The audiences can immediately understand the situation in a soccer game by observing the video sequence. The bird's-eye view video that enhances this overview feature is one of the advantages of free-viewpoint video. It is also important to investigate the virtual camera control methods using multi-touch gestures in the bird's-eye view video.

Figure 5. Example of bird's-eye view video

4.1.3 Instructions for subjects

In all evaluations, we asked subjects to imagine controlling the virtual camera to keep watching a ball using their favorite multi-touch gestures. It was possible to input the gestures whenever he/she wanted. In each trial, the attributes of the controllable camera parameters were fixed (e.g., gaze direction (orientation), camera position or zoom, so that we could investigate how the subjects used the different multi-touch gestures to control each attribute of the camera parameters. The subjects were given clear instructions. For example, the instruction for controlling the gaze direction (orientation) was "rotate the virtual camera as if you are shaking your head", for controlling the camera position, it was "shift the camera position" and for controlling zooming, it was "control the field of view of the virtual camera using zoom-in or zoom-out". Then, we explained the context of the target soccer scene. Table 1 shows the detailed instructions the subjects were given in each trial.

4.2 Environment for Subjective Evaluations

We conducted the subjective evaluations described in the previous section with 14 subjects (13 males and one female). They were from 22 years to 28 years old. We sent out questionnaires about their experience of 3D contents and to touch interfaces. The result is shown in Table 3. The evaluations were conducted with users who had various levels of the experience so that we could define the multi-touch gestures preferred by general users.

The equipment used in the evaluation is shown in Figure 6. The subjects sat down with a tablet PC. Although we did not tell the subjects how to hold the tablet PC, the subjects held it comfortably as they browsed free-viewpoint videos. In the experiment, we used ICONIA Tab-W500 PCs produced by Acer. The OS was Windows 7, 32bit version; the display size was 10.1 inches and the resolution was 1280 [pixels] x 800 [pixels]. The tablets had 2GB memory, 32GB HDD, AMD Dual-Core processer C-50 and AMD Radeon HD 6250 Graphic boards. The evaluated videos were generated and displayed at 30 fps.

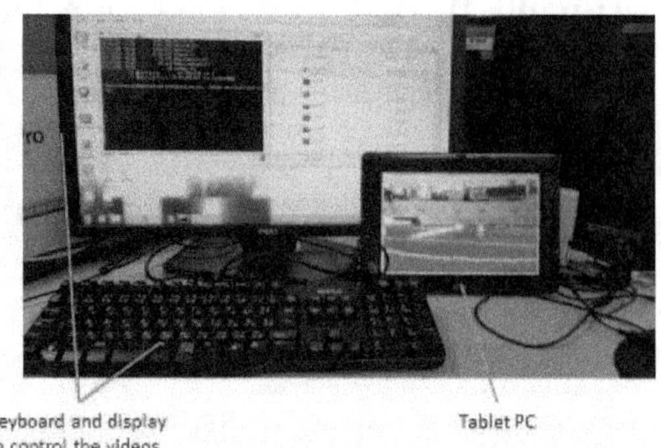

Figure 6. Environment for the experiment

4.3 Results and Discussion

We conducted subjective evaluations with 14 subjects by displaying 12 kinds of video sequences. For each video sequence, the subjects carried out their touch gestures three times. As a result, we extracted 504 multi-touch gesture data. In this section, we introduce the results of our analysis for the data set.

4.3.1 Tendency of the inputted multi-touch gestures

Table 4 summarizes the gestures inputted by the subjects. By monitoring the evaluation scenes, we confirmed that all given touch gestures were somehow meaningful for the virtual camera control in each video. We are curious that the results show almost all subjects choose stroke gestures using one or two fingers to move the virtual camera body (e.g., rotating the orientation and shifting the position), and to choose pinch-in/out gestures to control the zoom parameter.

4.3.2 Gestures for zoom-in/out

In both the field view videos and the bird's-eye view videos, 80-90% of the subjects input the pinch-in/ out gesture to control the zoom parameter (video 5, 6, 11, and 12). In some ordinary multi-touch-based applications such as photo albums, photographs are enlarged with the pitch-out gesture and the size is reduced with the pinch-in gesture. Thus it is reasonable to choose the pitch-in/out gestures to change the viewing size of the displayed videos.

For the other gestures, the subject must draw a circle. To zoom in, he or she draws a clockwise circle; to zoom out, the subject draws a counterclockwise circle. Schmalstieg et al. [20] found that the gesture of drawing a circle is suitable to observe a displayed object carefully; however, the gesture can be applied for a scene that has single object because it is difficult to set the focusing point of zoom-in/out. In our target scene, there were many objects to keep in focus. Thus, we consider that the pinch-in and pinch-out gestures are more suitable for controlling zoom-in and zoom-out functions of the virtual camera for capturing free-viewpoint videos.

4.3.3 Direction of stroke gestures

Figure 7 shows frequency of each stroke gesture inputted in the evaluation for 1-4 and 7-10, which we ask the subjects to move the virtual camera body. In the evaluation, in which the subjects were asked to input the direction of movement for rotating the virtual camera upward or shifting the position forward/right, the results were divided almost in half. One group moved their finger in the same direction as the ordered virtual camera motion. The other moved in the opposite direction (i.e., when the subjects wanted to move the virtual camera to the right, they moved their fingers from right to left on a tablet PC).

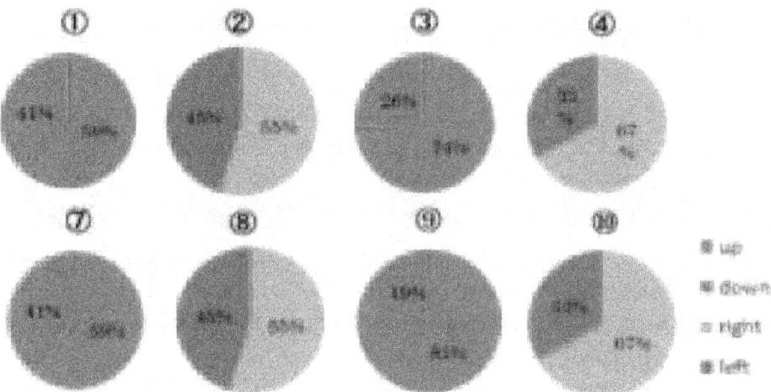

Figure 7. Frequency of stroke gesture inputs in evaluations for videos 1 - 4 and 7 - 10

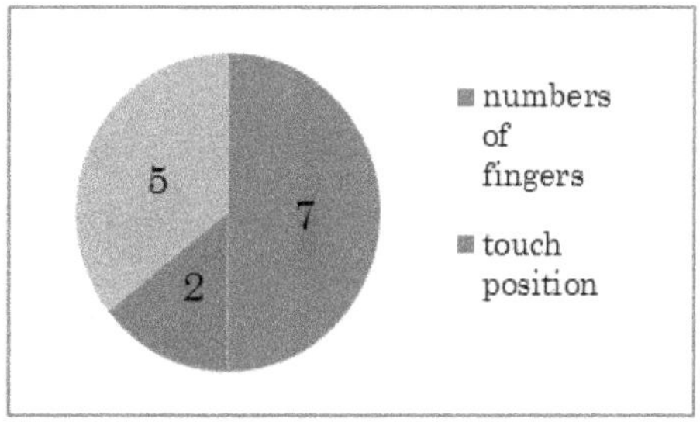

Figure 8. Frequency of each way of switching between the two types of camera control (rotating the orientation and shifting the position)

We investigated the reason by sending out short questionnaires to the subjects who chose the inverse direction stroke gesture. We found that they thought the ordered appearance (view) change could be realized by relatively moving the target space with fixing the virtual camera. For example, the interface of Google maps is designed such that when we want to see to the right of the map, we need to move our finger to left. This interface gives us the feeling given by daily activities such as flipping sheets of paper. We consider such natural behavior is important to designing intuitive multi-touch gestures. On the other hand, since the multi-touch interface should be accepted by general users, it is difficult to ignore that half of the subjects moved their finger in the same direction as the camera motion. There was no correlation between the direction of stroke gestures and the subjects' experience with using multi-touch interfaces. So, we think it is better to leave the choice to the users in a practical system.

4.3.4 Switching controlled parameters

In this evaluation, we found that many subjects switched the two types of camera control (i.e., rotating the camera to the right and shifting the camera position to right) by changing the number of the fingers they used to gesture. Almost all of them controlled the orientation of the camera with one finger and the viewpoint with two fingers. Since it is possible to more exactly point out the touching position, the subjects considered that it was possible to input precise gestures using one finger rather than two. Controlling the orientation of the virtual camera requires more precise operation than controlling the camera position. Even if we shift the view position a little while observing an object in the distance, the disparity is not so large. On the other hand, if we rotate the gaze direction a little in the same situation, the disparity becomes much larger. As a result, the subjects tended to be more careful when controlling the camera orientation and used a one-finger gesture.

Another principal way of switching the two types of camera control was changing the touching position without changing the number of touching fingers. Some subjects input gestures for controlling the camera orientation at the center of the tablet PC display, and for the camera position at the edges of the display. The reason for the difference is that they felt that controlling the camera orientation was an objective motion and controlling the camera position was a subjective one. Thus, they used input gestures like scrolling in a Web browser, which is suitable for changing the value, to control the camera position.

Figure 8 shows the frequency of each way of switching between the two types of camera control. Since half of the subjects chose the number of fingers as a switching method, it seems that switching by changing the number of touching fingers is more suitable for virtual camera control to capture the free-viewpoint videos.

4.3.5 Field View Video and Bird's-Eye View Video

In the evaluations, two types of videos (the field view videos and the bird's-eye view videos) were used. We investigated the difference in the input gestures by observing the two types of video sequences. Figure 10 shows the results of comparing the field view videos (videos 1 - 7) with the bird's-eye view videos (videos 2 - 8). Since the subjects chose almost all the same gestures, it seems that there was not a significant difference. In more detail, we calculated the similarity among the users' operations while using the two types of views. The similarity was given using the ratio of cosine similarity, which is the method to calculate the degree of similarity of the two vectors. The angle between two vectors is calculated with their inner product and magnitude. If the angle becomes small (i.e., the cosine value is close to 1), the two vectors have similar data. We generated 15 dimensional vectors for the 12 free-viewpoint videos used in the experiment, where the vector elements are the number of times the gestures were input by the subjects. Table 2 shows the results of the similarity between two free-viewpoint videos. We would like to note that the similarity between videos 1 and 7; 2 and 8; 3 and 9; 4 and 10; 5 and 11; and 6 and 12 is almost 1. In all these pairs, the subjects were asked to input multi-touch gestures for realizing a same virtual camera control when watching the two types of view (bird's-eye or field view) video.

4.4 Suitable Multi-touch Gestures for Capturing Free-viewpoint Video

To summarize our subjective evaluations, as illustrated in Figure 9, we proposed a multi-touch gesture suite for controlling a virtual camera to capture free-viewpoint videos. First of all, we utilized the extracted data indicating that users prefer to control the camera using simple strokes with one or two fingers. Since controlling the orientation of the camera requires more precise operation, we applied one-finger stroke gestures to controlling the orientation. Alternatively, a stroke gesture with two fingers was applied to the positional control of the virtual camera. For zoom-in or zoom-out control, we applied pinch-in/out gestures. The moving direction of the fingers depends on the user's preferences, so we have left that setting to users.

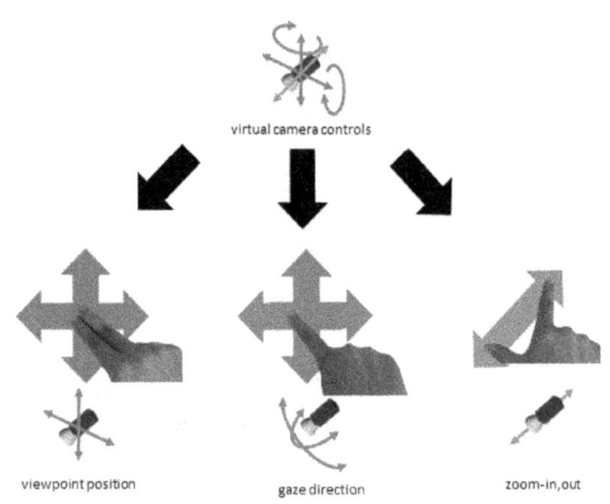

Figure 9. Our proposed multi-touch gestures for capturing free-viewpoint video

5. CONCLUSION

This paper introduced an intuitive and efficient 3D free-viewpoint video-browsing interface that applies multi-touch manipulation to virtual camera control. We focused on the multi-touch interface installed in tablet PCs as an input device. Subjective evaluations to define suitable gestures for virtual camera control were conducted. The results indicated the user's tendencies for virtual camera control and an example of the multi-touch gesture suite for controlling a virtual camera to capture free-viewpoint videos was proposed.

In future work, we will implement a few prototypes of the interfaces based on the extracted tendencies, and conduct user tests for evaluating the proposed multi-touch gestures.

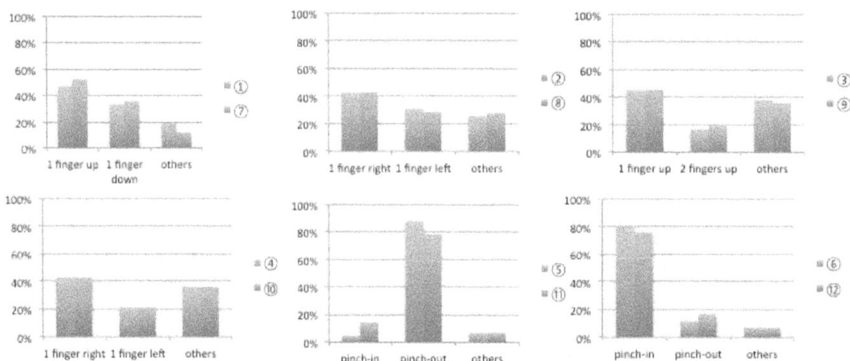

Figure 10. Results comparing the bird's-eye view to the field view videos

Table 1. Displayed videos in our evaluations

Bird's-eye view	Orientation	up	①	The ball is moving to upward. How do you control (rotate) the virtual camera capturing the ball as if you are shaking your head to change the orientation?
		right	②	The ball is moving to the right. How do you control (rotate) the virtual camera capturing the ball as if you are shaking your head to change the orientation?
	Viewpoint	forward	③	The ball is moving forward. How do you control (shift) the virtual camera capturing the ball with translation the camera position?
		right	④	The ball is moving rightward. How do you control (shift) the virtual camera capturing the ball with translation the camera position?
	Zoom	in	⑤	The ball is observed at the center. How do you control to control the field of view of the virtual camera to zoom in the ball?
		out	⑥	The ball is observed at the center. How do you control to control the field of view of the virtual camera to zoom out the ball?
Field view	Orientation	up	⑦	The ball is moving to upward. How do you control (rotate) the virtual camera capturing the ball as if you are shaking your head to change the orientation?
		right	⑧	The ball is moving to rightward. How do you control (rotate) the virtual camera capturing the ball as if you are shaking your head to change the orientation?
	Viewpoint	forward	⑨	The ball is moving forward. How do you control (shift) the virtual camera capturing the ball with translation the camera position?
		right	⑩	The ball is moving rightward. How do you control (shift) the virtual camera capturing the ball with translation the camera position?
	Zoom	in	⑪	The ball is observed at the center. How do you control to control the field of view of the virtual camera to zoom in the ball?
		out	⑫	The ball is observed at the center. How do you control to control the field of view of the virtual camera to zoom out the ball?

Table 2 Results of similarity of cosine

	①	②	③	④	⑤	⑥	⑦	⑧	⑨	⑩	⑪	⑫
①	–	–	–	–	–	–	–	–	–	–	–	–
②	0.026	–	–	–	–	–	–	–	–	–	–	–
③	0.863	0	–	–	–	–	–	–	–	–	–	–
④	0	0.933	0	–	–	–	–	–	–	–	–	–
⑤	0	0	0.056	0	–	–	–	–	–	–	–	–
⑥	0	0	0.141	0	0.197	–	–	–	–	–	–	–
⑦	0.993	0.01	0.871	0	0	0	–	–	–	–	–	–
⑧	0.027	0.998	0	0.946	0	0	0.01	–	–	–	–	–
⑨	0.872	0	0.979	0.019	0.011	0.135	0.88	0	–	–	–	–
⑩	0	0.933	0	1	0	0	0	0.946	0.019	–	–	–
⑪	0	0	0.073	0	0.98	0.359	0	0	0.029	0	–	–
⑫	0	0	0.146	0	0.237	0.992	0	0	0.139	0	0.382	–

Table 3. Experience and knowledge of the subjects

Questions								
Usually use a tablet PC?	Yes	8	No	6				
Familiar with for 3D contents?	Well	8	Sometimes	4	A little	2	Not at all	0

Table 4. The input gestures in the subjective evaluations

	1 finger up	1 finger down	1 finger right	1 finger left	2 fingers up	2 fingers down	2 fingers right	2 fingers left	pinch-in	pinch-out	circle(counter clockwise)	circle(clock wise)	1 finger arc	2 fingers arc	2tap
①	48%	33%	0%	0%	7%	0%	0%	0%	0%	0%	0%	0%	7%	5%	0%
②	0%	0%	43%	31%	0%	0%	7%	5%	0%	0%	0%	0%	7%	7%	0%
③	45%	12%	0%	0%	17%	14%	0%	0%	7%	2%	0%	2%	0%	0%	0%
④	0%	0%	43%	21%	0%	0%	14%	14%	0%	0%	0%	0%	0%	0%	7%
⑤	0%	0%	0%	0%	0%	0%	0%	0%	5%	88%	0%	7%	0%	0%	0%
⑥	0%	0%	0%	0%	0%	0%	0%	0%	81%	12%	7%	0%	0%	0%	0%
⑦	52%	36%	0%	0%	7%	0%	0%	0%	0%	0%	0%	0%	0%	5%	0%
⑧	0%	0%	43%	29%	0%	0%	7%	7%	0%	0%	0%	0%	7%	7%	0%
⑨	45%	12%	0%	0%	19%	7%	0%	0%	7%	0%	0%	2%	0%	0%	7%
⑩	0%	0%	43%	21%	0%	0%	14%	14%	0%	0%	0%	0%	0%	0%	7%
⑪	0%	0%	0%	0%	0%	0%	0%	0%	17%	76%	7%	0%	0%	0%	0%
⑫	0%	0%	0%	0%	0%	0%	0%	0%	79%	14%	0%	7%	0%	0%	0%

6. REFERENCES

[1] A. Cohe, M. Hachet, Understanding User Gestures for Manipulating 3D Objects from Touchscreen Inputs, In Proceedings of the 2012 Graphics Interface Conference (GI), 157-164, 2012.

[2] A. Cohe, F, Decle, M. Hachet, tBox: a 3d transformation widget designed for touch-screens, In Proceedings of the SIGCHI Conference on Human Factors in Computing Systems (CHI), 3005-3008,2011.

[3] J. Carranza, C. Theobalt, M. A. Mgnor, and H. P. Seidel, Free viewpoint Video of Human Actors, *ACM Transaction on Graphics*, Vol.22, No. 3, 569-577, 2003.

[4] F. Decle, M. Hachet, P. Guitton, ScrutiCam: Camera Manipulation Technique for 3D Objects Inspection, In Proceedings of the IEEE Symposium on 3D User Interfaces (3DUI), 19-22, 2009.

[5] J. Edelmann, A. Schilling, S. Fleck, THE DABR – A MULTITOUCH SYSTEM FOR INTUITIVE 3D SCENE NAVIGATION, In Proceedings of the 3DTV-Conference:The True Vision – capture, Transmission and Display of 3D Video (3DTV-CON), 1-4, 2009.

[6] M. Hachet, F. Decle, S. Knodel, P. Guitton, Navidget for Easy 3D Camera Positioning from 2D Inputs, In *Proceedings of IEEE Symposium on 3D User Interfaces* (3DUI), 83-89, 2008.

[7] M. Hancock, T. Cate, S. Carpendale, Sticky Tools: Full 6DOF Force-Based Interaction for Multitouch Tables, In Proceedings of ACM International Conference on Interactive Tabletops and Surfaces (ITS), 133-140, 2009.

[8] M. Hancock, S. Carpendale, A. Cockburn, Shallow-Depth 3D Interaction: Design and Evaluation of One-, Two-, and Three-Touch Techniques, In Proceedings of the SIGCHI Conference on Human Factors in Computing Systems (CHI), 1147-1156, 2007.

[9] U. Hinrichs, S. Carpendale, Gestures in the Wild: Studying Multi-Touch Gesture Sequences on Interactive Tabletop Exhibits, In *Proceedings of the SIGCHI Conference on Human Factors in Computing Systems* (CHI), ACM, 3023-3032, 2011.

[10] N. Inamoto, H. Saito, Immersive Observation of Virtualized Soccer Match at Real Stadium Model, In *Proceedings of International Symposium on Mixed and Augmented Reality* (ISMAR), 188-197, 2003.

[11] T. Kanade, P. Render, P. J. Narayanan, Virtualized Reality: Constructing Virtual Worlds from Real Scenes, *IEEE Multimedia*, Vol.4, No 1, 34-47, 1997.

[12] K. Kin, T. Miller, B. Bollensdorff, T. Derose, B. Hartmann, M. Agrawala, Eden: A professional Multitouch Tool for Constructing Virtual Organic Environments, In *Proceedings of the SIGCHI Conference on Human Factors in Computing Systems* (CHI), ACM, 1343 – 1352, 2011.

[13] T. Koyama, I. Kitahara, Y. Ohta, Live Mixed Reality 3D Video in Soccer Stadium, In *Proceedings of the 2nd IEEE/ACM*

International Symposium on Mixed and Augmented Reality (ISMAR), 178-187,2003.

[14] J. D. Mackinlay, S. K. Card, G. G. Robertson, Rapid Controlled Movement Through a Virtual 3D Workspace, In *Proceedings of the annual conference on Computer graphics and interactive techniques* (SIGGRAPH), 171-176, 1990.

[15] A. Martinet, G. Casiez, L. Grisoni, The Design and Evaluation of 3D Positioning Techniques for Multi-touch Displays, In *Proceedings of the IEEE Symposium on 3D User Interfaces* (3DUI), 115-118, 2011.

[16] W. Matusik, C. Buehler, R. Rasker, S. J. Gortler, L. McMillan, Image-based Visual Hulls, In *Proceedings of the 27th annual conference on Computer graphics and interactive techniques* (SIGGRAPH), 369-374, 2000.

[17] C. Moerman, D. Marchel, L. Frisoni, Drag'n Go: simple and fast navigation in virtual environment, In Proceedings of the IEEE Symposium on 3D User Interfaces (3DUI), 15-18, 2012.

[18] T. Ohnishi, R. Lindeman, K. Kiyokawa, Multiple Multi-Touch Touchpads for 3D Selection, In *Proceedings of the IEEE Symposium on 3D User Interfaces* (3DUI), 115-116, 2011.

[19] J. L. Reisman, P. L. Davidson, J. Y. Han, A Screen-Space Formulation for 2D and 3D Direct Manipulation, In Proceedings of the 22nd annual ACM symposium on User interface software and technology (UIST), 69-78, 2009

[20] D. Schmalstieg, L. M. Encarnacao, Z. Szalavari. Using transparent props for interaction with the virtual table, In *Proceedings of ACM Symposium on Interactive 3D Graphics* (I3D), ACM, 147-153, 1999.

[21] C. Ware, S. Osborne, Exploration and Virtual Camera Control in Virtual Three Dimensional Environments, In *Proceedings of ACM Symposium on Interactive 3D Graphics* (I3D), ACM, 175 – 183, 1990.

[22] T. Watanabe, I. Kitahara, Y. Kameda, Y. Ohta, 3D Free viewpoint Video Capturing Interface by Using Bimanual Operation, In Proceedings of the 3DTV-Conference:The True Vision – capture, Transmission and Display of 3D Video (3DTV-CON), 2010.

[23] J. O. Wobbrock, M. R. Morris, A. D. Wilson, User-defined gestures for surface computing, In Proceedings of the SIGCHI Conference on Human Factors in Computing Systems (CHI), 1083-1092, 2009.

[24] R. C. Zeleznik, A. S. Forsberg. Unicam – 2d gestural camera controls for 3D environments, In *Proceedings of ACM Symposium on Interactive 3D Graphics* (I3D), ACM, 169-173,1999.

Analysis of a Real-World HTTP Segment Streaming Case

Tomas Kupka, Carsten Griwodz, Pål Halvorsen
University of Oslo & Simula Research Laboratory
{tomasku,griff,paalh}@ifi.uio.no

Dag Johansen
University of Tromsø
dag@cs.uit.no

Torgeir Hovden
Comoyo
torgeir@comoyo.com

ABSTRACT

Today, adaptive HTTP segment streaming is a popular way to deliver video content to users. The benefits of HTTP segment streaming include its scalability, high performance and easy deployment, especially the possibility to reuse the already deployed HTTP infrastructure. However, current research focuses merely on client side statistics like for example achieved video qualities and adaption algorithms. To quantify the properties of such streaming systems from a service provider point of view, we have analyzed both sender and receiver side logging data provided by a popular Norwegian streaming provider. For example, we observe that more than 90% of live streaming clients send their requests for the same video segment with an inter-arrival time of only 10 seconds. Moreover, the logs indicate that the server sends substantially less data than is actually reported to be received by the clients, and the origin server streams data to less clients than there really are. Based on these facts, we conclude that HTTP segment streaming really makes use of the HTTP cache infrastructure without the need to change anything in the parts of Internet that are not controlled by the streaming provider.

Categories and Subject Descriptors

C.4 [**Performance of systems**]; D.2.8 [**Software Engineering**]: Metrics—*performance measures*; H.3.3 [**Information Search and Retrieval**]

General Terms

Measurement, Performance

Keywords

Live adaptive HTTP segment streaming, streaming, Microsoft Smooth streaming, streaming performance

EuroITV'13, June 24–26, 2013, Como, Italy.

1. INTRODUCTION

The number of video streaming services in the Internet is rapidly increasing. The number of videos streamed by these services is on the order of tens of billions per month where YouTube alone delivers more than four billion video views globally every day [21]. Furthermore, with only a few seconds delay, many major (sports) events like NBA basketball, NFL football and European soccer leagues are streamed *live*. For example, using the modern adaptive HTTP segment streaming technology, events like the 2010 Winter Olympics [22], 2010 FIFA World Cup [14] and NFL Super Bowl [14] have been streamed to millions of concurrent users over the Internet, supporting a wide range of devices ranging from mobile phones to HD displays.

As a video delivery technology, streaming over HTTP has become popular for various reasons. For example, it is scalable, runs on top of TCP, provides NAT friendliness and is allowed through most firewalls. The idea of splitting the original stream into segments and upload these to webservers in multiple qualities (bitrates) for adaptive delivery has been ratified for an international standard by ISO/IEC as MPEG Dynamic Adaptive Streaming over HTTP (MPEG-DASH) [19]. From the users' perspective, the video segment bitrate is selected based on observed resource availability and then downloaded like traditional web objects. Such an approach where video is streamed over TCP has been shown to be effective as long as congestion is avoided [20], and adaptive HTTP streaming has been proved to scale to millions of concurrent users [1]. Such a solution is therefore used by Microsoft's Smooth Streaming [22], Adobe's HTTP Dynamic Streaming [5] and Apple's live HTTP streaming [17] – technologies which are used in video streaming services from providers like Netflix, HBO, Hulu, TV2 sumo, Viaplay and Comoyo.

Current research has focused on various aspects of efficient client side delivery [6, 9, 15, 18]. To quantify the properties of such streaming systems from a service provider point of view, we have analyzed both sender and receiver side logging data provided by the Norwegian streaming provider Comoyo [7]. In this paper, we focus on live streaming events. In such scenarios, the segments are produced and made available for download periodically, i.e., a new segment becomes available as soon as it has been recorded, completely encoded and uploaded to a web-server, possibly located in a content distribution network (CDN) like Akamai or Level 3 [13]. The question that we answer in this paper is how adaptive HTTP segment streaming systems behave as a whole from a service provider point of view. For example, we observe that for live streaming, about 90% of clients download the same video segment within a 10 second period, i.e., 90% of the users are at most 10 seconds behind the most "live" client. Furthermore, we find that the clients report receiving much more video segments than the server actually sent. Based on the client IP distribution and the IP distribution as seen by the server, we conclude that HTTP cache proxies must be heavily used between the origin server and the clients. This proves for example that the existing HTTP infrastructure is efficiently used and offloads the origin server in real world networks.

The rest of this paper is organized as follows: In Section 2, we discuss examples of related work. Section 3 sketches the network setup used by Comoyo [7]. Section 4 through 6 analyse the provided logs. The analysis is split into client log analysis in Section 5 and into server log analysis in Section 6. We conclude the paper in Section 7, where we also discuss future work.

2. RELATED WORK

The popularity of HTTP segment streaming led to many real world studies for example, Müller et al. [15] collected data using different commercial streaming solutions as well as the emerging MPEG-DASH standard over 3G while driving a car. They compared the average bitrate, number of quality switches, the buffer level and the number of times the playout was paused because of re-buffering. A similar study was performed by Riiser et al. [18]. In their paper, they evaluated commercial players when on the move and streaming over 3G. They additionally proposed a location aware algorithm that pre-buffers segments if a network outage is expected on a certain part of a journey (based on previous historical records for that location). Houdaille et al. [9] focused on a fair sharing of network resources when multiple clients share the same home gateway and Akhshabi et al. [6] compared rate adaption schemes for HTTP segment streaming using synthetic workloads.

However, in general, studies, that we are aware of, focus all on the client side. In other words, they measure the user experience in one form or an other from the client side perspective. We believe this is mainly because the commercial content providers are protective of their data, and it is not easy to get any statistics or logs. However, the server-side performance is equally important in order to scale the service to 1000s of users. In this respect, we have a collaboration with a commercial streaming provider Comoyo [7], and we present the insights gained from both server and client side log analysis in this paper.

3. COMOYO'S STREAMING SETUP

Comoyo [7] is a Norwegian streaming provider, providing access to an extensive movie database of about 4000 movies. Additionally, Comoyo provides live streaming of various events like the Norwegian premier league. Most of the soccer games are played on Sundays and some on Wednesdays, and hence, these two days are the busiest times for football live streaming. The logs that are analysed in this paper are from a Wednesday, but first, we shall look at Comoyo's network infrastructure.

The network infrastructure for both on-demand and live services is illustrated in Figure 1. The difference is only in the source of the stream, which can be a live stream coming in over a satellite or a pre-recorded stream coming from the movie database. The on-demand movies are streamed either using the classic progressive streaming based on Windows Media Video (wmv) or via Microsoft's Smooth streaming [22] based on HTTP segment streaming, which is also used for live streaming. For live streaming, the video content is available in 0.5, 1, 2 and 4 MBit/s, and the audio in 96 Kbit/s.

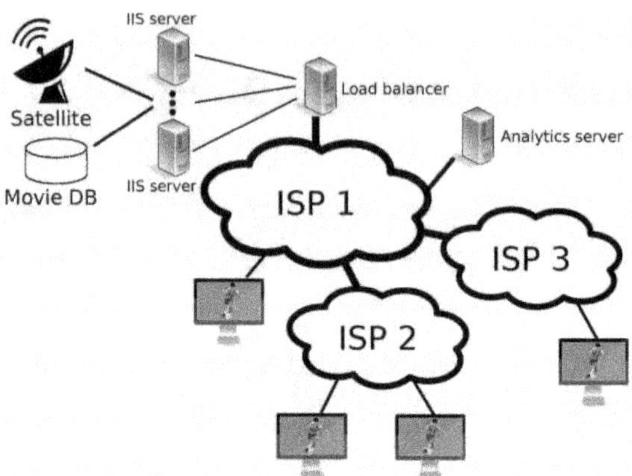

Figure 1: Streaming network infrastructure

Since, we focus only on Smooth streaming in this paper, we shall describe it briefly here. Smooth streaming [22] is a streaming technology by Microsoft. The file format used is based on ISO/IEC 14496-12 ISO Base Media File Format [2]. The reason for choosing it as its base is that it natively supports fragmentation. There are two formats specified, the disk file format and the wire file format. Each video (there is one for each bitrate) is stored as a single file in the disk file format. Additionally to the files containing the video data there are two more files on the server. One of these files (the *.ism* file) describes the relationship between the media tracks, bitrates and files on the disk. This file is only used by the server. The other file is for the client. It describes the available streams (codec, bitrates, resolutions, fragments, etc.) and it is the first file that the client requests to get information about the stream.

The Smooth streaming clients use a special URL structure to request a video fragment. For example, a client uses the following URL to request a SOCCER stream fragment that begins 123456789 time units from the content start[1] in a 1 Mbit/s bitrate from comoyo.com:

```
http://comoyo.com/SOCCER.ism/QualityLevels(1000000)/
Fragments(video=123456789)
```

The server parses the URL and extracts the requested fragment from the requested file. The official server that has these capabilities is the IIS server [16] with an installed Live Smooth Streaming extension [3]. It looks up in the *.ism* file the corresponding video stream and extracts from it the requested fragment. The fragment is then send to the client. Note that the fragments are cacheable since the request URLs of the same fragment (in the same bitrate) from two different clients look exactly the same and as such can be cached and delivered by the cache without consulting the origin server.

[1]The exact unit size is configurable, but is usually 100ns.

As Figure 1 shows, a number of IIS servers is deployed behind a load balancer that distributes the incoming requests. The load balancer is connected directly to the Internet, i.e., ISP 1 in the figure. However, content is not served exclusively to subscribers of ISP 1, but also to subscribers of other ISPs as illustrated in the figure. Besides the IIS servers; an analytics server is deployed and connected to the Internet. This server's role is to collect various information from the video clients. For example, the clients notify the analytics server when the video playout is started, stopped or when the quality changes.

4. LOG ANALYSIS

For our study, we were provided with two types of logs. We call these the server log and the client log based on where the information is collected and logged. We picked one server log and the corresponding client log from the same day, 23.05.2012. These logs were found representative for similar live streaming events. The logs are 24-hour logs collected for 8 soccer games. The server log contains information about every segment request as received by the load balancer. The most important information is the time of the request, the request URL, client's IP address, response size and streaming type, which can be progressive streaming, on-demand Smooth streaming or live Smooth streaming. Table 1 lists all the other important information that is logged about each request.

The client log is a collection of events collected from all the clients by the analytics server while the clients are streaming. Collected events are described in Table 2. The information logged for every event is described in Table 3. Specifically, every client event includes the information to which user in which session and at what time the event happened. Note that within a session a user can only stream one content, and a session can not be shared between multiple users. A typical sequence of events in one session is shown in Figure 2. In our analysis, we first analyse the logs separately and then point out the differences and inconsistencies and draw conclusions.

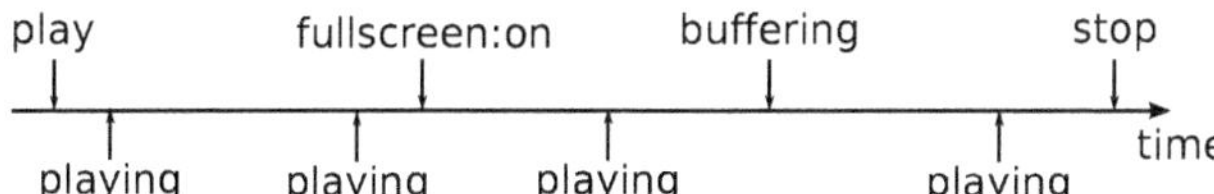

Figure 2: An example of events sent within a session

5. SERVER LOG ANALYSIS

Our analysis of the server log showed that there were 1328 unique client IP addresses in the log over the tracked 24-hour period. From all the IP addresses, 56% was tagged as live Smooth streaming, 6% as on-demand Smooth streaming and 44% as progressive streaming. 6% of all IP addresses was associated with more than one type of streaming, e.g., showed up one time as live Smooth streaming and another time as progressive streaming. In other words, if each user had a unique IP address there would had been 744 people watching the live stream.

Client IP	Client's IP address
HTTP method	HTTP method used (e.g. GET)
Request URL	The URL of the request
Response size	Response size in bytes
Useragent	Client's user agent
Cookie	Cookie information in the request
HTTP status	Status code of the response
Time taken	Time taken by the server in ms
Streaming type	The type of streaming. Possible values are progressive streaming, on-demand Smooth streaming, live Smooth streaming
Timestamp	Timestamp of the request in UNIX epoch format

Table 1: Information about every request in the server log file

play	The start of the playout
playing	Regularly sent every 60 seconds when the playout is in progress
position	Sent when the user seeks to a different position in the video
pause	Sent when the user pauses the playout
bitrate	Sent when the player switches to a different bitrate, i.e., resutling in a different video quality
buffer	Sent when an buffer underrun occurs on the client side
stop	Sent at the end of streaming
subtitles:on	Turn on subtitles
subtitles:off	Turn off subtitles
fullscreen:on	Switch to fullscreen
fullscreen:off	Quit fullscreen mode

Table 2: The reported client events

User device info.	This information includes the type of browser, OS, etc.
User ID	Identification of the user
Content ID	Identification of the content
Event ID	See Table 2
Event timestamp	Time of the event in seconds that elapsed since midnight
Event data	E.g. for bitrate event the bitrate the player switched to
Session ID	The session identification
Viewer	Always SilverlightPlayer v. 1.6.1.1
Client IP	Client's IP address as seen by the analytics server
Geographic info.	This information includes country, city etc.
ISP	Client's ISP

Table 3: Information about every client event

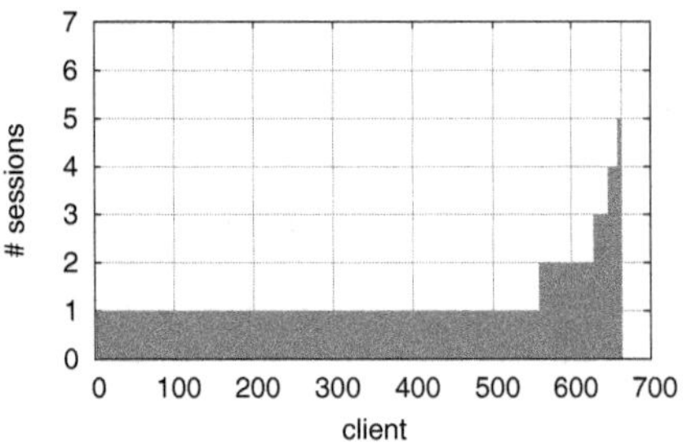

(a) Sessions per client IP

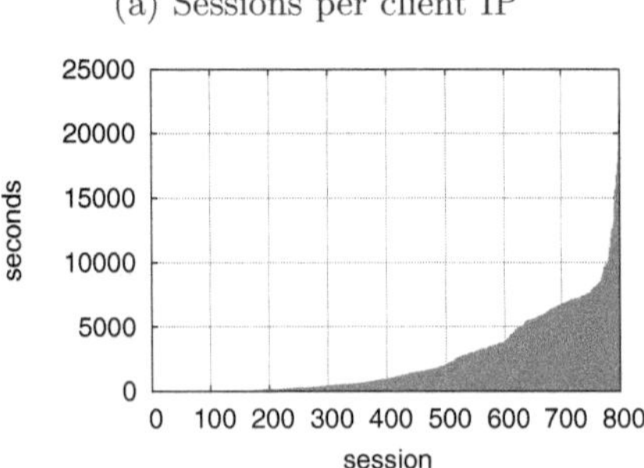

(b) Session duration

Figure 3: Sessions statistics based on the server log

Sessions.

Since the IIS server is basically a "dumb" HTTP server, it does not keep state about an ongoing session. It therefore does not log the user identification, content id or the session id (as is done by the analytics server in the client log, see Table 3). For this reason, we assumed (only for the server log) that a client is uniquely identified by its IP address, i.e., we assumed there is one client per IP address[2] (this proved to be a reasonable assumption, see Section 6).

The content streamed can be extracted from the URL that is included in the server log, since the URL has a fixed format for Smooth streaming as is described in Section 3. It is therefore possible to find out the content id, bitrate and the playout time in seconds of the segment within a video stream by parsing the URL. Unfortunately, it does not include the session id. We therefore approximated the session id with the combination of the client IP address (= client id) and the content id, e.g., if a user downloaded URLs with the content id A and B we say that the user had two sessions. Figure 3(a) shows that most of the clients had only one session. This is not what we find in the client logs as we will see later. It is also interesting to measure the live session duration. To do this, we calculated the time difference between the first and last request within a session. The session duration distribution is shown in Figure 3(b). The average session duration is 42 minutes (related to the break after 45 minutes in the game). Note that 107 clients are associated with multiple sessions, that is why there are more live sessions than live clients.

Byte transfers.

We noticed that some segments with the same playout time (included in URL, see Section 3) and within the same session were downloaded more than once, each time in a different bitrate. We conjecture this can happen because of two different reasons. First, multiple clients with different bandwidth limitations can be hiding behind the same IP (which can not be distinguished based on the available information). Second, the bitrate adaptation algorithm changes its decision and re-downloads the same segment in a different bitrate. Either way, bytes are unnecessarily downloaded, i.e., if there were multiple clients behind the same IP they could have downloaded the segment once and shared it via a cache; if the adaptation algorithm downloaded multiple bitrates of a segment it could have just downloaded the highest bitrate. We calculated the number of wasted bytes as the total number of bytes downloaded minus the bytes actually used. For this purpose, we assumed that only the highest bitrate segment for the same playout time within the same session is used by the client, i.e., if the same IP address downloads for the same playout time segments with 1 Mbit/s, 2 Mbit/s and 3 Mbit/s bitrate, only the 3 Mbit/s segment is played out and the other two segments are dis-

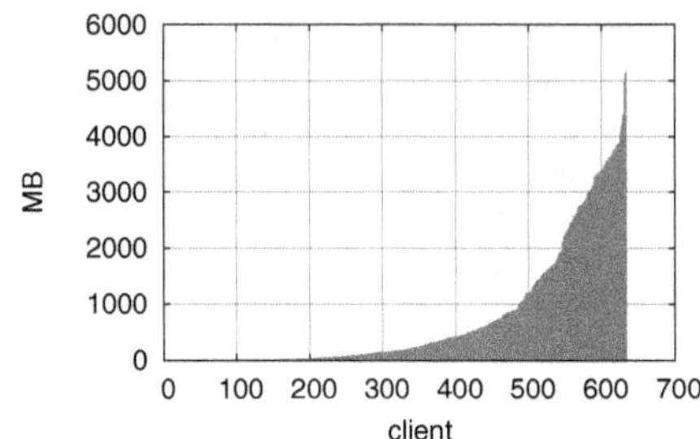

(a) Downloaded bytes

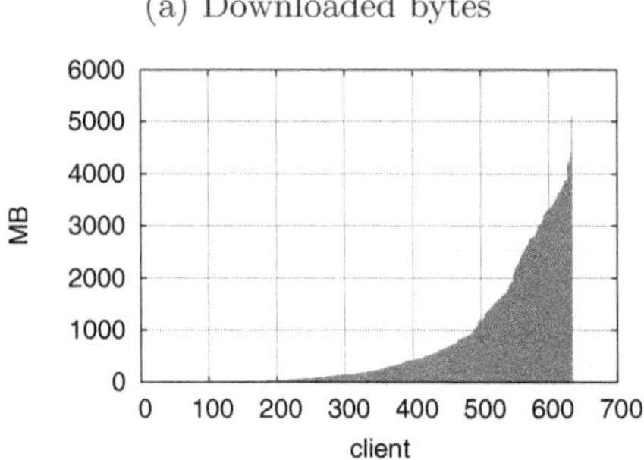

(b) Used bytes ($Downloaded - Wasted$)

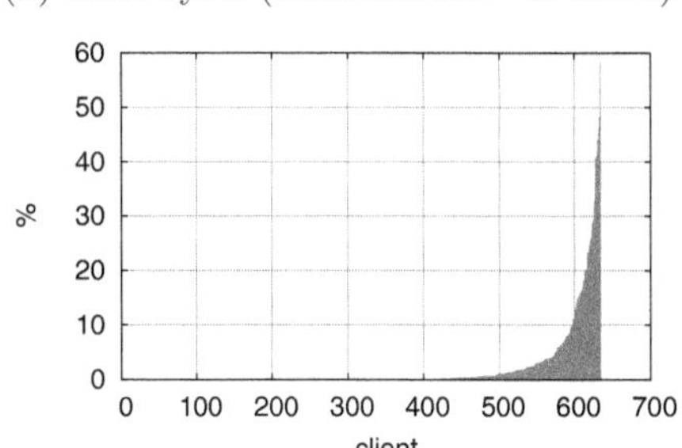

(c) Wasted bytes in percent of the total bytes downloaded

Figure 4: Per client bytes statistics

[2]We assume that, for example, there are no two clients sharing the same IP because they are behind the same NAT.

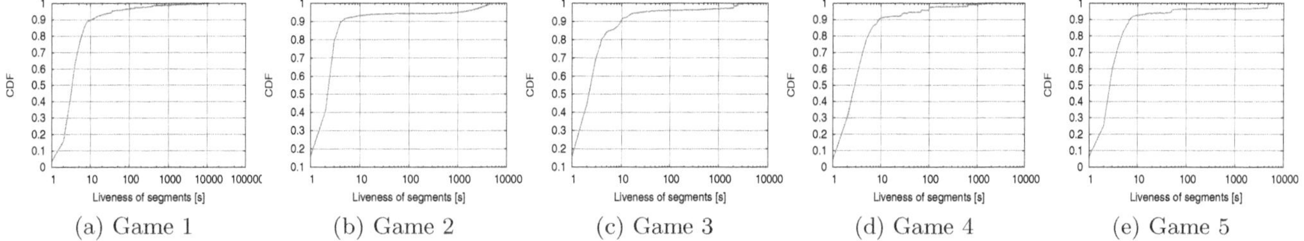

(a) Game 1 (b) Game 2 (c) Game 3 (d) Game 4 (e) Game 5

Figure 5: Liveness (absolute value) of segments based on the server log

carded (wasted). Based on the server log, approximately 13 GB out of the 467 GB downloaded in total were wasted this way. The 13 GB or 3% of the whole data could have been possibly saved if a different adaptation strategy or an HTTP streaming aware cache was used.

Figure 4 shows that there are clients that waste more than 50% of their downloaded bytes. Specifically, there is one client from Greece that contributes with 15.5% (2 GB) to the total amount of wasted bytes, even though its share on the total amount of downloaded bytes is only 0.85%. We think that this person might have recorded the stream in every quality, so there is also room for improvement on detecting and preventing behaviour like this (if stated in the service agreement).

Liveness.

Each log entry, i.e., request of segment i of a particular game, includes the server timestamp when the server served that particular download request, T_i^d. It also includes the timestamp of the playout time of the video segment, T_i^p. Since we do not know the relationship between T_i^d and T_i^p (we just know that they increase with the same rate), we find the minimum of $T_i^p - T_i^d$ over all segments i for each game and assume that this is the minimal liveness (time the client is behind the live stream) for that game. In other words, the video segment with the minimal $T^p - T^d$ has liveness 0, i.e., is as live as possible. In this respect, we compute the liveness of all other segments in each game. The results are plotted in Figure 5 for the 5 most popular games. We see that about 90% of all segments in each game was less than 10 seconds behind the live stream (the liveness measured relatively to the most live segment as explained above). This means that 90% of the requests for a particular segment came within a 10 seconds period. There are also some video segments that were sent with quite some delay. One plausible explanation is that these are from people that joined the stream later and played it from the start.

6. CLIENT LOG ANALYSIS

Every player that streams a live soccer game reports to the analytics server the events described in Table 3. The type of events reported is summarized in Table 2. The client log includes all reported events for a duration of 24-hours. The general statistics based on the client log are summarized in Table 4.

Client Location.

There were 6567 unique client IP addresses in the client log. This is significantly more than the corresponding 748 live streaming client addresses found in the corresponding server log. Not surprisingly most of the IP addresses was located in Norway (Figure 6), but there were also IP addresses from almost all corners of the world (Figure 7). The international IP addresses correlated with the areas that are

Number of IP addresses	6567
Number of users	6365
Number of sessions	20401 (or 3.21 per user on average)
Number of sessions with at least one buffer underrun	9495 (or 46% of total)
Number of sessions with at least one bitrate switch	20312 (or 99.5% of total)
Number of content ids	15
Number of countries	36
Number of cities	562
Number of ISPs	194
Number of users with multiple ISPs	113 (or 2% of all users)
Number of IPs with multiple users	31

Table 4: Statistics from the client log

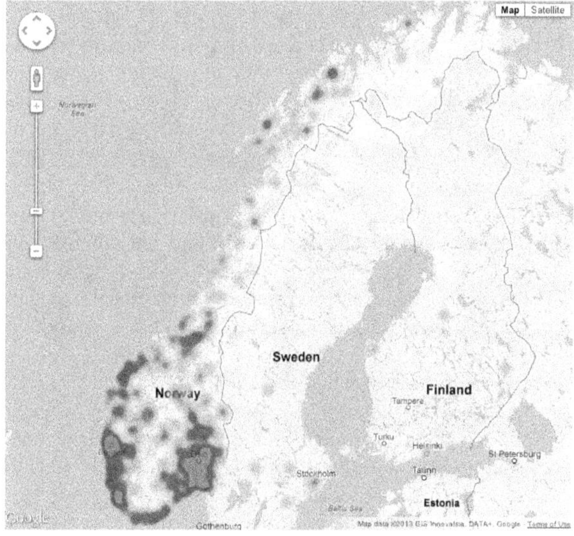

Figure 6: Geographical client distribution in Norway (the highest density of clients is in the red areas).

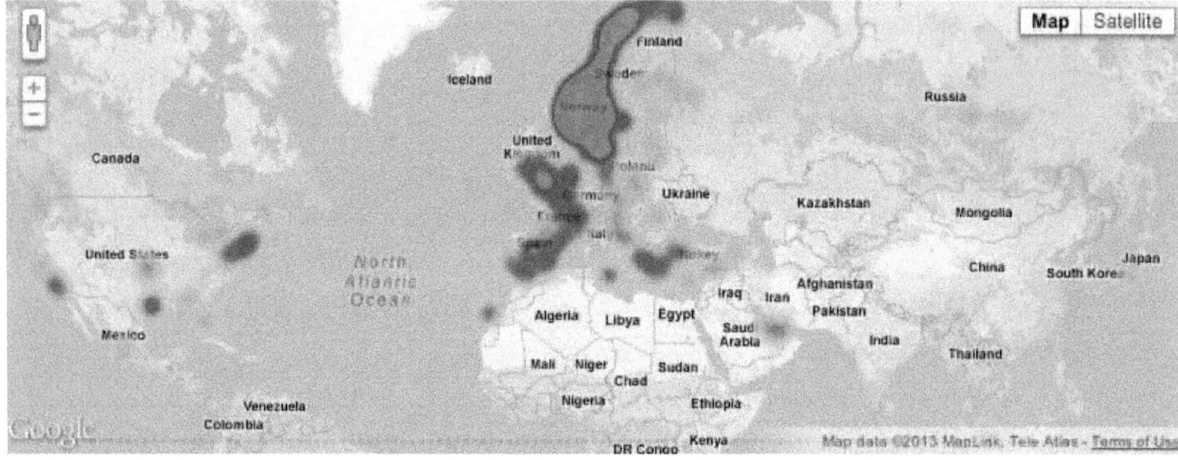

Figure 7: Geographical client distribution in the world (the highest density of clients is in the red areas).

know for high Norwegian population like Spain and England. Furthermore, the assumption in the server log analysis was that an IP address matches one user/client. Figure 8 shows that this assumption is quite reasonable. Only a very small number of users used more than one IP address, and an IP address was mostly used by only one user.

Furthermore, Figure 10 shows that most of the users used only one ISP (an IP address is hosted by one ISP only). Moreover, we saw that the majority of users used either Canal Digital Kabel TV AS (2560) or Telenor Norge AS (2454) as their ISPs. The other 1442 users were served by other providers (192 different network providers in total). We could see that the quality of the streaming in terms of the number of buffer underruns (periods the streaming is paused because of slow download) depends on the ISP, see Figure 9. For example, the majority of sessions at the Oslo Airport experienced buffer underruns (the airport provides a WiFi network).

Proxy Caching.

We hypothesise that the significant difference in the number of IP addresses in the server log and the client log is due to the use of cache proxies as known from traditional HTTP networks. The proxies reply to client requests before they get to the load balancer (Figure 1) and are logged. We expect the caches to be somehow related to ISPs, i.e., we would expect a cache on the ISP level to reply to all but the first request of the same segment, e.g., the open source proxy Squid [4] only sends the first request to the origin server all consecutive requests are served from the partially downloaded data in the cache.

Our hypothesis is supported by the IP to ISP ratio. The server log IP to ISP ratio for live streaming is 0.08 whereas the client log ratio is 0.03. In other words, we find 11% of the client log IP addresses in the server log, but find over 30% of ISPs from the client log in the server log (see Table 5 for exact numbers). We therefore hypothesise that HTTP caches located between the client and the server reply to a large number of client requests. This means that, as expected, one of the large benefits of HTTP segment streaming is the reuse of existing HTTP infrastructure.

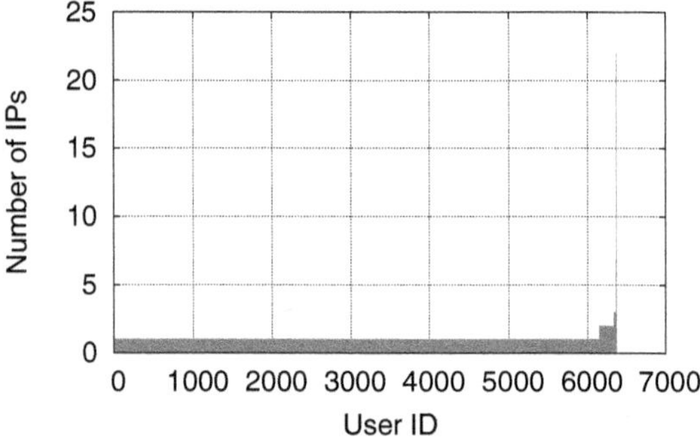

(a) Number of IP addresses per user

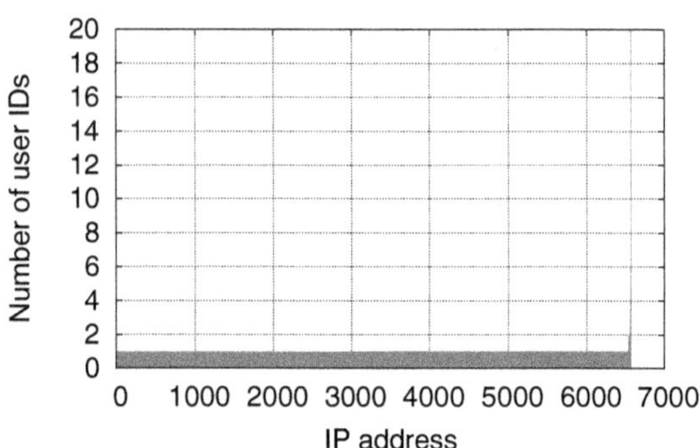

(b) Number of users per IP address

Figure 8: User to IP address mapping

Content and Sessions.

On the content side, we observe in Figure 11 that only about 2000 users streamed more than one content, i.e., one football game. The most popular content was game 1 with over 7500 sessions followed by 4 games with about 2000 to about 4000 sessions.

We also estimated the session durations based on the client log. This is not a straightforward task since only a small fraction of the sessions is started with a *play* event and terminated with a *stop* event (we suspect the user closes the browser window without first clicking on the stop button in

Type	Server log	Match in client log
live Smooth streaming	748 IPs	723 IPs (or 97%) hosted by 58 ISPs
on-demand Smooth streaming	74 IPs	5 IPs (or 7%) hosted by 3 ISPs
progressive streaming	581 IPs	390 IPs (or 67%) hosted by 32 ISPs

Table 5: IP to ISP statistics

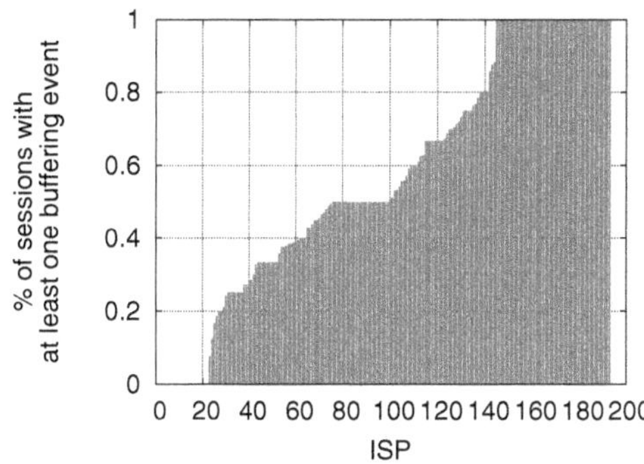

Figure 9: The percentage of sessions with at least one buffer underrun by ISP

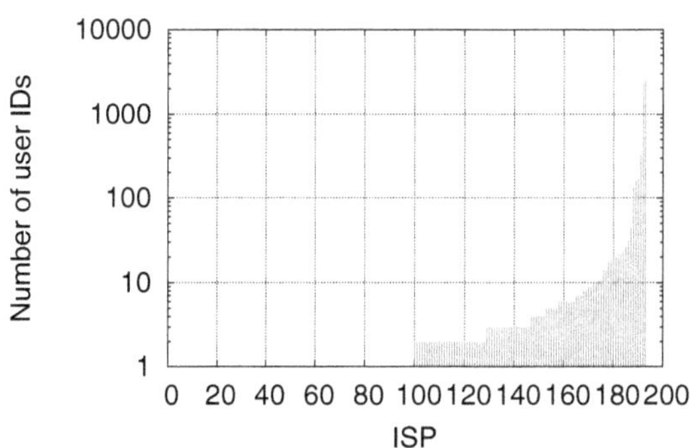

(a) Number of user IDs per ISP

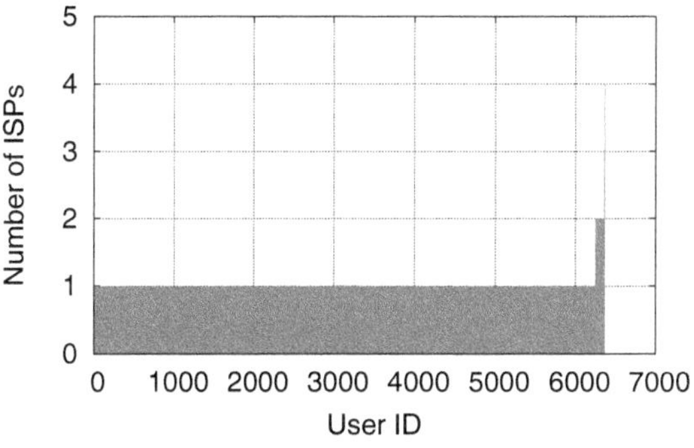

(b) Number of ISPs per user ID

Figure 10: ISP to user statistics

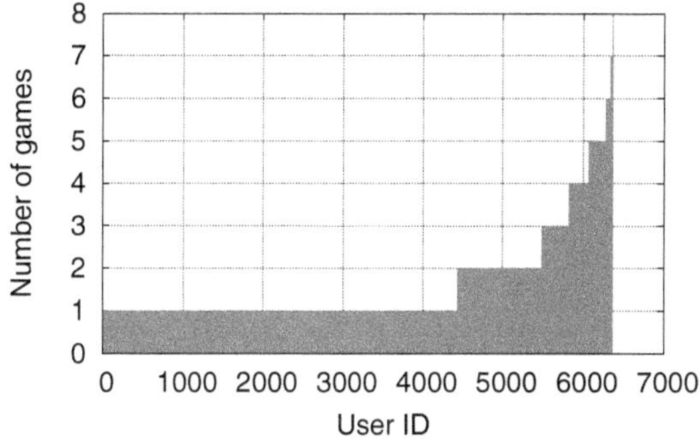

(a) Number of games streamed by a user

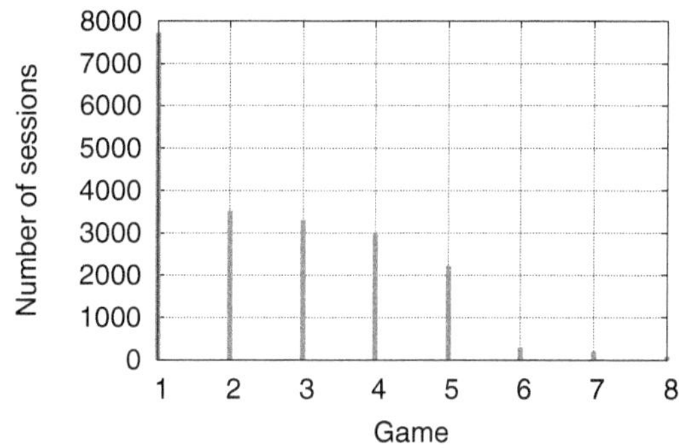

(b) Number of sessions per game (8 games were scheduled for that Sunday)

Figure 11: Content statistics

the player). We therefore calculated the session duration as the time difference between the first and last *playing* event[3]. Session durations calculated this way are, in general, shorter than they really were because they do not include the time before the first *playing* event and the time after the last *playing* event. Note also that the session duration represents the time the player was playing. Particularly, it does not represent the time the player window was open, but rather the time the player was playing a stream, e.g., the time when the stream was paused is excluded. Figure 12(b) shows that the calculated session durations ranged from 100 seconds to more than 3 hours. Please note that 6826 sessions did not contain a *playing* event and therefore are not represented in this graph (e.g., they were too short for a *playing* event to be sent).

Figure 12(c) shows that in the majority of sessions the bitrate was switched at least 3 times. However, Figure 12(d) shows that even with bitrate (quality) switching, the clients were not able to prevent buffer underruns. The logs report that more than a half of the sessions experienced at least one buffer underrun.

It is clear, from Figure 13, that the pressure on a streaming system like this is the biggest when a event begins. In this case all the games started at 5 PM UTC, and that is the time most of the clients joined. Figure 14 shows the number of active sessions over time.

[3]The *playing* event is sent regularly when the live stream is playing.

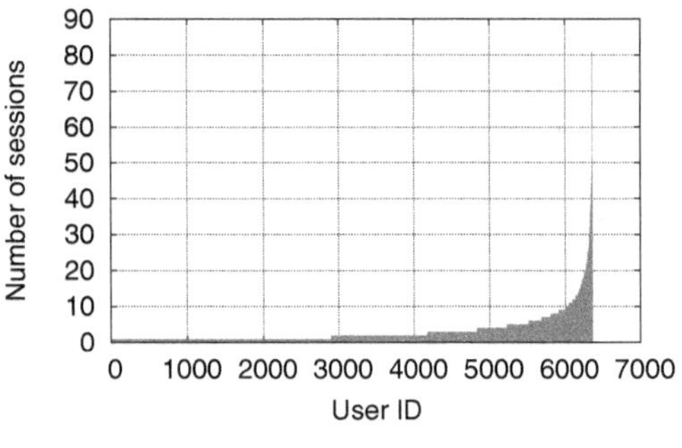

(a) Number of sessions per user

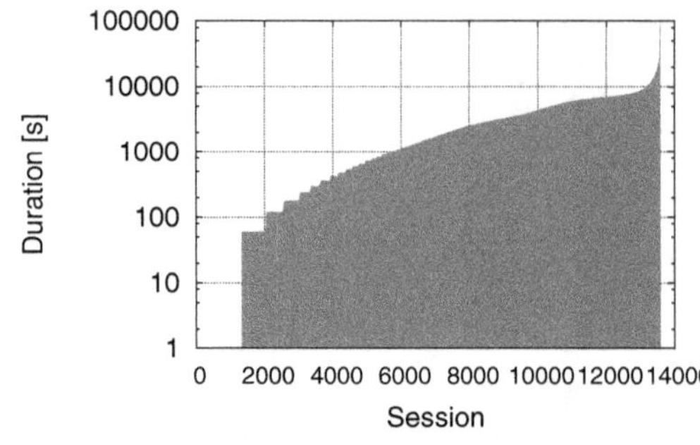

(b) Session duration based on the *playing* event

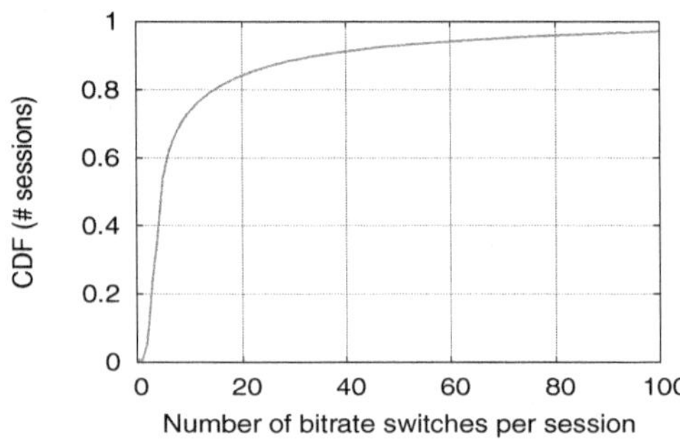

(c) Number of bitrate switches in a session

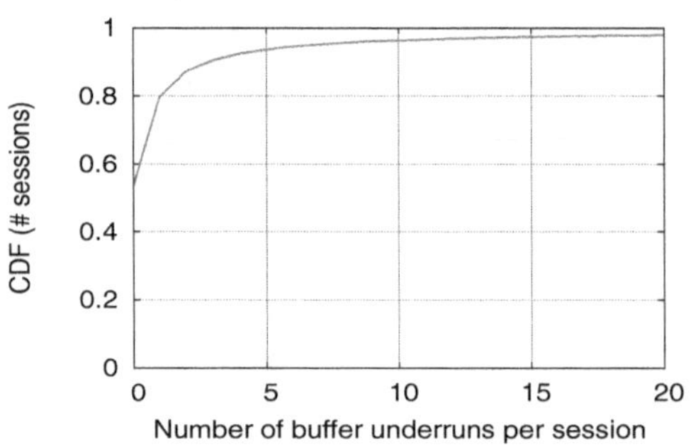

(d) Number of buffer underruns in a session

Figure 12: Session statistics based on the client log

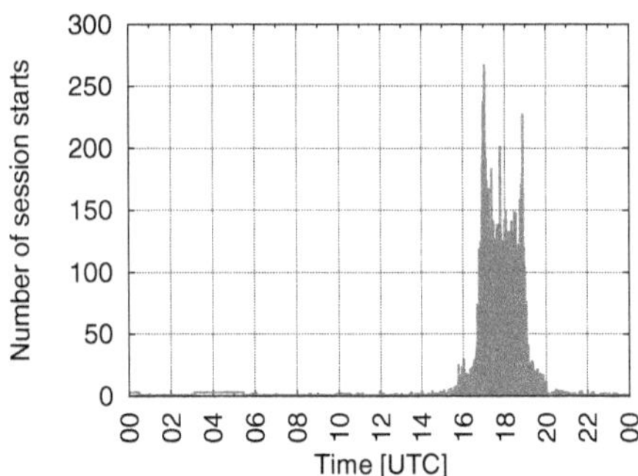

Figure 13: Number of sessions that were started per minute (all games started at 17:00 UTC)

Figure 14: Number of sessions at each point in time (minute resolution)

Byte transfers.

We also estimated the number of downloaded bytes based on the client log. We were very conservative and our estimations were very likely smaller than what the reality was. We split every session into periods bordered by *playing* events and/or *bitrate* events so that there was no *pause*, *stop*, *position*, *play* or *buffer* (buffer underrun) event inside a period. This ensures that the player was really playing during a period of time defined like this (session examples with bitrate switching are shown in Figure 15). A bitrate (the video content was available in 0.5, 1, 2 and 4 Mbit/s, and the audio in 96 KBit/s) was additionally assigned to each period based on the client *bitrate* event reports. The period duration multiplied by the corresponding bitrate gave us a good estimate of the bytes downloaded by the client in that period since Constant BitRate (CBR) encoding is used for the live streaming. The sum of bytes received in each period of a session gave us the total number of bytes downloaded in a session.

It is interesting that the final sum of bytes received by all sessions is many times higher than the number of bytes served by the server. The server log reports 551 GB in total whereas the estimate based on the client log is about 4.5 times higher. Even if only the smallest available bitrate is assigned to every period, the total amount of data received by the clients is 2.5 times higher than the value from the server log. This is another evidence for the presence of HTTP caches that respond to a significant number of requests.

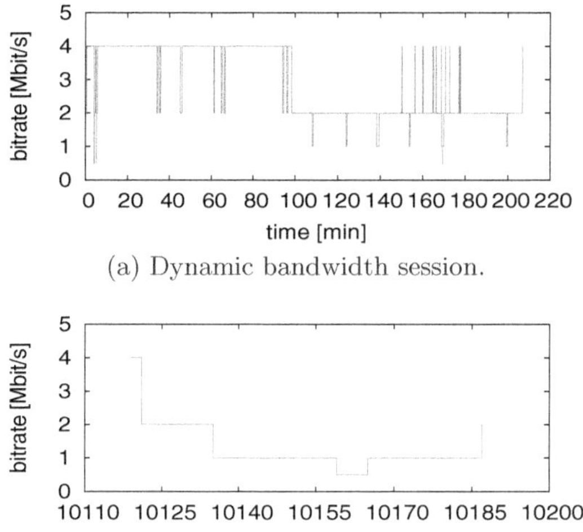

(a) Dynamic bandwidth session.

(b) Zoomed in to minutes 168 to 170.

Figure 15: Example of bitrate adaptation throughout a session based on client reports

7. CONCLUSION AND FUTURE WORK

In this paper, we analysed the logs of adaptive HTTP segment streaming provided by Comoyo [7], a Norwegian streaming provider. We analysed two types of logs; one from the origin server and one from the analytics server to which the clients report. The analysed data included all streaming sessions ranging from very dynamic to very static sessions, see example in Figure 15.

We observed that about 90% of the requests for the same segment in live streaming is sent within a period of 3 to 10 seconds depending on the content (analysed football game). This gives a great potential for caching. This also means that a HTTP proxy cache needs to cache a segment only very shortly.

We also deduced from the segments downloaded multiple times that at least 3% of the bytes downloaded could have been saved if more HTTP caches or a different adaptation strategy had been used. Moreover, based on the number of bytes downloaded and the IP address distribution in the server and the client log, we found evidence that segments must be delivered from other sources than from the examined servers. This suggests the presence of HTTP caching proxies in the Internet that had served the requests before they got to the load balancer. This means that the origin server is offloaded and that the idea of reusing the (unmodified) HTTP infrastructure really works in real scenarios.

Furthermore, since a significant portion of the data is distributed by HTTP caches that are most likely not aware of what they are serving, we conclude that it is important to look at how HTTP streaming unaware server performance can be increased by client side or very light server side modifications. In other words, it is important to look at the performance of HTTP servers that actually deal with HTTP segment streaming traffic.

The increase of the performance of such a server is, of course, only interesting if there is some kind of bottleneck. We can eliminate the disk as being the bottleneck for live streaming as the segments are small and clients are interested only in the most recent segments of the live stream. As such, the disk I/O poses no problem with current hardware. After the disk bottleneck elimination, we are left with the problem of a network bottleneck.

There are two types of bottlenecks - a client side and a server side bottleneck. The difference is that the client side bottleneck cannot be dealt with with a faster server, i.e., the server has the capacity to serve higher bitrate segments, but the client link is not fast enough to receive them in time. The obvious solution is to throw money at the problem and upgrade the client link. In our previous work we looked at how to deal with the client bottleneck using multiple links the client might have [8].

If the bottleneck is on the server side, the server capacity needs to be upgraded to deal with the bottleneck. This can be either done by using multiple servers or by increasing the capacity of the server already deployed. We looked at how to increase the capacity of a server without upgrading its physical link, i.e., by using different client request strategy, changing the TCP congestion control etc., in [11, 12].

As ongoing work, we are investigating new functionality provided to the user in order to further use the data after the large drop at 19:00 UTC in Figure 13. One idea is to allow the users to search for events and make their own video playlists on-the-fly [10] which again could be shared in social media etc. Such interactions will change the video access patterns, and new analysis will be needed.

Acknowledgment

This work is supported by the iAD (Information Access Disruptions) center for Research-based Innovation funded by Norwegian Research Council, project number 174867.

8. REFERENCES

[1] Move Networks. http://www.movenetworks.com, 2009.

[2] Coding of audio-visual objects: ISO base media file format, ISO/IEC 14496-12:2005. *ITU ISO/IEC*, 2010.

[3] IIS Smooth streaming extension. http://www.iis.net/downloads/microsoft/smooth-streaming, 2013.

[4] Squid. www.squid-cache.org/, 2013.

[5] Adobe Systems. HTTP dynamic streaming on the Adobe Flash platform. http://www.adobe.com/-products/httpdynamicstreaming/-pdfs/httpdynamicstreaming_wp_ue.pdf, 2010.

[6] S. Akhshabi, A. C. Begen, and C. Dovrolis. An experimental evaluation of rate-adaptation algorithms in adaptive streaming over http. In *Proc. of ACM MMSys*, pages 157–168, Feb. 2011.

[7] Comoyo. http://www.comoyo.com, 2012.

[8] K. R. Evensen, T. Kupka, D. Kaspar, P. Halvorsen, and C. Griwodz. Quality-adaptive scheduling for live streaming over multiple access networks. In *Proc. of NOSSDAV*, pages 21–26, 2010.

[9] R. Houdaille and S. Gouache. Shaping HTTP adaptive streams for a better user experience. In *Proc. of MMSys*, pages 1–9, 2012.

[10] D. Johansen, H. Johansen, T. Aarflot, J. Hurley, Å. Kvalnes, C. Gurrin, S. Sav, B. Olstad, E. Aaberg,

T. Endestad, H. Riiser, C. Griwodz, and P. Halvorsen. DAVVI: A prototype for the next generation multimedia entertainment platform. In *Proc. of ACM MM*, pages 989–990, 2009.

[11] T. Kupka, P. Halvorsen, and C. Griwodz. An evaluation of live adaptive HTTP segment streaming request strategies. In *Proc. of IEEE LCN*, pages 604–612, 2011.

[12] T. Kupka, P. Halvorsen, and C. Griwodz. Performance of On-Off Traffic Stemming From Live Adaptive Segment HTTP Video Streaming. In *Proc. of IEEE LCN*, pages 405–413, 2012.

[13] T. Lohmar, T. Einarsson, P. Frojdh, F. Gabin, and M. Kampmann. Dynamic adaptive HTTP streaming of live content. In *Proc. of WoWMoM*, pages 1–8, 2011.

[14] Move Networks. Internet television: Challenges and opportunities. Technical report, November 2008.

[15] C. Müller, S. Lederer, and C. Timmerer. An evaluation of dynamic adaptive streaming over http in vehicular environments. In *Proc. of MoVid*, pages 37–42, 2012.

[16] D. Nelson. Internet Information Services. http://www.iis.net/learn/media/live-smooth-streaming, 2012.

[17] R. Pantos (ed). HTTP Live Streaming. http://www.ietf.org/internet-drafts/draft-pantos-http-live-streaming-10.txt, 2013.

[18] H. Riiser, H. S. Bergsaker, P. Vigmostad, P. Halvorsen, and C. Griwodz. A comparison of quality scheduling in commercial adaptive http streaming solutions on a 3g network. In *Proc. of MoVid*, pages 25–30, 2012.

[19] T. Stockhammer. Dynamic adaptive streaming over HTTP - standards and design principles. In *Proc. of MMSys*, pages 133–144, 2011.

[20] B. Wang, J. Kurose, P. Shenoy, and D. Towsley. Multimedia streaming via TCP: an analytic performance study. In *Proc. of ACM MM*, pages 16:1–16:22, 2004.

[21] YouTube statistics. http://youtube.com/t/press_statistics, 2012.

[22] A. Zambelli. Smooth streaming technical overview. http://learn.iis.net/page.aspx/626/smooth-streaming-technical-overview/, 2009.

User Information Needs for Environmental Opinion-forming and Decision-making in Link-enriched Video

Ana Carina Palumbo *
University of Amsterdam
Science Park 904, 1098 XH Amsterdam, NL
a.c.palumbo@tue.nl

Lynda Hardman †
Centrum Wiskunde & Informatica (CWI)
Science Park 123, 1098 XG Amsterdam, NL
Lynda.Hardman@cwi.nl

ABSTRACT

Link-enriched video can support users in informative processes of environmental opinion-forming and decision-making. To enable this, we need to specify the information that should be captured in an annotation schema for describing the video. We conducted expert interviews to elicit users' potential information needs. We carried out a user survey to assess the relevance of the identified information types. Finally, we observed users' behaviour and needs when presented with a selection of video segments. Our results indicate that certain types of information about the environmental problem, the opinions expressed, the people expressing them and the sources are more relevant for users.

Categories and Subject Descriptors

H.5.4 [**Hypertext/Hypermedia**]: User issues

Keywords

User information needs, link-enriched video, opinion-making, decision-making, interactive TV, environmental issues

1. INTRODUCTION

The deterioration of the ecosystems on which we depend is currently one of the most important international concerns we face. The participation of informed citizens in the debate on decisions affecting sustainable development of our society is fundamental to addressing these issues [2]. Television plays a key role in providing audiovisual information, which enables the communication of extensive amounts of information quickly. It also enables conveying feelings, emotions and abstract interactions, which are difficult to express through other media [5]. By combining the potential of audiovisual information to engage and entertain with the capacity of digital hypermedia to connect ideas, link-enriched video can enhance the processes of opinion-forming and decision-making, e.g. in the context of environmental issues.

Our challenge is to identify the information that should be captured in video annotations that will support the enrichment of TV broadcasts with information that is pertinent to users. The main problem is the gap between what users' need and what broadcasters can do to meet these needs within their time and budget constraints. In this paper, we specify the information that should be captured in an annotation schema for link-enriched video that is able to support users' information requirements in the processes of opinion-forming and decision-making on environmental issues. Specifically, we identify information that can be provided and prioritise users' information requirements.

We conducted expert interviews (section 3) to consolidate an inventory of types of information that users need to form opinions and make decisions on these issues. From these we selected specific user requirements by conducting: a user survey (section 4) to gain insights into the information and tools that users claim they need; and a user experiment (section 5), to directly observe the types of information that participants used.

2. RELATED WORK

Environmental issues are notoriously complex, and require the understanding of different types of information from many perspectives [6, 7]. Opinion-forming and decision-making require "a complex search for information, full of detours, enriched by feedback from casting about in all directions, gathering and discarding information, fueled by fluctuating uncertainty, indistinct and conflicting concepts" [7](p.86). Link-enriched video has the potential for rapid information location within collections of video material by allowing users to select content segments, jump to related information during playback, and return to material earlier in the sequence or a previous segment [3].

Through a literature review, we identified and selected a number of information types that describe aspects of environmental videos. These provide an initial framework for the study of information requirements in this domain. The types and their references can be seen in Table 1.

We use this groundwork on information to form opinions to guide the series of interviews with experts on the topic.

When environmental problems are under debate, people express their opinions. Therefore, considering argumentation and rhetoric in the annotation of videos in this domain

*currently affiliated with Eindhoven University of Technology, Eindhoven, NL

†also affiliated with Institute of Informatics, University of Amsterdam, NL

EuroITV'13, June 24–26, 2013, Como, Italy.

Table 1: Information that could be captured in an annotation schema. The items were selected from literature (L) and/or interviews (I).

Types of information	Sources
Details of the environmental problem	
Basic description: Type - Subject - Location - Date - Physical, chemical and biological processes.	L[4] and I4
Impact: scale (individual, local, regional, national, global) - Affect on human communities, health and species	L[4], I2, I4 and I5
Temporality: Background of the issue (lessons learned, causes) - Outcomes (short-term) - Future scenarios (long-term)	L[4], I1, I2, I4 and I5
Personal implications: Responsibilities - Instruments to take action	I4
People involved: Types of actors - Objectives	L[4] and I5
Position/Opinion	
Argument: Direction (favorable, unfavorable, neutral) - Degree - Saliency - Rethorics (Ethos,pathos/intensity, logos) - Temporality - Advantages - Relation to other arguments (opposite, similar) - Dimensions (Scientific, social, cultural, economic, political, ethical, technical, legal, safety and security, historical)	L[4], I1, I2, I3, I4 and I5
Public opinion: Stage of attention cycle (saliency) - Distribution - Saliency - Extent of consensus	L[4]
Person	
Type of actor - Details (age, name, culture, values, profession and occupation, location, educational background, biography, related organisations and affiliations) - Arguments expressed - Personal benefits	L[4], I1, I3 and I4
Sources	
Name - Level of trustworthiness	I2, I3 and I5
Type (Wikipedia - Books, - Magazines - Scientific papers and reports - Newspapers - Documentaries - Videos - TV programs - Websites - Social sites - Radio broadcasts)	

is fundamental. We base the information types related to points of views and positions on the model proposed by Bocconi et al.[1].

3. EXPERT INTERVIEWS

Expert interviews were conducted to inventorise the spectrum of types of information users need.

3.1 Method

Interviewees. Five experts were interviewed: a project manager who works for a broadcaster (interviewee 1, I1); a social communicator who works with environmental risks and disasters in a governmental organization (I2); a researcher and developer of videos and documentaries in a research institute (I3); a project manager who designs environmental education programs in a non-governmental organisation (I4); and an expert who works for environmental and water consultancy in a governmental organisation (I5).

Interview. Two types of semistructured interviews were conducted: one for experts from the field of video production and broadcasting (I1, I3) and another for environmental governance experts (I2, I4, I5). The specific questions asked can be found in [4].

3.2 Results and discussion

According to the experts, users require sufficient information to form a complete, unbiased and informed opinion or decision (I4 and I5). The types of information that experts mentioned are summarized in Table 1.

Both literature and experts indicate that environmental video content can be enriched from multiple perspectives. All information previously elicited was organized in types and subtypes to configure a set of data to be used in the next stages of the study (Table 1). Annotating all these types of information in video content might not be viable and feasible for broadcasters. A prioritisation is needed and thus we ask potential users.

4. USER SURVEY

We conducted a user survey to assess the types of information that could be captured in an annotation schema. Users were asked to reflect on their information needs when forming their environmental opinions or decisions.

4.1 Method

Survey. An online survey of 19 questions was carried out from May 26th to June 3rd 2012. It included general questions and specific ones on shale gas drilling. Users had to imagine that there was a project aiming to build a shale gas drilling pad in their own region. They were asked to suppose they were informing themselves on the topic. Two video segments were shown (see questions in [4]).

Measurements. The types of information in Table 1 were included in the survey. For most of the questions Likert scales of 5 items were used. In other cases, users were asked to rank variables using a drag and drop function (options were randomly ordered to reduce bias). Text fields for comments and open questions were included.

Participants. The sample comprised 213 participants. Respondents varied in age, gender, country of residence, level of education and frequency of time spent watching TV, online videos and reading the newspaper.

4.2 Results and discussion

The main results of the survey are shown in Figure 1.

Whereas around 90% of the respondents agreed that identifying whether a statement is a fact or an opinion is important (M = 4.3, SD = 0.7), only 73% stated that it is possible to differentiate them in a discourse (M = 3.8 , SD = 0.8).

Even though users claimed that factual information is more relevant than emotional and ethical arguments, when being asked about the usefulness of two video segments they had seen (one classified by them as an opinion and the other as a blend of facts and opinions) no difference was made. There is a contradiction between what they claimed that is relevant and the level of usefulness assigned to the video segments they watched.

The 3 most useful sources of information for users are: scientific papers and reports (M = 9.5, SD = 2.6), documentaries (M = 8.4, SD = 2.4); and newspapers (M = 6.6, SD = 3.2).

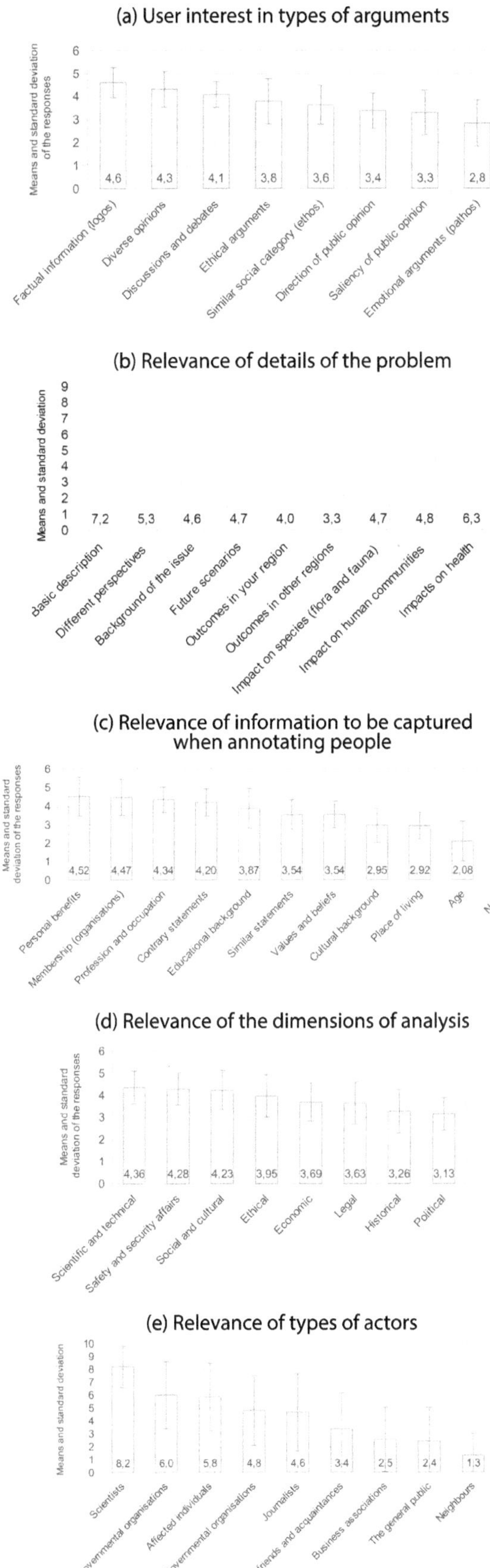

Figure 1: User survey results.

The most frequent need expressed was to have access to evidence, facts, objective and unbiased information, exact data and results from research. Opinions from "independent" people (not affiliated or part of a company) holding a "neutral" and "honest" position were highlighted.

Users prefer complete information over pieces of information. They want benefits and disadvantages, pros and cons of the issue and costs and benefits. Risks involved and other aspects related to the safety, security and health implications, and the consequences in the short and long term were mentioned by many participants. Some users find the scientific and technical perspective more relevant than the security and safety perspective, whereas in open questions users' expressed their interest for the latter.

A frequent information need is to know about possible alternatives to achieve the same goals. Users want to compare benefits and disadvantages of these to choose the best cost-benefit balance. They are also interested in knowing about the level of community need for the benefits of the issue under discussion. Several participants mentioned the need to know who benefits, who is making money out of it, who support the initiatives and who are against.

5. USER EXPERIMENT

To assess the level of priority of the types of information needed for forming opinions and making decisions, we observed the way users perform information tasks on a specific environmental issue. We compared the results with those obtained in the user survey.

5.1 Method

We created a website including meaningful titles for 25 segments on shale gas drilling depicting a single speaker's or organisation's opinion. Each represented a type of information from our initial set (Table 1). 6 participants, of which 5 female, had to select a video segment using its title, from which we inferred the type of information they wanted.

Procedure. Users watched a short video introducing the issue and were then asked to imagine that they had to decide whether to sign a petition for or against shale gas extraction in their region. To inform themselves, they chose 3 video segments. Before and after watching each segment they were asked about their expectations and the reasons for their choice. After each segment, they were given 3 minutes to consult online resources. They had to decide whether they agreed with or believed the ideas expressed; or they had to check any other additional information need triggered by the segment. Finally, users were asked to rank and explain the usefulness of the sources they used.

5.2 Results and discussion

Three participants (P1, P5 and P6) decided to watch the "Arguments supporting shale gas drilling". Other segments were chosen by two participants: "Impact on human communities" (P1, P3), "Arguments against shale gas drilling" (P4, P5), "Potential outcomes" (P4, P6), "Diverse opinions" (P2, P6), "Health implications" (P3, P5). Users chose to access the video segments using a dimension of analysis in 4 cases, the direction of the opinion in 5 cases and using other categories in 9 cases. None of the participants chose a segment titled by the type of actor giving arguments. A detailed analysis of the results can be found in [4]. The most frequently expressed needs were safety, security and health implications, outcomes and future scenarios.

Words and phrases such as "impacts", "effects", "risks" "disadvantages" and "health implications" were used in users' queries. When searching for information online after watching each segment, some participants found difficulties in specifying terms for their queries. One participant expressed her/his desire to access an explanation in "easy words". 3 participants were interested in checking alternatives to shale gas drilling or solutions implying less negative impacts for human beings (P2, P3, P6).

The need to consider opinions from people that are more neutral, objective or that do not have particular interests or benefits on the issue was highlighted.

Even though most of the participants stated their intention to be informed from many points of view, they tended to choose all the segments following one topic or direction. 5 participants decided to be against shale gas drilling almost from the beginning of the experiment (P4 was the exception). When a segment contained supporting arguments they considered them as less useful and biased. This confirms what persuasion theories state: users tend to give more weight to arguments that support their own opinion.

Two participants (P3 and P4) had chosen 3 segments in advance and later decided to switch them. This could indicate that information needs evolve while forming an opinion and are renewed each time new information is processed.

All participants used the Google search engine, 3 accessed Wikipedia, 2 scanned Google results lists, 1 accessed Google Scholar (P2), 1 used YouTube (P5). Other sources used were news sites and web sites specialized in the topic.

The use of Wikipedia in the experiment suggests its relevance, even though in the survey it was ranked in the 6th position out of 9. Users did not always pay attention to the name of the source or its credibility. Many of the sources they opened were not familiar to them. Only one participant reflected on the importance of the trustworthiness of the source (P2). Participant 3 considered whether the site contents were reviewed by experts or not to rank them. In many cases users scanned scientific papers and reports.

Even though on several occasions participants mentioned they wanted to listen to an expert, none chose to access a segment representing the opinion of a particular actor. To determine whether to believe a speaker or not, users searched for concepts mentioned by the speaker. A participant (P4) expressed s/he wanted to know the name of the speaker to look for information about his background.

6. CONCLUSION

From the broad spectrum of information that could be captured in video annotation we identified the most meaningful types for users:

- Details about the problem: location and date; advantages and disadvantages of the main environmental processes involved, the level of community need for the resulting benefits, possible alternatives and outcomes for the community.

- Position or opinion: direction of arguments, relations between the main topic and other topics and arguments about the future; safety, security and health implications; scientific and technical perspective; level of subjectivity (pathos) and objectivity (logos) identified in a discourse; direction of opinions (favourable, neutral or unfavourable) and relation between them (opposite or similar).

- Person: how the speakers benefit from the position they hold, the organisations and affiliations s/he belongs to, his/her name, profession and occupation; links to different arguments expressed by a speaker and relation between those arguments (opposite or similar); experts' and scientists' explanations.

- Sources: the name and level of trustworthiness is relevant; users prefer scientific papers and reports and Wikipedia to enrich the video content.

7. ACKNOWLEDGMENTS

This work was partially supported by LinkedTV, funded by the European Commission through the 7th Framework Programme (FP7-287911). We would like to thank: Frank Nack, University of Amsterdam, NL; Daphne Willems, Daphnia: vision on rivers, NL; Lotte Belice Baltussen, Research & Development Department, Netherlands Institute for Sound and Vision, NL; Nicolas de Abreu Pereira, Rundfunk Berlin-Brandenburg, Germany; Santiago Gaitan, Delft University of Technology, NL; Veronica Viduzzi, Emmanuelle Beauxis-Aussalet, Mieke Leyssen and Stefano Bocconi.

8. REFERENCES

[1] S. Bocconi, F.-M. Nack, and L. Hardman. Automatic Generation Of Matter-Of-Opinion Video Documentaries. *Journal of Web Semantics*, 6(2):139 – 150, April 2008.

[2] R. Declaration. Rio declaration on environment and development. In *Report of the United Nations conference on environment and development, Rio de Janeiro*, pages 3 – 14, 1992.

[3] A. Girgensohn, L. Wilcox, F. Shipman, and S. Bly. Designing affordances for the navigation of detail-on-demand hypervideo. In *Proceedings of the working conference on advanced visual interfaces*, pages 290 – 297, 2004.

[4] A. C. Palumbo and L. Hardman. Investigation Towards Link-Enriched Video: User Information Needs For Environmental Opinion-Forming And Decision-Making. CWI Technical Report INS-1301, URI: http://persistent-identifier.org/?identifier=urn: nbn:nl:ui:18-21150, CWI, January 2013.

[5] F. Shipman, A. Girgensohn, and L. Wilcox. Authoring, viewing, and generating hypervideo: An overview of hyper-hitchcock. *ACM Trans. Multimedia Comput. Commun. Appl.*, 5(2):1 – 19, Nov. 2008.

[6] C. L. Spash. Informing and forming preferences in environmental valuation: Coral reef biodiversity. *Economic Psychology*, 23(5):665–687, Oct. 2002.

[7] M. Zeleny. *Multiple criteria decision making*, volume 25. McGraw-Hill New York, 1982.

NEXT - Graphical Editor for Authoring NCL Documents Supporting Composite Templates

Douglas Paulo de Mattos
MídiaCom Lab, Computer Science Institute
Fluminense Federal University
Niterói, RJ, Brazil
douglas@midiacom.uff.br

Júlia Varanda da Silva
MídiaCom Lab, Computer Science Institute
Fluminense Federal University
Niterói, RJ, Brazil
julia@midiacom.uff.br

Débora Christina Muchaluat-Saade
MídiaCom Lab, Computer Science Institute
Fluminense Federal University
Niterói, RJ, Brazil
debora@midiacom.uff.br

ABSTRACT

Using the Ginga-NCL middleware, interactive multimedia applications for the Brazilian digital TV system are written in NCL (Nested Context Language). Although programming skills are not required when using a declarative authoring language, authors need to have at least a basic knowledge of the language in order to develop an application. Aiming at facilitating and spreading the use of NCL, this paper presents a graphical editor that allows the development of NCL documents for authors with no knowledge of the language. The proposed editor is called NEXT (NCL Editor Supporting XTemplate). To provide that facility, the editor uses hypermedia composite templates, which represent generic structures for NCL programs. Those templates are specified in the XTemplate 3.0 language. In addition, NEXT offers other functionalities, such as creation and editing NCL documents in different views, which facilitate the development of digital TV applications. Those functionalities are provided as a set of plugins, which makes the tool extensible and adaptable to different author skills.

Categories and Subject Descriptors

H.5.2 [**Information Interfaces and Presentation**]: User Interfaces—*graphical user interfaces (GUI), interaction styles*; D.2.2 [**Software Engineering**]: Programming environments—*graphical environments, integrated environments*

Keywords

NCL, Digital TV, authoring tool, plugins, XTemplate, graphical editor, NEXT, hypermedia composite templates

1. INTRODUCTION

Digital TV applications can be represented as multimedia documents containing different media content types, such as text, image, audio and video. In addition, these applications allow interactivity, offering a richer environment for the viewer, once they provide different kinds of content besides traditional audio and video in a television system.

Interactive multimedia applications can be developed using declarative or imperative languages. In the declarative environment of the Brazilian digital television system [4] and in the H.761 standard for IPTV services [6], applications are written in NCL (Nested Context Language). This language is based on the NCM (Nested Context Model) [18] hypermedia conceptual model. It allows structuring hypermedia documents logically using compositions and specifying spatio-temporal relationships as connectors and links using the event-based paradigm. Although it is very expressive for interactive multimedia document authoring, using connectors and links for expressing synchronization relations requires good knowledge of NCM and NCL.

On the other hand, the current digital TV and IPTV scenarios require simple and quick methodologies for creating interactive multimedia applications. Therefore, authors with different skills should contribute to their development, including the ones that are not programmers.

In order to allow the development of NCL documents for authors with no knowledge of the language, this paper proposes a graphical editor that provides the use of generic structures named hypermedia composite templates for authoring NCL documents. The proposed editor is called NEXT (NCL Editor Supporting XTemplate). Those templates specify the document generic structure, in other words, they define synchronization relationships among document media objects. Hypermedia composite templates are specified in XTemplate 3.0 [17], an XML-based declarative language that can be used together with NCL.

When composite templates are used, NCL documents just need to define specific media content, since all synchronization relationships can be defined in the template. This way, NCL authors do not have to define connectors and links and document authoring becomes much simpler. NEXT creates template graphical representation from XTemplate XML specifications and allows using them for developing NCL documents. It allows authors that do not have basic knowledge of NCL to develop interactive applications for the Brazilian Digital TV System and the ITU H.761 standard using the declarative paradigm.

Moreover, the development of NCL documents is facilitated using other NEXT's functionalities, such as creation and edition of NCL documents in different views. They are

EuroITV'13, June 24–26, 2013, Como, Italy.

provided as a set of plugins, which makes NEXT extensible and adaptable to different author skills. NEXT is the unique graphical editor for creating NCL documents using an extensible template library. Any template written in XTemplate 3.0 can be used by NEXT. To facilitate the creation of templates, a graphical editor named EDITEC [10] is available and will be discussed in the related work section.

The rest of the paper is structured as follows. Section 2 presents related work. In Section 3, template usage for multimedia document authoring and XTemplate are discussed. Section 4 presents the NEXT editor. Section 5 presents NEXT's architecture and other functionalities. Conclusion and future work are given in Section 6.

2. RELATED WORK

EDITEC [10] is a graphical editor for creating multimedia composite templates based on XTemplate 3.0. The tool offers structural, layout and textual views, allowing the user to have better template understanding during the authoring process. The structural view allows creating elements and defining relations that will be part of the template. The layout view presents screen regions in which media content can be displayed. The text view is divided into two parts, one to display the template XML code and another to display region, descriptor and rule bases of the template. In addition, EDITEC provides iteration options for creating sets of links relating generic document components. EDITEC allows creating composite templates but it does not provide the creation of complete multimedia documents using composite templates.

Composer 3 [11] is intended to offer a single authoring environment suitable for different author profiles, from home users to content producers. Its architectural pattern is based on a micro-kernel, a core model and extensions. The micro-kernel groups minimum system functionality and coordinates how extensions should collaborate. Extensions are made through plugins, which are programs that interact with the micro-kernel and add new features/resources to the tool. Although it facilitates multimedia document authoring, Composer 3 does not allow the use of templates for document authoring. It neither performs document simulation, nor checks the consistency of document presentation behavior. Such features could be added to the tool through the development of new plugins.

In LAMP [8], the main goal is the automatic generation of multimedia presentations modeling template schemes. The idea is to offer the user the possibility of defining the presentation layout and behavior, characteristics and attributes of its objects, without worrying about the instances that will be used for fulfilling the template. That work focused on coordination and synchronization of continuous media objects. The automatic generation of presentation is divided into two parts: first, the template is defined and, second, data that will be used in presentation is dragged from a data repository and inserted into the template. Then, the presentation can be started. The tool consists of an editor, a document presentation simulator and a generator that considers the template and media nodes chosen by the author. It also provides a formatter, which presents the final document. Although LAMP provides several features for multimedia document authoring, it does not check the consistency of document presentation behavior.

Mobile Multimedia Presentation Editor [12] enables the creation of multimedia presentations with several types of media content on a mobile device. The user interface consists of different views and menu options, which change according to the currently used view. It provides play (run), edit, preview and temporal views. Created presentations are represented using SMIL [5] and temporal relations are based on a page sequence. The temporal view presents two timelines that appear on the screen top. A set of predefined templates is offered by the tool and the user may create a presentation without using them. The tool fails to provide all SMIL language functionalities and does not allow the use of all media types such as video for document creation.

T-MAESTRO [13] is a system that aims at selecting personalized content, which relates entertainment with learning, to be displayed to the user of an interactive digital television system. Thus, it is necessary to provide interactive contents to the system. In order to adapt SCORM [3] learning objects (created by the RELOAD editor) to the system, the A-SCORM Course Creator Tool was developed. The authoring tool allows creating adaptive courses with a minimal knowledge of the system architecture. A-SCORM enriches the functionality of RELOAD by adding new features that allows specifying different organizations for SCORM courses. For doing so, two sub-modules were added: the adaptation rules editor and SCO (Sharable Content Object) repository. The first one, SCO creator tool, allows authors to specify characteristics such as content appearance and configuration, as well as adaptation of SCOs. Thus, authors simply select a template, among those provided by the tool, then specify the options that vary according to the rules and finally create adaptation rules. The second sub-module allows including SCOs, which were created by the first sub-module, in the course that is being developed.

Template-Based MHP Authoring Tool [9] aims at generating compatible codes with MHP (Multimedia Home Platform) [7], the European standard for interactive digital television middleware, where interactive applications are written in Java. The authoring tool was designed to allow adding new features or new user interface components. It employs a metaprogram generator, which is responsible for creating the real program generator for each module that composes the MHP application according to the XML-based template description of the module. A real program generator produces the Java source code for a module according to its XML description. Thus, to add new functionalities to the tool, it is only necessary to change the metaprogram generator and the template description of the real program generator.

3. DOCUMENT AUTHORING USING TEMPLATES

In order to minimize difficulties related to developing applications, multimedia authoring systems are offered. They are designed to facilitate authoring work providing a graphical environment for editing multimedia documents by authors with little knowledge about the authoring language used. One way to facilitate further multimedia application development is providing graphical editors that offer templates.

A template is a document structure without specific content. It defines essential parts of a declarative multimedia document, such as how media content should behave during its execution. Thus, the authors task is to indicate which media content should be used in order to complete the template information and create a final document. Figure 1 shows the concept of an editor that offers templates for creating multimedia documents.

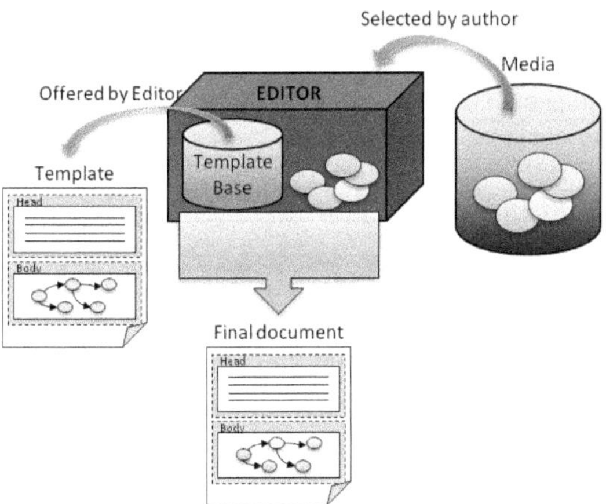

Figure 1: Editor for creating multimedia documents using templates

As the template is responsible for specifying the multimedia document structure, or for defining its nodes and links, where nodes represent media content and links, relationships among them, it is interesting that the template reuses predefined structures, which define generic relations in the document, making its authoring easier. A declarative language that allows hypermedia composite template definition is XTemplate 3.0 [17].

XTemplate allows the definition of hypermedia composite templates using XML, where templates specify generic components and their relationships, whose detailed specification is responsibility of the document that will use the template. XTemplate can be used together with NCL 3.0. Moreover, composite templates can be created graphically using EDITEC [10], which generates the XML template code. In its third version, XTemplate defines a template through four main parts: head, vocabulary, body and constraints.

The head is responsible for importing <descriptorBase> and <connectorBase> elements from an NCL document, which will be used for specifying presentation characteristics and spatio-temporal relations, respectively. The vocabulary specifies template components, which may have interface points (ports) or nested components. It is also responsible for defining connectors that will be used for creating links. The body defines interface points (ports), vocabulary component instances and relationships among those components. Constraints allow adding restrictions on vocabulary elements. Each constraint has an expression in the XPath language [1].

When a document uses a template, that document should refer to the template being used. Therefore, in order to allow using templates in NCL documents, NCL was extended including a few elements and attributes. In the NCL document head, a <templateBase> element is declared indicating the template bases used. Each NCL context (including the NCL body) may declare a xtemplate attribute indicating the specific template to be used. Different templates can be used by different context nodes in the same NCL document. When a context node uses a template, its component nodes (media, context and switch nodes) should declare an xlabel attribute identifying the template component roles those nodes will play.

When an NCL document uses templates, it must be processed in order to generate a final complete standard NCL document including the template information. Figure 2 shows the processing of a document that uses a composite template. The processor uses as inputs the extended NCL document and the composite template, and outputs the final standard NCL document, ready to be deployed.

The template processor analyzes the template head to get additional documents to be used, generates a new document including new anchors, nodes and links specified in the template and validates template constraints. The final document can then be run on a standard GINGA-NCL middleware implementation.

4. NEXT

In order to allow the graphical creation of NCL documents using hypermedia composite templates, this paper proposes NEXT. It is a graphical editor that is able to understand a template specification in XML and generate its graphical representation, allowing the author to complete the document definition indicating media objects that will play template component roles.

When an NCL document uses templates, there is no need for specifying the application presentation behavior with connectors and links, as templates will already do that. Therefore, NCL authoring becomes much simpler. NEXT allows authors with no knowledge about NCL to build an NCL multimedia document using templates.

Any hypermedia composite template created with the XTemplate 3.0 language can be used by NEXT. Besides allowing its use by authors without NCL knowledge, NEXT also facilitates the creation of applications that require large amounts of code, often repetitive in its specification.

An example of how a template may help is the following. Consider a video that has 100 subtitles. To synchronize them with the video, it is necessary to define a video anchor (video fragment) for each subtitle and synchronization links relating them. Thus, typing this application code becomes monotonous and error-prone. If a template to synchronize a video with subtitles is available and the author uses NEXT, he only has to define the video content, its anchors and subtitle texts. All connectors and links for temporal synchronization will be automatically generated by the editor considering the template specification.

Another important issue is helping authors to understand what the template specifies, as the author will probably use a library with several different templates available. Each template specified using XTemplate has an RDF [2] file, which explains its functionality.

In addition, NEXT creates a template graphical representation, which is based on a group of screens, showing how final media content will be displayed on the TV screen. The screen abstraction is not part of a template specification in XTemplate. Different screens are generated by NEXT as an interpretation of a template components and links.

A template screen can be understood as a group of nodes that will be presented at the same time and, possibly, in different positions on the screen. Every time a new media

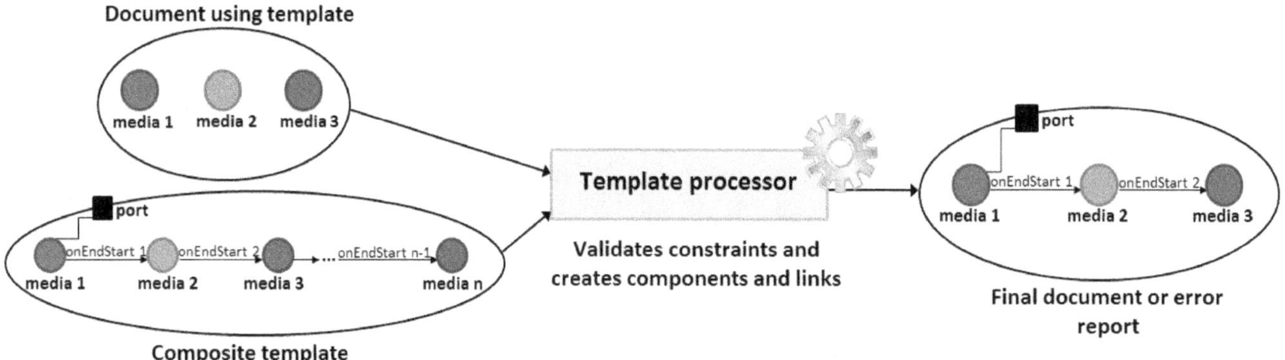

Figure 2: Document processing using composite templates

node is presented or removed, according to the template link specification, a new screen for this template is created by NEXT. Figure 3 illustrates the screen concept. Screens 1 and 2 are different because the blue button is not displayed on Screen 2.

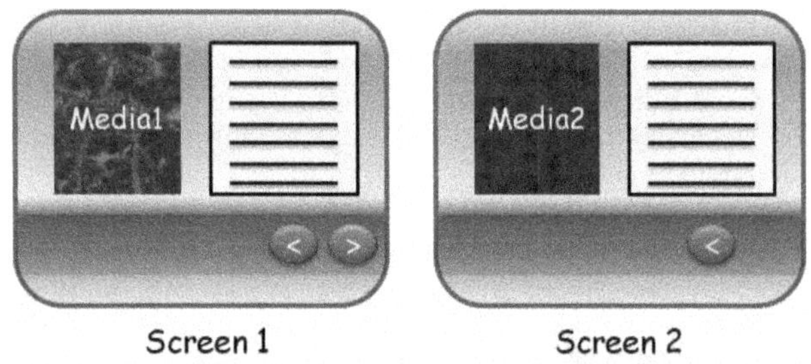

Figure 3: NEXT's template screen example

NEXT uses visual symbols to present the template specification graphically to users. The graphical symbols used by NEXT are shown in Table 1.

Figure 4 shows how NEXT would graphically represent the template screens shown in Figure 3.

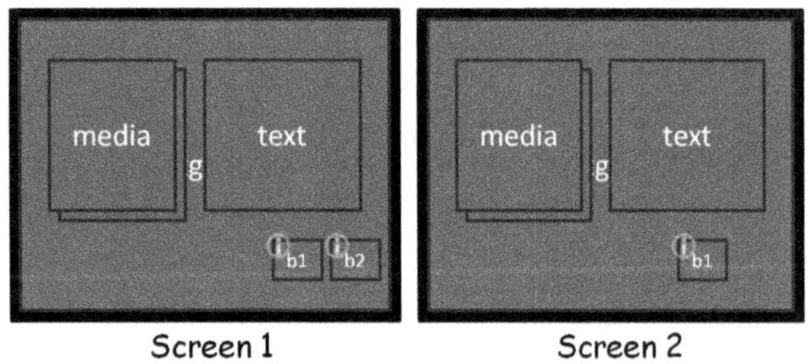

Figure 4: Example of screen graphical representation in NEXT

In order to create the template graphical representation, NEXT reads template XML files found in the template base using an API called aXT, which is responsible for creating a Java object that contains information about the template components. A template may specify, in its head, an NCL document containing layout specification (region and the descriptor bases). The NCL layout document is read with an API called aNaa [15], which creates another Java object, this

Table 1: Symbols used for representing template screens

Symbol	Meaning
fullScreen	A blue rectangle represents a node whose content should be chosen by the author.
logo	A red rectangle indicates a specific NCL node that is already defined in the template and cannot be chosen by the author.
fullScreen	A shaded rectangle represents a set of nodes, i.e., multiple nodes may appear in the same screen region at different times. In this case, the author can specify different content.
fullScreen	The "i" icon on a component indicates that it enables interactivity, that is, the viewer can use the remote control to interact with the application.
	A musical note represents an audio component and analogous to rectangles, the note color may be blue, red or shadowed.

time to provide the necessary information about the layout of each template component.

The next step is to consider the information in the template vocabulary. Using it, it is possible to know the template components and their interface points. For each component, a Java object is created, which stores information about the component layout, size and position on the screen. After knowing all template components, NEXT begins to read the template body. Template links allow NEXT to know the relationships among its components and to find out which elements may be selected using the remote control.

The template graphical representation may contain one or more screens. This quantity will be defined according to the template links. To define the first screen, it is necessary to verify the template components that are referenced by <port> elements. After that, to complete the first screen, it is necessary to check whether any link starts the presentation of any other template component when port element components start. After determining the first screen, NEXT verifies all links that were not checked to found out if new screens are needed. For example, suppose a document that starts with a video and, when the video stops, an image is presented before the document ends. In that case, the first screen will present only the video node, and a second screen will be needed to present the image node. NEXT is capable of determining the second screen as it scans all template links and finds out that an image node will be presented after the video. If new nodes are presented or finished according to template links, new screens will be created by NEXT.

After creating the graphical representation for each template in the template base, those representations are displayed to the author so he can select a template to use. Figure 5 shows the templates found in a template base. On the left side, a screen list shows the first screen for each template. When one of those screens is selected by the user, all different screens that compose the selected template are displayed on the right side, as well as its explaining text.

In that example, the template base provides four templates. The first one, *Slide Presentation Template*, is used for creating a slide presentation. The navigation is given by the remote control red and green keys that present it in progressive and regressive order respectively, and the yellow key ends the presentation. The *Advertisement Template* synchronizes an audio announcement with the corresponding product and price images and a button. Products and prices are displayed over a background image and the blue key allows the viewer to select the desired products. When the advertisement ends, the template displays the products that were selected by the user. The third template, named *Quiz*, is used for presenting a quiz when the yellow key is pressed, while a video is playing. Each screen presents four answers, which are chosen by one of the color keys of the remote control. This template must be used together with another one that defines the quiz screens. The last one, called *Subtitles*, presents a video with synchronized subtitles.

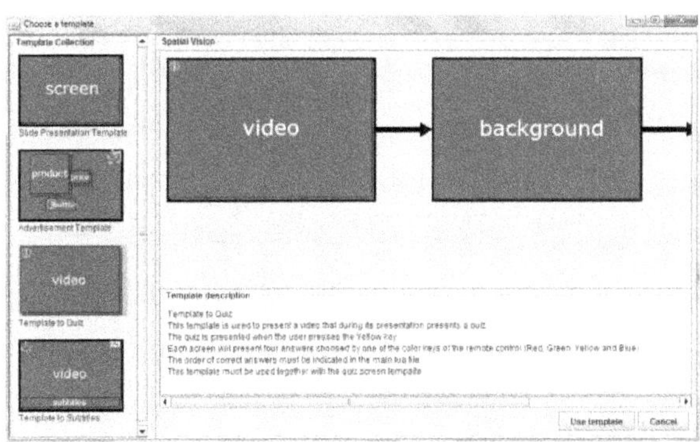

Figure 5: NEXT window for choosing a template

When a template is selected, a new window allows the creation of an NCL document filling the template with specific content. On the left side, each template screen in shown and on the right side the user can complete the content for a screen. If there are components representing sets of nodes in the same screen, that screen can be duplicated by the author in order to define different content nodes. In the first version of NEXT, template components were filled using a table according to Figure 6. The graphical editor was tested by ten computer science students, who did not have knowledge of NCL, and two students that had, in order to assess its practical usability and functionality. After briefly explaining about NCL and the concept of templates, all users were able to create NCL documents using templates, even the ones with no knowledge about NCL. The tool usage was not explained. The students could just use the help menu. After those tests, they answered a questionnaire about using the editor. The questions were objective and the choices ranged on a scale from zero to five, where zero was the worst and five the best option. Table 2 shows the result.

Table 2: Questionnaire and results

Question	Average Score	Conclusion
How do you evaluate your knowledge about NCL?	1	Very low
How much do you understand the concept of template and its purpose?	4	Good
Regarding the facility to understand the tool, how would you rate it?	4	Easy
How much did you need to use the help menu?	1	Very little
Could you develop the final NCL application with the tool?	-	Everyone succeeded
How do you evaluate the graphical representation of template screens?	4	Good
Could you understand the template functionality through the screens displayed and its explaining text?	4,5	Good understanding
How do you evaluate the table to fill in template components?	4,5	Good
How easy is to create media anchors?	4	Easy

Besides that, they gave suggestions for improving the graphical editor. The main suggestions that were given are: drag media directly from a repository to a template screen, display the picture of the media chosen instead of rectangles and improve indication of when a media anchor can be created. These suggestions were implemented and a second version of the tool is available. Therefore, that new version is discussed hereinafter.

Instead of using a table to fill in template components, the own template screen is now used. In other words, when the author selects one of the template screens on the left side of the editor window, the selected screen is displayed in a bigger size on the right side to allow users to fill in template components using drag-and-drop from the media repository, which is also provided by NEXT. In addition, the editor shows the description of each template component when the mouse is over it.

When the author wants to fill in a template component, he opens the media repository, selects a media file and drags

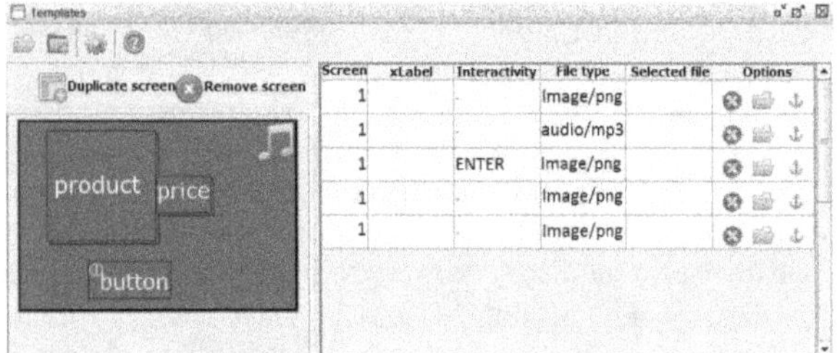

Figure 6: NEXT's first version user interface

and drops it on the template component. Figure 7 shows how the media repository and the new version of the editor work together.

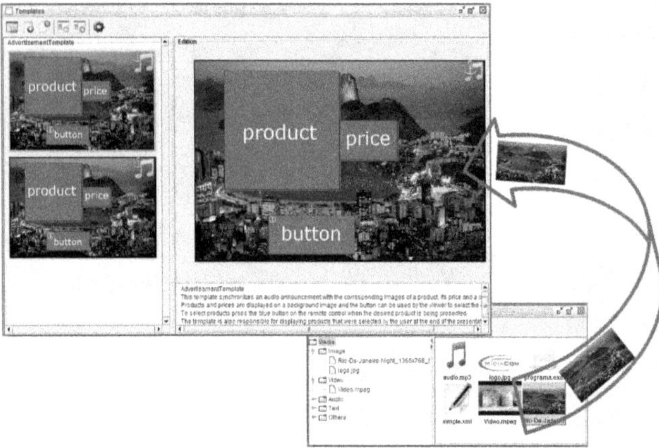

Figure 7: Drag and drop

Instead of using red and blue rectangles as in the NEXT's first version, components are visually changed according to their type. Figure 8 shows how each visual component is modified after being filled by the user. If it is an image or a video, its content is displayed; if it is a text, its color becomes green and its content is displayed; if it is an audio, its color becomes green; and for other media types, a special icon is displayed.

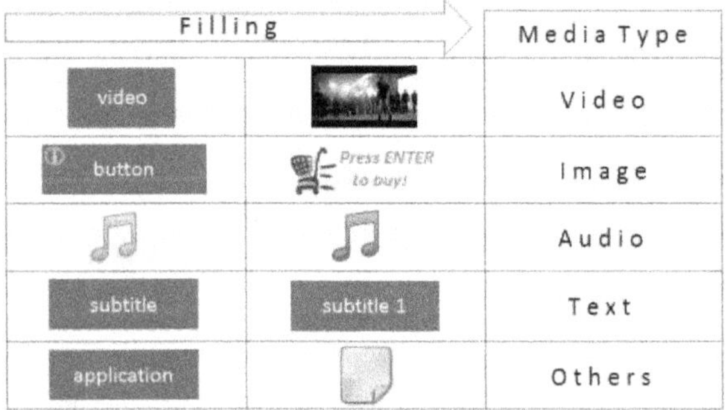

Figure 8: How template components change after being filled

When media anchors should be created, an anchor icon is shown over the component in the corresponding template screen. In order to create an anchor, the author must click on the anchor icon and another window for anchor creation will be shown, as illustrated in Figure 9. Continuous media content (video or audio) is displayed using JMF (Java Media Framewok). Begin and end time instants for each anchor creation can be easily set by clicking on available buttons and defining new anchors.

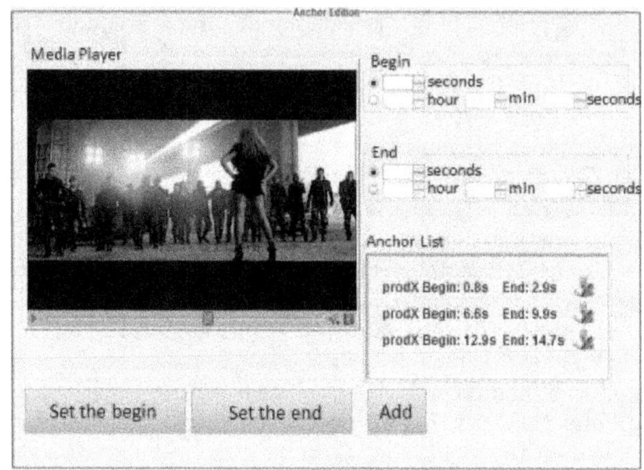

Figure 9: Window for anchor creation and edition

5. IMPLEMENTATION

The proposed editor is based on a modular architecture composed of a main core and a set of plugins, which allow new features to be easily added to the tool.

In order to maintain the editor features independently of available plugins, a program core capable of dealing with the NCL document and also of communicating with plugins was developed. NEXT architecture can be seen in Figure 10. The editor was developed in Java to provide portability and to allow the use of other tools already implemented in Java, such as aNaa (API for NCL Authoring and Analysis) [15]. NEXT core is composed by main, mainTree, common, repository and plugin packages, and also by aNaa, which is added to the library. As it can be seen in Figure 10, NEXT is completely independent of its plugins. Notice that NEXT graphical editing functionalities are developed as plugins, including the creation of NCL documents using templates.

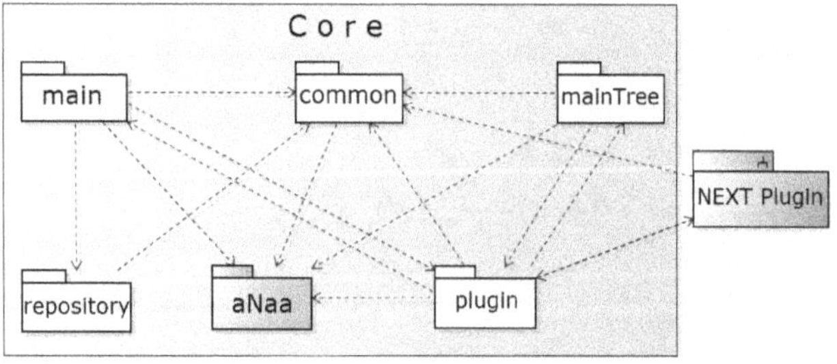

Figure 10: NEXT architecture

The aNaa API creates Java objects for representing each NCL document element. Using its parse method, it is possible to create a new NCL document and, using its load method, it is possible to import an existent NCL document into the editor. That API also offers methods for creating and editing NCL elements, which must be used by plugins in order to perform modifications in the document. Moreover, it is also responsible for the document structural and behavioral analysis. The created NCLDoc object is analyzed and, if any error is found, the author is warned. aNaa is able to find syntax and reference errors and also spatio-temporal undesired behaviors in the document definition [16].

With the NCLDoc object generated by aNaa, the mainTree package creates a tree representation of the NCL document, which facilitates element location and visualization in the editor. The main package is responsible for the editor graphical view. It implements buttons for creating and editing NCL documents and allows adding and removing plugins. The common package allows the editor use in different languages (English and Portuguese) and implements some auxiliary methods that are used by other tool packages. The plugin package has methods for exchanging information between the core and the plugins. Thus, it is possible to collect information from the plugins, such as the list of NCL elements of interest, as well as to keep them informed about changes in the document. It is also responsible for creating the editor XML configuration file, which has information about the installed plugins that must be offered to the user, when the program starts. The repository package implements the media repository.

NEXT plugins are responsible for making changes in the NCL document being created or edited. To develop a plugin for NEXT, the programmer must create a program in Java (with jar extension), which will be incorporated into NEXT. As it is necessary to provide communication between the plugin and the core, the jar file must necessarily have a package called myPlugin, which must implement StartPlugin and PluginInfo classes.

The *StartPlugin* class is responsible for initiating the plugin and will be called by the NEXT core, when the author wishes to use it. This class must also implement the *getUpdateAtNCLDocument* method to be notified about changes in the document. Therefore, it is possible for the author to use different plugins at the same time. The *PluginInfo* class is responsible for providing plugin information to the core. It consists of *getAlias* and *getNCLInterest methods.* The first method reports the alias of the plugin, that is, the menu name that will be displayed by the editor to represent the plugin. The second method should return a list with the NCL elements of interest for the plugin. Therefore, whenever one of the NCL elements of the list suffers changes that have not been caused by the plugin, it will receive a notification of adding, removing or editing the element.

To make changes in the NCL document, the plugin must use the *HandleNCLDocument* object, passed as a parameter in the constructor of the *StartPlugin* class. By applying the *getNCLDocument* method to the object, the program receives an *NCLDoc* object, which represents the NCL document in accordance to aNaa. Owning this object, changes must be applied using the aNaa API methods.

5.1 Other Functionalities

Aiming at adding other important functionalities to the NEXT, some plugins were developed. Those plugins provide the creation and editing of NCL documents in different views. It allows NEXT to be used by different author profiles, including those that already know NCL. For doing so, there is a need for enabling the document creation with or without the use of templates. Therefore, three plugins were developed. The first one provides a Connector Editor allowing the graphical creation of hypermedia connectors. The second one offers a Layout View for graphical creation of regions and descriptors. The third one provides a Structural View for creation of context nodes and its internal elements. NCL knowledge is required for using these three plugins. Each plugin will be briefly presented in the following paragraphs.

The Layout View allows the creation of NCL region bases, regions and descriptors. Therefore, its list of interest includes only those three elements. It provides a simple interface where spatial regions can be easily manipulated using the mouse. For descriptor creation, it is necessary to have a basic NCL knowledge, as the author must fill in some descriptor element attributes. It can also be used as a standalone application, where the author can create an NCL document that contains only region and descriptor bases. This document may be used by other NCL documents, which can import its bases. Figure 11 presents the Layout View plugin interface.

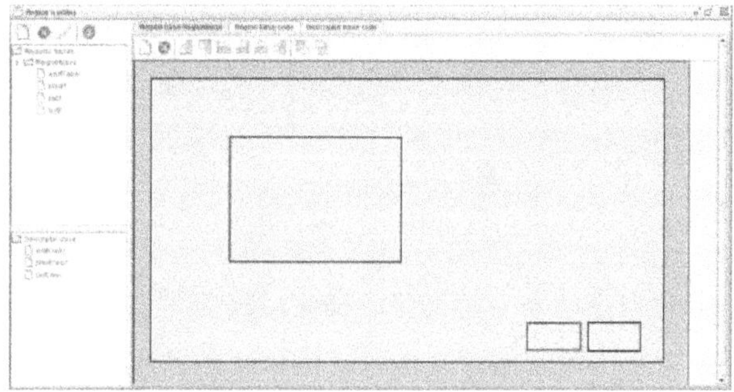

Figure 11: Plugin for region and descriptor edition

The Structural View plugin allows graphical editing of NCL context nodes. It provides authoring of the document body or context node components, as well as its links. The user interface is based on drag-and-drop in order to facilitate authoring work. However, a basic NCL knowledge about the document body and its components is required for using this plugin, since some NCL elements such as anchors (media fragment or property) and switches, may not be so intuitive for a beginning author. Figure 12 shows its user interface.

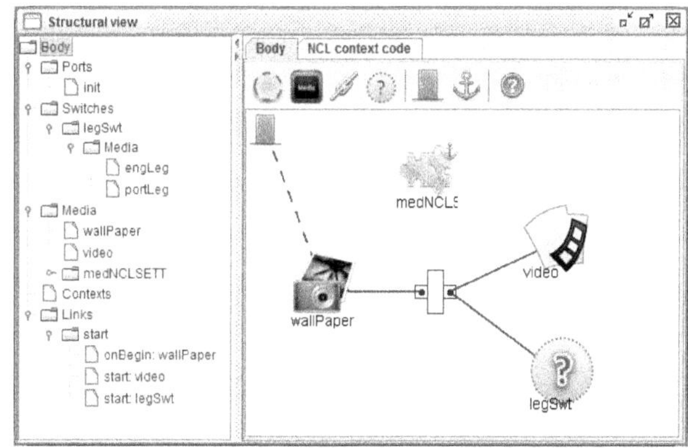

Figure 12: Structural View Plugin

The Structural View plugin receives information of some head elements, although it does not allow changes on them.

However, changes involving those head elements, in some cases, may cause changes in body elements. In that case, the necessary changes are made by the plugin in order to maintain document consistency.

The Connector Editor provides the graphical authoring of hypermedia connectors [14], which define spatio-temporal relations for NCL documents. It offers a drag and drop user interface and displays messages to guide the author during connector creation. As the Layout View, this plugin can also be used as a standalone application, allowing the document to contain just a connector base. Figure 13 shows the Connector Editor plugin interface.

Since the Connector Editor plugin only handles hypermedia connectors, its list of interest comprises only the connector base and connectors.

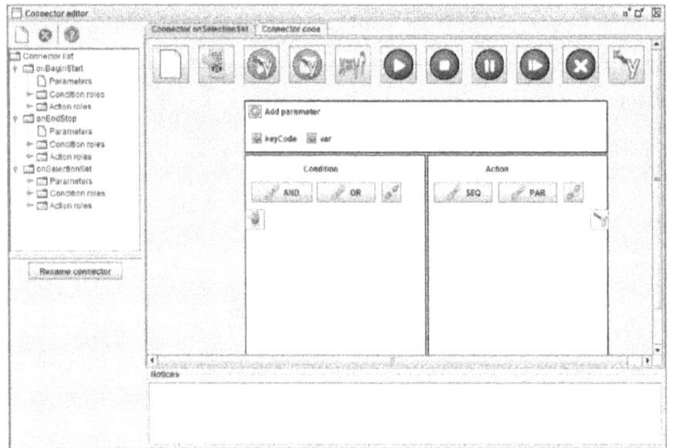

Figure 13: Plugin for connector edition

NEXT allows showing the NCL document structure as a tree, providing a neat document view. Document elements occupy different branches in the tree according to their types. NCL XML code visualization is also possible. In addition, NEXT offers two language options, English and Portuguese. Figure 14 presents the graphical interface of the NCL textual view. Notice that those functionalities are implemented within NEXT core, they are not plugins.

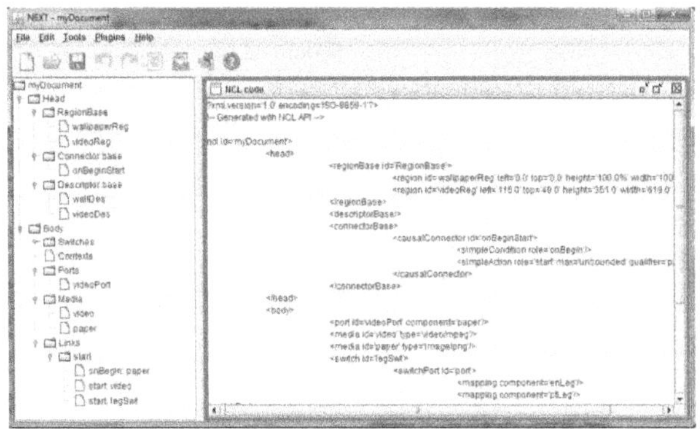

Figure 14: Textual view

Besides allowing the creation of new NCL documents, it is also possible to edit standard NCL documents. It is possible to open and modify any NCL file, regardless of where it has been created.

NEXT presents a list of already installed plugins, where the user can choose to use any of them. The selected plugin will be displayed on the central area, which in Figure 14 is showing the NCL code. Besides, it is possible to add and remove plugins dynamically.

NEXT allows authors with no NCL knowledge to create applications using any composite templates specified in XTemplate. Generic templates are supported by other tools discussed in the related work section. However, there is no NCL authoring tool, which supports template usage to create interactive multimedia applications. NEXT is extensible including new features through plugins, enabling its use by different author profiles. Extensibility and adaptability are provided by some related works including Composer, which allows creating NCL documents. NEXT provides different views of the multimedia application as other discussed works, except A-SCORM. NEXT allows document analysis, that is, it indicates to the author possible incorrect definitions or inconsistent specifications, considering both syntactic and semantic language rules. Multimedia spatio-temporal consistency should also be verified, notifying the author about inconsistencies that can probably cause unexpected application presentation behavior. Although Composer is able to analyze applications, it does not analyze a document spatio-temporal behavior.

6. CONCLUSIONS

The main contribution of this paper was the proposal of a graphical editor for creating NCL documents, called NEXT, which supports the use of composite templates created with the XTemplate 3.0 language. This work aimed at allowing authors with no knowledge of NCL to develop applications for digital television interactive services and facilitating and accelerating the creation of NCL documents. No other NCL graphical editor provides the possibility of using a flexible set of templates. NEXT provides the use of any XTemplate 3.0 template to build NCL documents.

Initial usability tests showed that users with no knowledge of NCL are able to create documents using templates with NEXT.

The editor allows extending its functionalities through the inclusion of plugins and is available in two languages. Besides graphical editing of NCL document using templates, three other plugins have been developed allowing the use of NEXT by authors with basic NCL knowledge.

As future work, new usability tests will be carried. Other plugins will be developed to extend NEXT, offering more facility for document creation. EDITEC, a template graphical editor, will be adapted as a new plugin, integrating the facility of creating new templates in the graphical environment. Moreover, a plugin that allows the document presentation simulation is another important future work. A Temporal View plugin will be very helpful for developing applications by providing a better understanding about temporal events. An editable Textual View will also be very useful for authors who already know NCL.

7. ACKNOWLEDGMENTS

This work was partially supported by CAPES, CNPq and FAPERJ.

8. REFERENCES

[1] XML Path Language - XPath. http://www.w3.org/TR/xpath, 1999.

[2] W3C Semantic Web Activity. http://www.w3.org/2001/sw/, 2001.
[3] SCORM 2004 3rd Edition. http://www.adlnet.gov/capabilities/scorm/scorm-2004 3rd, 2006.
[4] *Digital terrestrial television - Data coding and transmission specification for digital broadcasting Part 2: Ginga-NCL for fixed and mobile receivers - XML application language for application coding.* ABNT 15606-2, 2007.
[5] Synchronized Multimedia Integration Language - SMIL. http://www.w3.org/TR/SMIL/, 2007.
[6] *Nested Context Language (NCL) and Ginga-NCL for IPTV.* ITU H.761 recommendation, 2009.
[7] Multimedia Home Plataform. http://www.mhp.org/, 2010.
[8] A. Celentano and G. O. Template-based Generation of Multimedia Presentations. *International Journal of Software Engineering and Knowledge Engineering*, 13:419–445.
[9] H. Chiao, K. Hsu, Y. Chen, and S. Yuan. A Template-based MHP Authoring Tool. In *Proceedings of the Sixth IEEE International Conference on Computer and Information Technology*, page 138, 2006.
[10] J. R. Damasceno, J. A. F. Santos, and D. C. Muchaluat-Saade. Editec: Hypermedia Composite Template Graphical Editor for Interactive tv Authoring. *ACM Symposium on Document Engineering*, pages 77–80, 2011.
[11] R. L. Guimarães, R. Costa, and L. F. Soares. Composer: Authoring Tool for iTV Programs. *European Interactive TV Conference - EuroITV*, 2008.
[12] T. Jokela, J. T. Lehikoinen, and H. Korhonen. Mobile Multimedia Presentation Editor: Enabling Creation of Audio-visual Stories on Mobile Devices. In *Proceedings of 26th ACM SIGCHI*, pages 63–72, 2008.
[13] M. R. López, R. P. D. Redondo, A. F. Vilas, J. J. P. Arias, M. L. Nores, J. G. Duque, A. G. Solla, and M. R. Cabrer. T-MAESTRO and its authoring tool: using adaptation to integrate entertainment into personalized t-learning. *Multimedia Tools and Applications*, pages 409–451, 2008.
[14] D. C. Muchaluat-Saade and L. F. G. Soares. XConnector & XTemplate: Improving the Expressiveness and Reuse in Web Authoring Languages. *New Review of Hypermedia and Multimedia*, 8:139–169, 2002.
[15] J. A. F. Santos. Multimedia and Hypermedia Document Validation and Verification Using a Model Driven Approach. Masters thesis, Federal Fluminense University, 2012.
[16] J. A. F. Santos, C. Braga, and D. C. Muchaluat-Saade. A model-driven Approach for the Analysis of Multimedia Documents. *5th International Conference on Software Language Engineering - SLE 2012/Doctoral Symposium*, 2012.
[17] J. A. F. Santos and D. C. Muchaluat-Saade. XTemplate 3.0: Spatio-temporal Semantics and Structure Reuse for Hypermedia Compositions. *Multimedia Tools and Applications*, 61(3):645–673, 2012.
[18] L. Soares, R. Rodrigues, and D. C. Muchaluat-Saade. Modeling Authoring and Formatting Hypermedia Documents in The Hyperprop System. *ACM Multimedia Systems Journal*, 8(2):118–134, 2000.

A Hybrid Architecture for Delivery of Panoramic Video

Martin J. Prins, Omar A. Niamut, and Ray van Brandenburg
TNO
Brassersplein 2
Delft, Netherlands
{martin.prins,omar.niamut,ray.vanbrandenburg}@tno.nl

Jean-François Macq, Patrice Rondao Alface, and Nico Verzijp
Alcatel-Lucent Bell Labs
Copernicuslaan 50
2018 Antwerp, Belgium
{jean-francois.macq, patrice.rondao_alface, nico.verzijp}@alcatel-lucent.com

ABSTRACT

The media industry is being pulled in the often-opposing directions of increased realism (high resolution, stereoscopic, large screen) and personalisation (selection and control of content, availability on many devices). Within the EU FP7 project FascinatE, a capture, production and delivery system capable of allowing end-users to interactively view and navigate around an ultra-high resolution video panorama showing a live event is being developed. In this paper we report on the latest developments of the FascinatE delivery network. We build upon an initial version of this delivery network architecture and its constituent functional components and propose a hybrid element to combine the two underlying delivery mechanisms that have previously been reported on. This hybrid aspect enables the delivery network to function in an end-to-end live delivery scenario.

Categories and Subject Descriptors

H.5.1 [Information Interfaces And Presentation]: Multimedia Information Systems – *video, immersive media, interactive media*

General Terms

Performance, Design, Experimentation, Standardization.

Keywords

Immersive media, ultra-high definition, panoramic video, media aware networking, tiled streaming, MPEG-DASH.

1. INTRODUCTION

New kinds of ultra-high resolution sensors and ultra large displays are generally considered to be the logical next step in providing a more immersive visual experience to end users. This notion of immersive media with ultra-high definition TV (UHDTV) and displays, highlighted by the NHK work on 8K Super Hi-Vision video [1] and the Fraunhofer HHI 6K OMNICAM system [2] seems contradictory with the explosive growth of device diversity. That is, having the content available on an increasing number of mobile devices, such as smartphones and tablets, each with its own characteristics, facilitates the user in selecting and controlling content. In contrast, UHDTV still assumes a more or less passive behaviour on the end user's side.

Within the EU FP7 project FascinatE [3] a capture, production and delivery system capable of supporting interaction, such as pan/tilt/zoom (PTZ) navigation, with immersive media has been developed by a consortium of 11 European partners from the broadcast, film, telecoms and academic sectors. This system allows end-users to interactively view and navigate around an ultra-high resolution video panorama showing a live event, with the accompanying audio automatically changing to match the selected view. The output is adapted to the end-user device, ranging from a mobile handset to an immersive panoramic display. At the production side, an audio and video capture system is developed that delivers a so-called Layered Scene Representation (LSR), i.e. a multi-resolution, multi-source representation of the audiovisual environment [4]. In addition, content analysis and scripting systems are employed to control the shot framing options presented to the viewer. Intelligent networks with processing components are used to repurpose the content to suit different device types and framing selections, and user terminals supporting innovative gesture-based interaction methods allow viewers to control and display content suited to their needs.

The business rationale for such a system was previously pointed out in [5]. A good overview of the entire FascinatE system and all its subcomponents and achievements is given in [6]. In particular, a more detailed look at the FascinatE delivery network and its initial achievements is given in [7]. There, the initial FascinatE reference delivery network architecture is defined, as well as its functional components. Also, an initial look at two novel delivery mechanisms and their implementations is provided.

Now in its fourth and final year, the FascinatE project aims to conclude its developments with a demonstration where live capture of ultra-high resolution panoramic video content is processed and delivered in real-time to a variety of end-terminals. This paper reports on the latest developments of the FascinatE delivery network, that enable live delivery of panoramic video. In this paper we discuss the final delivery network architecture, present a new hybrid network component, which combines the two delivery mechanisms described in [7]. The paper is organized as follows; section 2 reconsiders the delivery network architecture and its functional components. In section 3 we present latest developments for two delivery mechanisms. In section 4 we address the rationale and implementation of a hybrid distribution method. We conclude this paper with section 5, which discusses our findings and outlook for the future.

EuroITV'13, June 24-26, 2013, Como, Italy.

2. DELIVERY NETWORK

In order to deliver the LSR to end-devices, the delivery network needs to ingest the whole set of audiovisual (A/V) data produced to support immersive and personalized applications. This typically translates into very demanding bandwidth requirements. As an example, the live delivery of the immersive A/V material in an LSR consisting of an OMNICAM and three HD image sequences would require an uncompressed data rate of more than 16 Gbps. In situations where the full LSR is to be received by an end-user terminal, say in the case of a theatre with large-scale immersive rendering conditions, the delivery requires massive end-to-end bandwidth provisioning, even when using mezzanine or broadcast video compression. But FascinatE also aims at delivering immersive video services to terminal devices with lower bandwidth access or less processing power. In particular, a high-end home set-up capable of processing the full LSR for interactive rendering, but with typical residential network access, may be unable to receive the data rate of the complete LSR. Finally in case of low-powered devices, such as mobile phones or tablets, FascinatE introduces media proxies, capable of performing some or all rendering functionality on behalf of the end-client.

2.1 Delivery Network Functional Components

In order to be able to a) ingest audiovisual content in real time with data-rates up to 16Gbps b) accommodate delivery of the LSR to devices with different characteristics, e.g. processing power, supported resolution, network bandwidth, while maintaining interactivity and PTZ-control and c) to have a solution which is scalable, the delivery network contains three functional components: the A/V Ingest, A/V Proxies and A/V Relays. The A/V Ingest and A/V Proxy components have already been described in [7], whilst the A/V Relay is a new component, which will be discussed in more detail. For the sake of completeness this section continues with a short summary of the A/V Ingest and A/V Proxy and their roles.

2.1.1 A/V Ingest

The A/V Ingest, as its name suggests, provides ingestion of A/V data coming from the production domain at the left-hand-side of the delivery network. A/V data in the production domain is typically provided in uncompressed form and thus leads to high data rates, which typically cannot be transported over most delivery networks. This means that the data needs to be preprocessed before delivery. Furthermore the data needs to be inserted in the delivery network.

To be able to transmit a 7Kx2K video panorama over the network, the A/V Ingest uses a form of multi-resolution tiling before encoding the panoramic video. Therefore the source video frame in its original form and one or more downscaled versions are split into tiles, such that the tiles can be accessed and thus distributed independently. Multi-rate encoding is performed to a). further reduce the required bandwidth to transmit the content by using standard compression techniques and b). to offer the content in different representations, distinguished by bitrate, such that the delivery network can make use of rate-adaptation mechanisms such as adaptive-bitrate streaming or a bandwidth-optimal selection of tiles [8]. The multi-rate encoding process results in media delivery units that we refer to as segments. The segments are subsequently ingested into the delivery network and transmitted to the A/V Proxies by a Segment Transport Server.

2.1.2 A/V Proxy

At the right-hand side of the network, the A/V Proxy component is responsible for ensuring that A/V segments required by one user, or a local set of end-users, are delivered and reassembled according to their activity requests. The A/V Proxy can also perform in-network A/V processing using a rendering node to adapt to personalized requests and/or personalized delivery conditions, such as access bandwidth and device capabilities. For devices with limited processing power, these in-network rendering operations can range from 2D reframing and

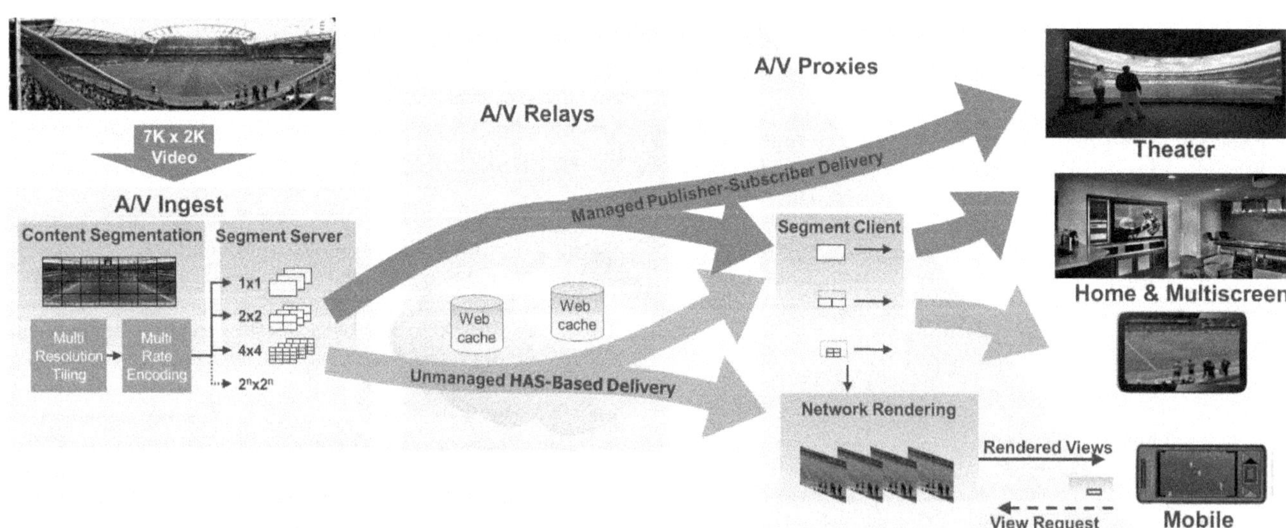

Figure 1- FascinatE network functionality and delivery mechanisms.

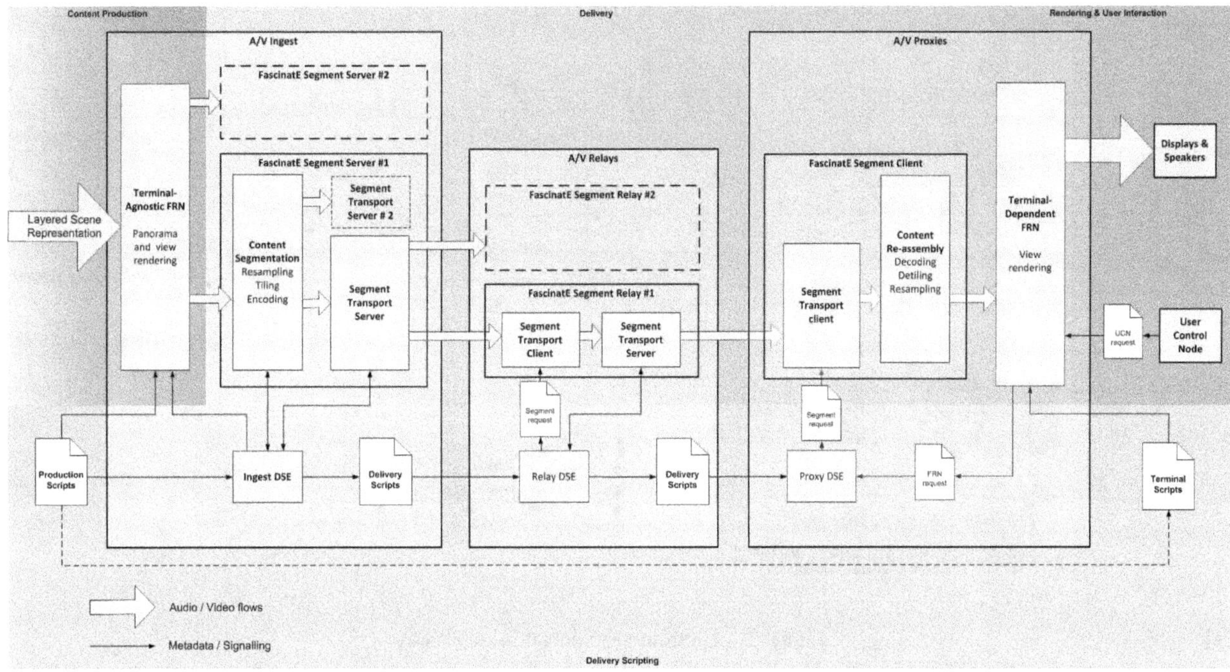

Figure 2 - Delivery network architecture, showcasing the A/V Relay component

downscaling to full 3D rendering, compensating the cylindrical or spherical distortion of the panorama [9].

Figure 1 gives an overview of the FascinatE network functionality and delivery mechanisms, essentially highlighting the functions of the A/V Ingest and A/V Proxy. As previously described in [6] and [7], this architecture has been designed to support content delivery over both managed networks, e.g. walled-garden IPTV, and unmanaged video delivery networks, over the public Internet . As indicated in Figure 1, two types of end-to-end delivery mechanisms have been mapped onto this architecture; a publisher-subscriber (Pub/Sub) mechanisms to support interactive video delivery (see section 3.3) and extensions of HTTP Adaptive Streaming (HAS) to spatially segmented content (see section 3.2).

In this paper, we highlight the third component of the delivery network, i.e. the A/V Relay, that allows for a more flexible combination of the two different transport mechanisms over the end-to-end delivery chain.

2.1.3 A/V Relay

When looking at the delivery path between the two ends of the FascinatE delivery network, one can see that there are at least two sets of transport requirements. At the A/V Ingest side, the entirety of the content has to be efficiently pushed to the delivery network. At the A/V Proxy side, less bandwidth is usually available and only a fraction of the content will be requested on behalf of the served end-users and these requests may have a high dynamicity depending on the level of interactivity of end-users. To support the connection between a 'push' source at the A/V Ingest and a multitude of 'pull' A/V Proxies, we propose to introduce an intermediate tier of nodes, called A/V Relays. On one hand, each A/V relay is required to be located sufficiently deep in the network to support the reception of the entire content from the A/V Ingest. On the other hand, it must serve the aggregated sets of interaction requests from the A/V Proxies.

In addition, an A/V Relay can contain functions to aggregate, cache and/or relay segment requests to other nodes. Importantly, the A/V Relay can be also be used as a demarcation point between delivery modes for the downstream A/V flows. As illustrated in Figure 2, the A/V Relay contains two functions back-to-back, referred to as the Segment Transport Client and Server. Note that the Segment Transport Server is similar to the one located in A/V Proxies. They allow terminating the transport from the A/V Ingest and initiating the transport towards the A/V Proxies. In section 3, we explain this role when using either HTTP or Pub/Sub based transport end-to-end. In section 4, we focus on a hybrid scenario, where we use Pub/Sub delivery on the Ingest – Relay path and HTTP transport on the Relay-Proxy one. With this scenario we tend to further reduce bandwidth usage in a heterogeneous delivery network setting, as the content then only needs to be transmitted once between an A/V Ingest and the AV/Relay while being used by both Pub/Sub and Tiled HAS delivery mechanisms.

3. DELIVERY MECHANISMS FOR INTERACTIVE IMMERSIVE MEDIA

As discussed in section 2, the actual transport of A/V segments takes place between Segment Transport Servers and Clients. Two specific delivery mechanisms have been developed between the A/V Ingest and the A/V proxy, catering for different usage scenarios and network deployments: Tiled HAS (section 3.2) and Pub/Sub (section 3.3). Compatibility of the media being distributed by the mechanisms is achieved by defining a common tiling format, as described below.

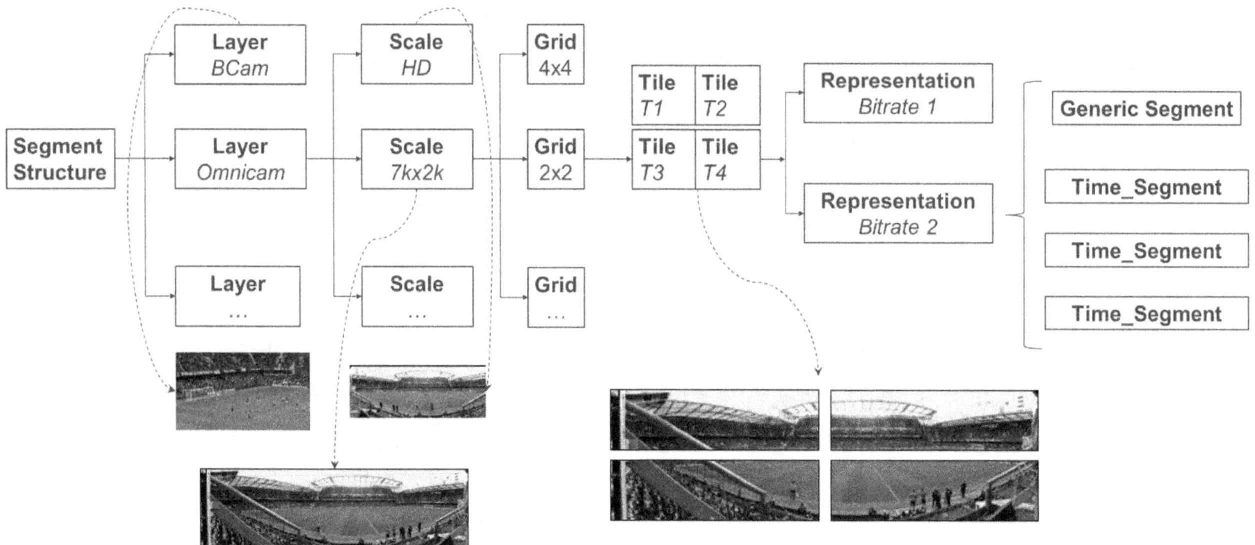

Figure 3 – Content segmentation hierarchy

3.1 Content Segmentation

By having a common tiling format we ensure that A/V content needed to be processed, tiled and encoded only once at the A/V Ingest; the resulting tiles can then be used by both tiled HAS and Pub/Sub. As discussed in section 2 the A/V Ingest encompasses both multi-resolution tiling and multi-rate encoding, thereby enabling the offering of panoramic video in a range of presentations.

A content addressing schema has been created which describes the hierarchy of segmented content. This hierarchy of segmented content and some example segments is shown in Figure 3. The hierarchy consists of the following structure:

- For every layer in the LSR, one or more resolution scales are created;
- For every scale, one or more MxN tiling grids are created, leading to a set of tiled video streams per scale;
- For every tile, one or more representations at different quality settings are created;
- For every representation, the associated tiled video stream can be temporally segmented.

3.2 Tiled HTTP Adaptive Streaming

Tiled HAS combines the flexibility of HTTP Adaptive Streaming solutions such as MPEG-DASH [10]. with the concept of tiled and zoomable content [11,12]. HAS solutions typically combine multi-rate encoding with temporally segmented delivery of video streams via HTTP, enabling adaptive delivery of video by providing a video in different representations (e.g. bitrates), allowing a client to seamlessly and adaptively switch between segments from different representations, based on changing network conditions. By combining HAS with the concept of tiling, a delivery mechanism is created which allows for the adaptive and selective delivery of parts of the content the user expresses interest in.

Within the delivery network a tiled HAS framework was developed, consisting of a Segment Transport Server (STS), a Segment Transport Client (STC), a video rendering engine and a user control node (UCN). The STS provides the tiled HAS segments which are requested by the STC. The video rendering engine determines, through the use of a tiled HAS manifest (see next subsection), which segments are needed to recreate a certain view or region-of-interest (ROI) as requested by the user via the UCN, and uses the STC to request those segments. The segments that have been retrieved by the STC are recombined and presented on a display.

3.2.1 Tiled HAS Manifest

Content tiling can be seen as the spatial counterpart of temporal segmentation in current HAS delivery mechanisms. This means that existing HAS manifests such as an HTTP Live Streaming manifest or MPEG-DASH Media Presentation Description (MPD), can easily be extended for tiled HAS, as previously noted by [12].

The content segmentation hierarchy presented in Section 3.1 is used as a basis for our Tiled HAS manifest. The differences (blue) and similarities (purple) between the Tiled-HAS manifest and the MPEG-DASH Media Presentation Description (MPD) and its hierarchal structure are depicted in Figure 4. The signalling component (green) will be discussed in the next subsection.

3.2.2 Signalling framework

Within HAS based mechanisms like MPEG-DASH, data and control path are tightly coupled. That is, HAS is typically purely client-driven, which means that session updates can only be signaled by updates in the manifest file. The client therefore needs to poll the server providing the manifest to see if the session has been updated, i.e. by using the HTTP *If-Modified-Since* header parameter.

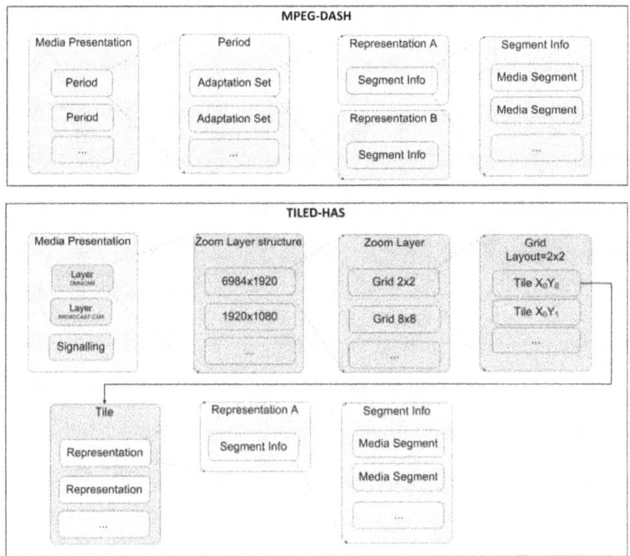

Figure 4 – Comparison of MPEG-DASH and Tiled-HAS manifest structures

Some efforts have been made within the DASH-Manifest to guide how often a client needs to check for updates, by means of the *minimumUpdatePeriod* attribute, which specifies the smallest period between potential changes to the manifest.

A disadvantage of these solutions is that asynchronous real-time communication as triggered by the server, for instance for session notifications or real-time events, is not possible. Furthermore the delivery network may need to provide other control information out-of scope of the Tiled HAS media presentation, for instance for signaling the availability of other content, such as alternate camera feeds, or production scripts, specifying specific ROI information which could be used to automatically control what the user will see, resulting in a virtual camera feed.

To be able to fulfill both requirements the Tiled HAS Framework is extended with a signaling framework (Figure 5) based on the WebSocket standard, a web-technology which allows for low-latency two-way communications over a single TCP connection [14]. Websockets is part of the HTML 5 standard [15], but also used outside of the web-domain. With the signalling framework, the Tiled HAS delivery components can provide signaling to each other with three types of signaling:

1. *Segment Server - Video Renderer;* relates to the delivery function and providing real-time updates regarding the LSR. It is primarily used by the STS to signal updates and session information to the video renderer, such as an updated manifest, alternative Segment Transport Servers, or the availability of new layers of the LSR (e.g. a broadcast camera) or so-called production scripts. These scripts contain production information which can be used to assist or automate PTZ controls or provide the user with predefined virtual camera feeds. An example is the tracking of a specific player during a soccer match.
2. *Video Renderer - User Control Node;* enables user interaction with the video renderer and allows for providing information that can enhance the presentation and interaction on the UCN. Examples are requesting session information, controlling the session (PTZ operations or traditional Trick Play) and subscription to session related events, such as additional video streams.
3. *Video Renderer Pass-through;* enables pass-through of signaling messages: events originated by the STS can be forwarded to the UCN as well. Example is the subscription to specific live events, such as additional video streams, or (chat)-messages that should be displayed on the UCN instead of the traditional display.

The signaling framework supports two types of communication, i.e. request-based and subscription-based. The former provides the means to retrieve session information or alter a session, the latter allows for subscribing to specific events transmitted by the video renderer, akin to the Pub/Sub mechanism. The messages are defined in JavaScript Object Notation (JSON) [16].

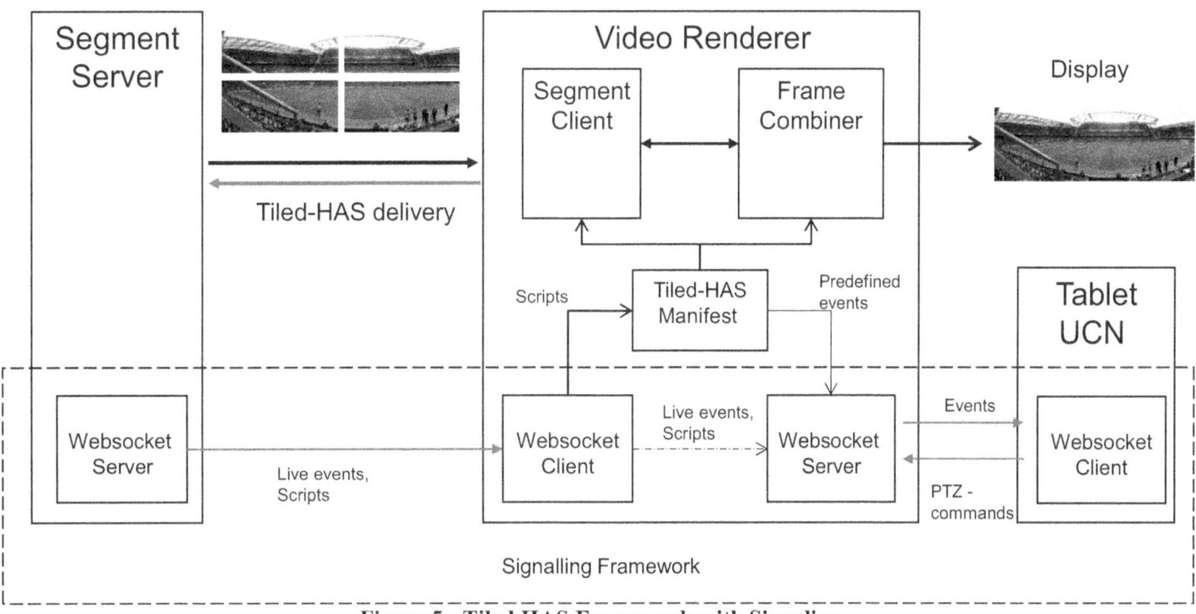

Figure 5 - Tiled HAS Framework with Signaling

3.3 Pub/Sub based delivery

Another category of transport mechanisms studied in the FascinatE project is based on a publisher-subscriber (Pub/Sub) architecture. This is inspired from traditional large-scale publish-subscribe systems where a large set of information flows need to be delivered to a large set of users. Usually each user is interested in only a subset of the information streams, and therefore only subscribes to a subset of the publisher. A characteristic of most Pub/Sub implementations is that they can seamlessly combine the use of unicast and multicast channels. As such they allow the system to selectively and adaptively transmit audio/visual data. This is of particular interest for managed networks, where combinations of IP multicast and unicast transport can be performed. They can be optimized depending on the bandwidth resources and content popularity. In the following sections, we explain the rationale of using Pub/Sub for video streaming and indicate where the mechanisms are used in the overall FascinatE delivery network.

3.3.1 Mapping of Video Tile Streaming to Pub/Sub

By interpreting a video stream as a sequence of messages, e.g. at the granularity of individual frames or of short subsequences, Pub/Sub mechanisms can be extended to support scalable delivery of a large number of video sources. In the case of tiled panoramic content, the proposed approach consists in assigning a given tile (characterized in terms of spatial location, resolution, quality, … as depicted in the hierarchy of Figure 1) to a unique publisher. At the server side, all the available content is thus offered by jointly running a set of publisher threads. At the other end, each client is responsible for subscribing to a subset of the video tiles, depending on their need for sub-regions of the entire panoramic scene. In our previous work [8], we have studied techniques to optimally select a subset of tiles within a multi-resolution and multi-quality hierarchy, under bandwidth constraints.

A second means to optimize the usage of network resources is to efficiently select the type of transports (unicast or multicast) allowed by Pub/Sub. The general idea is to publish tiles requested by many end-users on multicast channels and publish the less popular on unicast channels. The combination of multicast and unicast to optimally distribute content under capacity constraints has been studied in [17]. The application to the transport of tiled content has been studied in [18] in the case of using a shared medium (such as Wireless LAN).

4. HYBRID A/V RELAY

The delivery configurations studied and demonstrated in FascinatE include full end-to-end usage of either HAS or Pub/Sub –based delivery. This has already been discussed in our earlier work to full extent. In this section we focus on a new third delivery configuration which combines both Pub/Sub and Tiled HAS transport for a hybrid delivery mechanism.

4.1 Architecture

As explained in section 2.1.3, the A/V Relay is introduced as a demarcation point in the end-to-end delivery path. The Ingest-Relay portion is expected to support the delivery of the full set of video tiles. In contrast, the Relay-Proxy portion is expected to optimally transport the content, taking into account the interactivity requests. As depicted in Figure 6, we have developed a set of Pub/Sub mechanisms that cover the need of both portions of the network. From the Ingest, the entire content can be multicast to all Relays, whereas from each Relay a managed combinations of multicast and unicast transport is used to transport the requested subset of tiles to the connected Proxies. Figure 6 also indicates that the A/V Relay can be implemented as a hybrid node, receiving on one side the content from the Ingest using Pub/Sub and re-offering the content for HAS-based transport as described in section 3.2.

As the two delivery mechanisms use a common tiling format the two methods can also easily be combined two offer a single integrated solution to provide a panoramic video over both managed and unmanaged networks. In a typical example, a service provider may already have a managed delivery in its core network delivery part, thereby leveraging high bandwidth conditions and efficient delivery via both unicast and multicast. New functionality is added by the hybrid A/V relay function, by converting the streams to Tiled HAS which is more suited to best-effort based delivery, e.g. unmanaged broadband and mobile Internet connections. In the next section, we cover the implementation of this hybrid usage scenario.

4.2 Implementation

The Hybrid A/V Relay implementation consists of three components: a *Pub/Sub Segment Transport Client*, a *Tiled-HAS Segmenter*, and a HAS Segment Transport Server.

The *Pub/Sub Segment Transport Client* subscribes to the Pub/Sub Segment Transport Server at the A/V Ingest, which serves the Panoramic Video with a video resolution of 6976 by 1920 pixels at 25 fps. The Panoramic Video is provided in four different resolutions, with a dyadic tiling approach ranging from 1x1 up to 8x8 (see Table 1).

Tile layout	Frame resolution	Number of streams
1x1	872x240	1
2x2	1744x480	4
4x4	3488x960	16
8x8	6976x1920	64

Table 1: Content tiling settings for the Hybrid A/V Relay implementation

The 1x1 and 8x8 tiling structures are used by both the Pub/Sub A/V Proxy as the Tiled-HAS based Proxy implementations, whilst the 2x2 and 4x4 layouts are used for the Tiled-HAS based Proxy only, as this implementation uses more intermediate resolution layers. The resulting 85 streams (85 tiles per video frame) are offered to the Segment Transport Server, to which the Hybrid Segment Relay subscribes; the Pub/Sub Segment Relay only subscribes to 65 streams (8x8 and 1x1). The streams are encoded into the H264 format with a GOP size of 1 second (or 25 frames).

For each tile a separate Pub/Sub session is opened, leading to the reception of 85 independent video streams by the *Pub/Sub Segment Transport Client.*

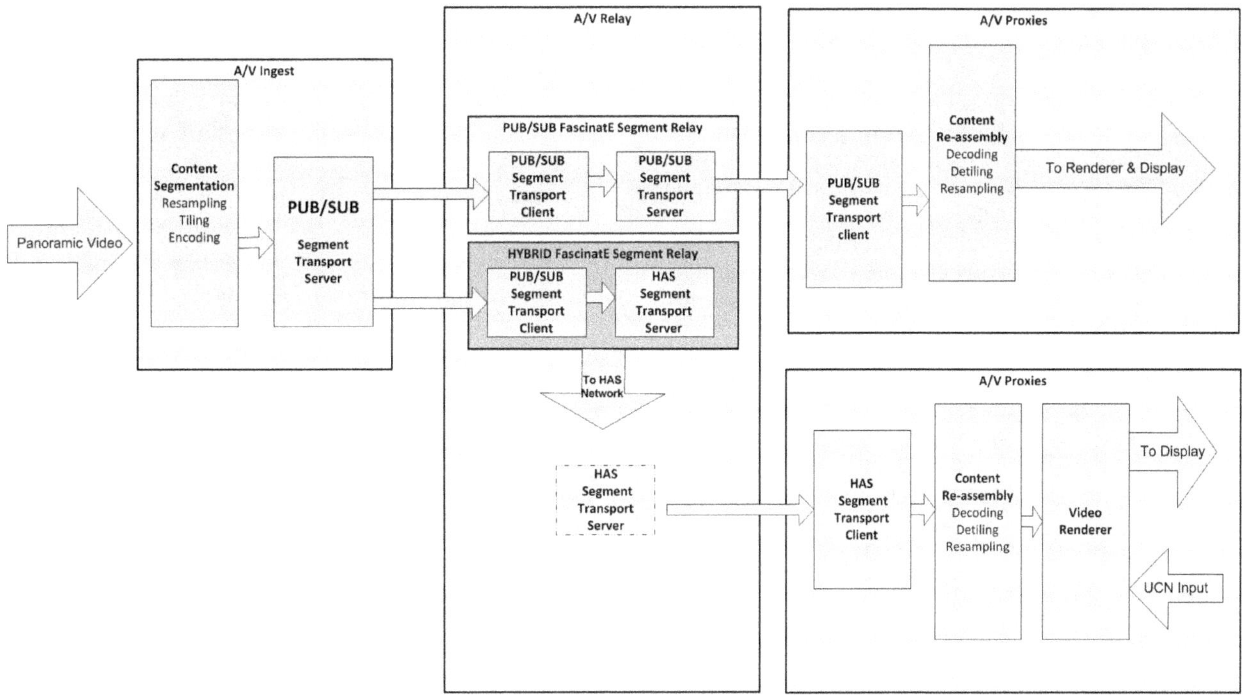

Figure 6- Hybrid A/V Relay

Once the *Tiled-HAS Segmenter* retrieves the 85 corresponding video streams it can start segmenting each stream to its corresponding Tiled-HAS segments. To start creating Tiled-HAS segments for each stream, two constraints have to be fulfilled:

- Every segment needs to start with an independently decodable video frame (an I-frame).
- For each respective stream the initial segment must start with the same frame number of PTS, otherwise the content does not remain synchronized.

This means that the Tiled-HAS Segmenter, when for every stream the same I-frame has been received, can start writing the first segments. In the case that for some streams preceding I-frames have been received, they will be discarded.

Once the first segments have been written and have been moved to the HAS Segment Transport server (which can be in the same system, or reside on another network (for instance a CDN), the Tiled-HAS manifest is generated. Once this Tiled-HAS manifest has been signaled to the Tiled-HAS Proxy implementation (see next section), the transmission of the HAS-segment can start, after which the Panoramic Video can be displayed on a HD screen, whilst offering the user control of the view by means of the Tablet UCN. As earlier described in [7] the Tablet UCN can also be used to provide trick play, picture-in-picture through multi-ROI rendering and predetermined ROI selection. Also, additional layers from the LSR, e.g. from broadcast cameras can be made available on the companion screen. This last feature is however not yet supported in the current implementation.

4.3 Hybrid A/V Signalling

An important aspect of configuring and bootstrapping for the A/V Relay Hybrid implementation is the signaling of the control plane information, such that the A/V Relay can be dynamically instantiated or adjusted. Subsequently the Tiled-HAS based A/V proxy needs to be notified when the Hybrid A/V relay has started.

In the next subsections we elaborate on the signaling functions of the Pub/Sub and the Tiled-HAS framework which are used in the Hybrid Relay implementation.

4.3.1 Pub/Sub signalling

In addition to being used for transporting the video data, the Pub/Sub mechanism can also be used to push signaling data. In this section, we focus on the specific case of using Pub/Sub on the Ingest-Relay portion of the network. The signaling data are required by the Relay (and then by the Proxy), in order to correctly request tiles and handle the received video data. We develop a specific signaling format that provides the subscribers with the following information.

- Description the tiling pattern (grid resolution and scale)
- Unique Identification of content segment spatially and temporally
- Resource locators allowing the subscription to each segment

Based on this information the Hybrid A/V Relay is able to retrieve the segments for the Panoramic Video, but also configure and start the Tiled-HAS Segmenter: it requires the information to start segmentation but also to be able to create the Tiled HAS manifest with the correct configuration parameters.

4.3.2 Tiled-HAS signaling

As we discussed in section 3.2.2 we developed a signaling framework to amongst others be able to transmit live events to our Video Renderer. For the Hybrid A/V Relay Implementation we had the Video Renderer connect to the WebSocket Server in the Segment Server, such that it could receive the manifest in real-time once the first segments where created. This prevented needing to poll the server for the manifest and thus also introducing unnecessary delays.

5. CONCLUSIONS

In this paper we presented the latest developments of the Fascinate delivery network. The delivery network architecture has been extended with A/V Relay functionality, which allows for more scalable and fine-grained delivery functions. Another advantage of this A/V Relay function is the ability to create hybrid delivery functions, which combine the two delivery mechanisms developed within FascinatE, i.e. Publisher/Subscriber and tiled HTTP Adaptive Streaming. Due to a common tiling format, the existing implementations presented in our earlier work, are now easily combined to create a hybrid implementation which allows for real-time conversion of Pub/Sub transport to Tiled-HAS based transport, allowing the FascinatE delivery network to provide high resolution panoramic video over both managed networks towards high-end devices, as well as over unmanaged, over the top delivery, to mobile and fixed devices connected to the Internet. With this hybrid implementation we further have reduced bandwidth usage as the content only needs to be transmitted once between an A/V Ingest and the AV/Relay while being used by both Pub/Sub and Tiled HAS delivery mechanisms.

5.1 Future work

The feasibility of the updated delivery network will be presented during a final demonstration, where live capture of ultra-high resolution panoramic video content is processed and delivered in real-time to a variety of end-terminals. This demonstration will take place at MediaCityUK, in May 2013, and will include additional FascinatE system components, showcasing a variety of novel aspects for interactive and immersive media delivery. In particular, this demonstration allows us to further evaluate the hybrid A/V relay, in which multicast transmission is used first, after which the A/V relay filters the tile layouts that not needed by the Pub/Sub A/V proxies. A demonstration will also take place at the EuroITV 2013 conference. Additionally, we will investigate if the hybrid A/V Relay can also be used in a novel way for temporally storing A/V content in the form of tiled HAS segments, which could then be reused for Pub/Sub distribution for content replay scenarios. Lastly, we will investigate an alternative content tiling scheme that is particularly suited for tiled HAS delivery towards mobile devices only.

6. ACKNOWLEDGMENTS

The research leading to these results has received funding from the European Union's Seventh Framework Programme (FP7/2007-2013) under grant agreement no. 248138. Thanks to SIS Live, the Premier League, Compagnie Sasha Waltz & Guests, the Berlin Philharmonic Orchestra, KUK Filmproduktion and the BBC for their assistance during the test shoots and permission to use images. The authors would like to thank the FascinatE project partners for the insightful discussions and collaboration.

7. REFERENCES

[1] M. Maeda, Y. Shishikui, F. Suginoshita, Y. Takiguchi, T. Nakatogawa, M. Kanazawa, K. Mitani, K. Hamasaki, M. Iwaki and Y. Nojiri. "Steps Toward the Practical Use of Super Hi-Vision". NAB2006, Las Vegas, USA, Apr. 2006.

[2] R. Schäfer, P. Kauff, and C. Weissig. "Ultra-high resolution video production and display as basis of a format agnostic production system", IBC2010, Amsterdam, Netherlands, Sep. 2010.

[3] Fascinate Project, Official FascinatE Website. http://www.fascinate-project.com, 2013

[4] G.A. Thomas, O. Schreer, B. Shirley, J. Spille. "Combining panoramic image and 3D audio capture with conventional coverage for immersive and interactive content production", IBC 2011, Amsterdam, The Netherlands, 11th Sept., 2011.

[5] O. Niamut, T. Bachet, and S. Limonard. "High-resolution video, more is more?" Proc. of the 9th international interactive conference on Interactive television. ACM, 2011.

[6] O. A. Niamut *et al*, "Towards A Format-agnostic Approach for Production, Delivery and Rendering of Immersive Media", MMSYS 2013 Proc., Feb. 2013.

[7] O. Niamut, J.-F. Macq, M. Prins, R. Van Brandenburg, N. Verzijp, P. Rondao Alface, "Towards Scalable And Interactive Delivery of Immersive Media", NEM Summit 2012, Oct. 2012.

[8] P. Rondao Alface, J.-F. Macq, and N. Verzijp, "Interactive Omnidirectional Video Delivery: A Bandwidth-Effective Approach", Bell Labs Tech. Journal 16(4): 135-147, 2012.

[9] J.-F. Macq, N. Verzijp, M. Aerts, F. Vandeputte, E. Six, "Demo: Omnidirectional video navigation on a tablet PC using a camera-based orientation tracker", in Proc. of the Fifth ACM/IEEE ICDSC 2011, Ghent, Belgium, Aug. 2011.

[10] ISO/IEC JTC 1/SC 29, "Information technology -- Dynamic adaptive streaming over HTTP (DASH) -- Part 1: Media presentation description and segment formats". ISO/IEC 23009-1:2012. April 2012.

[11] O.A. Niamut, M.J. Prins, R. van Brandenburg, A. Havekes "Spatial Tiling And Streaming In An Immersive Media Delivery Network", in Adjunct Proc. of EuroITV 2011, Lisbon, Portugal, June 2011

[12] Khiem, N., Ravindra, G., Carlier, A., and Ooi., W. 2010. Supporting zoomable video streams with dynamic region-of-interest cropping. In Proceedings of the first annual ACM MMSys '10. ACM, New York, NY, USA, 259-270.

[13] N. Khiem, Q. Minh, G. Ravindra, and W. T. Ooi. "Adaptive encoding of zoomable video streams based on user access pattern." Proc. of the second annual ACM MMSys, 2011.

[14] I. Fette, A. Melnikov, "The WebSocket Protocol", IETF RFC 6455. http://tools.ietf.org/html/rfc6455. December 2011

[15] World Wide Web Consortium (W3C), "HTML 5.1 - A vocabulary and associated APIs for HTML and XHTML". http://www.w3.org/TR/html51/. W3C Working Draft 17 December 2012

[16] JSON, Javascript Object Notation, http://www.json.org.

[17] D. Bickson, E. N. Hoch, N. Naaman, and Y. Tock, „A hybrid multicast-unicast infrastructure for efficient publish-subscribe in enterprise networks", In Proc. of the 3rd Annual SYSTOR'10, pp. 1–7, Haifa, Israel, 2010.

[18] G. Ravindra and W. T. Ooi, "On Tile Assignment for Region-of-Interest Video Streaming in a Wireless LAN", In Proc. of the 22nd International Workshop on NOSSDAV'12, Toronto, Canada, 7-8 June 2012

Windy Sight Surfers: Sensing and Awareness of 360º Immersive Videos on the Move

João Ramalho
LaSIGE, Faculty of Sciences
University of Lisbon
1749-016 Lisboa, Portugal
+351217500087
joaoramalho90@gmail.com

Teresa Chambel
LaSIGE, Faculty of Sciences
University of Lisbon
1749-016 Lisboa, Portugal
+351217500087
tc@di.fc.ul.pt

ABSTRACT
Immersion in video has a strong impact on the viewers' emotions, their sense of presence and engagement. In this paper, we present the design and user evaluation of Windy Sight Surfers, an interactive mobile application, to be used on its own or as a TV second screen, providing an approach to an immersive environment, for the capture, visualization and navigation of georeferenced 360º and high definition videos. In this approach, wind is used as a means to enhance immersion, by giving a more realistic feel of wind, speed and orientation to the user. Usability results revealed increased immersion and appreciation for both experience sensing and context awareness features, which are the two main categories of the designed features.

Categories and Subject Descriptors
H.5.1 [Information Interfaces and Presentation (I.7)]: Multimedia Information Systems – *video, hypertext navigation and maps*; H.5.2 [Information Interfaces and Presentation (I.7)]: User Interfaces – *interaction styles, evaluation*;

General Terms
Design, Experimentation, Human Factors.

Keywords
Interactive Video, Immersion, 360º, Presence, Sensing, Awareness, Mobility, Second Screen, Speed, Wind.

1. INTRODUCTION
Video allows to capture and present events and scenarios with great authenticity, realism, and emotional impact, and it is becoming pervasive in our lives, in personal capturing and display devices, over the Internet, in social media, and through video on demand services on iTV [13,14]. Immersion in video has a strong impact on the viewers' emotions, their sense of presence and engagement [5,21]. 360º videos could be highly immersive, by allowing the user the experience of being surrounded by the video. Immersion may be defined as a feature of display technology determined by inclusion, surround effect, sensory modalities and vividness through resolution [19]. Slater & Wilbur proposed that the Immersion property is associated with Presence, which relates to the viewer's conscious feeling of being inside the virtual world. Wirth et al. [22] added to Slater and Wilbur's definition of presence that this state includes perceived self-location in the virtual world.

In today's world, mobile devices proliferate and represent, by the wide range of sensors and actuators they are increasingly incorporating, a wide range of opportunities to capture and display 360º and HD video and metadata (e.g. geo-location and speed) with the potential to allow augmenting human-computer interaction and supporting more powerful and immersive video user experiences. But there are challenges to the design of effective environments that make use of this immersion potential.

Wide screens and CAVES, or domes - with varying angles of projection, possibly towards full immersion, in cylinder or spherical (e.g. allosphere.ucsb.edu) environments - provide privileged display conditions for immersive video view for their dimensions, but they are not very handy, and especially CAVES are not widely available. Whereas, the flexibility associated with mobile devices could allow users to actually turn around, as if they hold in their hands a window to the video where they are immersed in, and bring this with them everywhere. As second screens, mobile devices may also be used to help navigation in a video that is projected or displayed outside in a wider screen (from TV to CAVE), allowing e.g. to display and interact with additional info and navigational aids, clearing the video display of additional extraneous info, for increased sense of immersion and flexibility, and even to decide to catch the current video on TV and go on watching it on the move.

Moreover, this is the first time an entire new generation has grown up watching content on demand, which is characterized by the Internet, Mobile and Social environments. Nielsen (www.nielsen. com/us/en/newswire/2012/introducing-generation-c.html) defines this group by the connected behavior, referring to it as the Generation-C. In this context, video is one of the most pervasive content types, and its consumption is ever increasing. Recently, YouTube announced one billion active users every month (http://youtube-global.blogspot.pt/2013/03/onebillion-strong.html) which translates to one out of every two people on the Internet using YouTube, and therefore consuming video content. Addressing Generation-C, Google (http://adwordsagency .blogspot .pt/2013/03/how-does-gen-c-watch-youtube-on-all.html) states that YouTube usage on smartphones mirrors usage on PCs,

EuroITV'13, June 24–26, 2013, Como, Italy.

and around 70% of the population watch YouTube on two or more devices, which confirms the significance of multiscreen computing.

In this paper, we present the design and first user evaluation of Windy Sight Surfers (Windy SS), an interactive mobile application for the capture, visualization and navigation of 360° immersive videos, designed to empower users in their immersive video experiences, both accessing other users' videos and sharing their own. Videos are captured by the users as 360° videos with several metadata associated to them, and published afterwards to an online community, where they can be searched and viewed in novel and immersive ways. The designed wind prototype allows users to get a more realistic feel of movement and speed in a mobile environment.

Section 2 highlights the main challenges and most related work. Section 3 presents Windy Sight Surfers in terms of video and metadata capturing, sharing and designed features to increase immersion during viewing, while section 4 addresses and discusses its user evaluation. The paper ends in section 5 with main conclusions and directions for future work.

2. RELATED WORK

This work builds on our previous work on 360° hypervideo [13], an interactive web application, which evolved to allow capturing, sharing and navigating georeferenced 360° videos and movies, synchronized with maps [14].

Visualizing and interacting with 360° hypervideo is a quite new area of research. Challenges include providing users with: an appropriate interface capable to explore 360° contents, with the feeling of looking around and an increased sense of immersion; appropriate affordances to understand the hypervideo structure and to navigate it effectively in a 360° hypervideo space, dealing with disorientation and cognitive load, in this richer and more challenging 360° context, where video represents rich information that changes in time and that may be out of sight in 360°. In the scenario of georeferencing and sharing of 360° videos, other challenges include orientation and navigation in videos and maps and support to find and access large amounts of videos and users.

Briefly, most relevant related work concerns to hypervideo and immersive environments (mainly Virtual Reality (VR) and Augmented Reality (AR), images, like Google Street View, seldom video), georeferencing and maps, orientation, cognitive load and filtering, and navigation and off-screen content in mobile devices [13,14,6,15,20,7,24,2]. Recent PanoramaGL library allows the visualization of 360° and spherical photos by using the mobile device as a window to the photo. Courtois et al. [4] evaluated interest and readiness of users to use mobile devices as second screens when watching television. Bleumers et al. [1] found: that certain genre-specific content elements are more suitable for omnidirectional (360°) video than others from a user perspective (such as hobby/sport programs that trigger interest in mastering the skills that are displayed); and more circumstantial elements, such as when there is little progress in the program and users are prone to seek temporary distraction, could potentially also make omnidirectional video viewing worthwhile.

Regarding the use of sensors and actuators, e.g. Mendes et al. [11] presented art galleries where user's movements influence video being displayed. Moon and Kim [12] showed that the use of wind as output increased the sense of presence in VR, and their application did not allow user interaction, as the movement occurred with a pre-defined animation path. Cardin et al. [3] presented a head mounted wind display targeted at a VR flight simulator application. In the experience, participants should determine the wind direction, which they did with a variation of 8.5 degrees. Kojima et al [8] developed a wearable device for presenting the sensation of localized wind. The prototype focused on the ears and used a combination of sound and small fans. Conclusions stated that ears can perceive wind with high spatial resolution, and a local wind sensation can increase immersion. Lehmann et al. [9] evaluated the differences between merely visual, stationary and head mounted wind, with considerable increase in presence when using stationary and head mounted wind prototypes. But these approaches target VR - not video, nor mobile environments - as they require heavy and very specific equipment. Furthermore, none of them presents methods to capture wind metadata and couple it with video as a way to increase realism and immersion.

3. WINDY SIGHT SURFERS

This section presents Windy Sight Surfers, with a special focus on the video and metadata capture and publishing, the designed features to increase immersion during viewing, which divide into two main categories: Experience Sensing and Context Awareness, the interaction with TV screens and its searching methods, supported by the system architecture presented on the last sub-section.

3.1 Video & Metadata Capture

Videos are captured in two categories: 360° and High Definition videos. For High Definition videos, we used a GoPro Hero 2, a 'wearable' camera mostly targeted at adventure video. For 360° videos, we used a Sony Bloggie Handy Cam1, a 360° camera available to the general consumer that allows converting the captured video into a cylindrical projection. Concurrently, we captured the geo-references along the video trajectories with a smartphone, using the GPS (latitude, longitude, and speed every half second) and digital compass (orientation angle, whenever direction changes are > 45°) sensors. Also, a 3G Internet connection was used to acquire data from the OpenWeatherMap's weather forecast web service (temperature, weather status, rain, wind speed and orientation). All metadata is stored in an XML file associated with its related video.

The reason to not integrate these functionalities (video & metadata capture) in the same device is related to the hardware availability. Currently, the affordable 360° video cameras do not provide very good image quality, and most cameras do not have the sensors that are common to find in any smartphone. Ideally, this process will be integrated in the same device.

With the aim of creating a social platform, users can submit and view each other's routes and experiences.

3.2 Video & Metadata Publishing

Wind Surfers has its ultimate objective in providing a social platform where users can submit and view each other's routes and experiences. Therefore, in the application's video publishing page, the user imports the video and the file containing the video's metadata information. Some key information, such as the city or administrative area and the country where the video was captured, or the date when the video was recorded, is automatically inferenced to help the indexing in server side. The videos may also be indexed by their popularity, most recent and by themes, very much similar to other web applications like Youtube.

When publishing videos, there is the possibility to override some of the automatically generated values. Namely, the weather information given by the OpenWeatherMap web service may be corrected for a more accurate value. Users are given this possibility because the OpenWeatherMap web service collects the weather information from the nearest weather station to the real GPS coordinates. This station can be far away from the real location, and therefore information might not be as accurate as desired. Also, if there was no Internet connection by the time the video was captured, the user can introduce some missing metadata, such as the local where the video was recorded (through a long press in a map), the recording date, and weather information.

3.3 Video Viewing: Experience Sensing Features

Videos are displayed in full screen on an Android tablet. 360° videos are mapped onto a transitional canvas that is in turn rendered around a cylinder, to represent the 360° view and allow the feeling of being "partially" surrounded by the video. Taking advantage of the compass within the tablet, by moving the tablet around, the user can continuously pan around the 360° video in both left and right directions, as if it was a window to the 360° video surrounding the user. Additionally, users can pan around the video without having to move, as the entire screen consists of a drag interface.

As a mean to increase immersion through *sensing the experience*, in addition to watching the video by moving around, a Wind Accessory was developed (Fig.1). This prototype is based on the Arduino Mega ADK; it is mounted on the back of the tablet and controls two fans generating a combined air flow of 180CFM (maximum). While the videos are being reproduced, the fans blow wind to the viewer taking into account the video's metadata. Particularly, the RPM of the fans is relative to the speed of the movement, and the wind's speed and orientation when video was captured, so that, while the camera is facing against the wind orientation, the fans' RPM (rotations per minute) are much higher than when the camera is orientated towards the wind orientation.

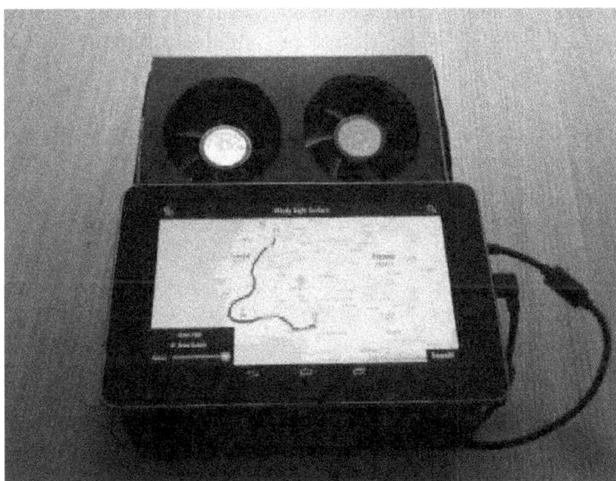

Figure 1. Video Sensing: Wind Accessory connected to the Android Tablet

3.4 Video Viewing: Context Awareness Features

To increase immersion through *context awareness*, when the video is being reproduced, an overlay is always present above it, which enriches the video with some information from the video's metadata (Fig.2). This strives to increase immersion through context awareness. Such information is divided in two categories:

1) where the information is *always present* on the screen throughout the video reproduction, which includes a circular "pie-chart"-like interface that uses a red angle to highlight where the user is looking at (Fig.2, top right corner). In addition, it also serves the purpose of quick angle navigation, by dragging the red area to the desired position [14]. There is also a speedometer showing the speed of the movement in the video (Fig.2, bottom right corner) and a G-Force meter (Fig.2, bottom left corner);

2) related to information that appears *momentarily* and is related to specific portions of the video. This includes visual and vibratory notices to notify the user about relevant happenings on specific moments, such as the maximum G-Force value in the video. This feature can be activated in two modes:

2.1) presenting *all the momentary information* available;

2.2) presenting momentary information to the users taking into account their *preferences*.

During the video search process, some information is retained, about the user's choices, filters and keywords, and while the video is being reproduced this data is taken into account. So, when the user searches for 'fast' or 'speedy' videos, a maximum speed notice is presented, or when the user searches mainly for routes instead of adventure experiences, no information about speed or G-Forces is shown. This feature is described in more detail in a further section of this article, which is specifically related to the video search process.

Figure 2. Video Context Awareness: Video being reproduced with the overlay, displaying the orientation (top right), speed (bottom right) and G-Force (bottom left)

Navigation between videos and related information, like points of interest, is accomplished through touchable hyperlinks that can be defined in space (all around the image) and time (along the video) [14]. These hyperlinks are touchable by the user and, for the moment, there are 2 types of hyperlinks: *Crossing Trajectories* (in blue), to mark and allow navigating to other videos with a trajectory that crosses the current one in the current position; and

Experience Markers (in red), that mark other videos with experiences in the current location.

To support this type of navigation and orientation, besides the possibility to Link (navigate) to the link destination, Link Awareness is vital since most of the content is outside the viewport, and links change along time. For this purpose:

1) Hotspots are highlighted through labeled colored balloons on the video, providing additional info about the video object and link;

2) Hotspot Indicators appear on the Lateral edges of the window to provide awareness about the location (vertical position) and distance of the links that are not in the viewport (the closer the hotspot, the bigger the hotspot indicator).

By taking into account the user's preferences, each of the described features throughout the section can be disabled at any time.

3.5 Searching Videos

The application's initial page displays a zoomed out map view with bubbles that represent clusters of videos in specific areas. As the user zooms in, the results shown are adapted to the zoom level, until the zoom reaches the level where each video can be shown in the map as a single marker (Fig.4). There are two types of videos: the ones which represent routes, and therefore are represented as paths in the map with a blue marker (Fig.4 upper blue marker); and the ones which represent experiences, which are represented as simple red markers in the map (Fig.4 bottom red markers).

Users can start a new search at any point in the application by clicking in the magnifying glass icon (Fig. 3 in the upper right corner). Videos can be searched directly on a map, using their fingers to draw paths that represent routes in which he is interested, or drawing bubbles that represent areas in which they are is interested (Fig.3); also, the user can use a more conventional search method, consisting of a set of keywords to conduct search, as well as a set of filters related to the nature of the application (slow/fast, rain/sunny, day/night, year span, experience/route, etc.).

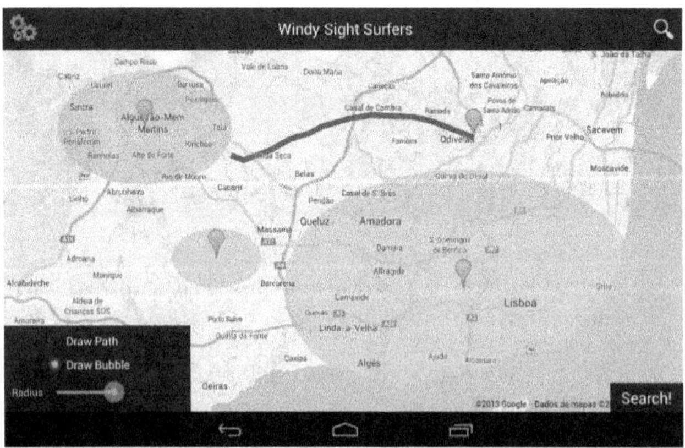

Figure 3. Video search through the map in Windy Sight Surfers.

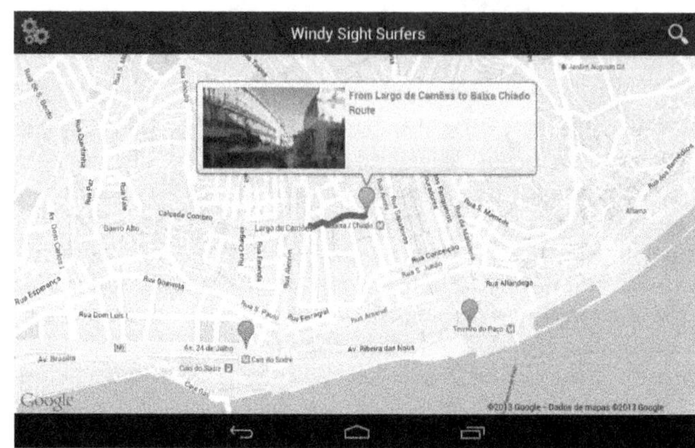

Figure 4. Search Results Displayed on the Map.

The process to retrieve the search results depends on which of the methods used to conduct it. If the user performed a search by directly drawing a route on the map, the search follows an algorithm that divides the path into several bounding boxes and retrieves videos that are also located in the majority of those bounding boxes. The reason for this is that if the search related to simple match between any of the coordinates (just one bounding box), the result could be erroneous (e.g. a search of an horizontal path retrieves as result a vertical path perpendicular to the searched path). In the case the search was conducted by drawing a circle on the map, the results are simply based on videos contained on the drawn bubble. Lastly, if the search was conducted using the filters system, the result is based on a simple filtering of the movies that comply with the user's demands.

The presentation of the search results can be done in two different ways: the first and default one, is the presentation of the videos in the map, following the same approach as the application's initial page. The user can select one of the videos by touching its marker on the map, which pops up an info-box, containing a picture as the title of the video (Fig.4); the second, is a more conventional list of videos, which consists in a cover-flow containing a picture and title of each of the videos.

3.6 Interaction with TV & Wider Screens

As an additional feature, seen as a mean to increase immersion, and considering the context of this work, the application has the capability to interact with a wider screens like TVs, and take advantage of their size. Therefore, an application designed to "future TV sets" was developed. Actually, this is a web application running in a standard web browser supporting WebGL. In this case, the computer acts as a TV, and no mouse or keyboard interaction is used to control the web application. All the interaction with the computer is achieved via the mobile application, which can be seen as an enhanced remote control. As the application thrives to be as seamless to the user as possible, the connection between the mobile device and the TV has as its only requirements that the application is installed in the mobile device, the computer (acting as a TV) has a web-browser with the named web application running, and there is a wireless network to which both devices have access.

Whenever the mobile application detects a TV in a short range the user is given the possibility to extend the application to the TV screen. If the user accepts this option, the mobile application rearranges itself so that the videos are reproduced in the TV screen, and the mobile device's screen is used for second-screen purposes. In this situation, as the TV becomes responsible for reproducing the videos, the mobile application has two main roles in the environment: control the TV, and displaying additional metadata, related videos, etc. Next there are presented some of the application functionalities.

Hyperlinks. The approach for presenting hyperlinks in the TV screen is the same used to present them in the standalone mobile application (hotspots and hotspot indicators in section 3.4). However in order to make it easier for the user to navigate through the hyperlinks shown in the video on the TV screen, they also appear in the mobile application's minimap (Fig.5). This feature serves another purpose: once the minimap displays the 360º of the video at any time, the user has easy access to hyperlinks that are out of sight on the angle being viewed on the TV screen. In order for the user to be aware of those hyperlinks, whenever a hyperlink first appears outside the TV viewport, the mobile device emits a short vibrating alert. To select a hyperlink the user must long press its hotspot, as a simple press could possibly create false positives, for example when the user is panning around the video through the minimap.

Minimap. When a 360º video is being reproduced, a minimap with a projection of the 360º video may be shown to provide increased perception and control of the whole video image (Fig.5). As an overlay on the minimap, a red rectangle highlights the angle being reproduced in the video, and this angle can be changed by dragging the rectangle or by clicking any other part of the minimap. When viewing the video on the TV, or another wider screen, and the minimap on the second screen on the mobile application, selection and dragging may be done through touch, and in addition, the view angle can be changed by turning the mobile device as if it was a steering wheel.

Figure 5. Viewing 360º video: Top) The viewport of the video; Bottom) The video minimap.

Geographical Navigation & Orientation in the 360º Videos and Maps. While videos are being viewed on the TV, the mobile application has the ability to present a map, which identifies video trajectories. The path corresponding to the current video is shown in green (color that humans are more sensible to, and associated with growth, life and action), while other trajectories in the area are shown in blue (color that humans are less sensible to, and associated with calmness). When a trajectory has been viewed it is painted red, to help users track which ones were already viewed. By touching or dragging the trajectories, the user can navigate to the corresponding video and time. The map and video are synchronized, with a View Area Marker similar to the circular "pie-chart"-like interface described in section 3.4 for localization (moving along the trajectory) and orientation (rotating in accordance with the 360º video rotation by the user).

Related Videos. While videos are being viewed in the TV, the mobile application also shows a list of related videos, clickable by the user. This list consists not only of videos that relate to the video being viewed, but it also contains videos that match the user preferences. Therefore, the first videos of the list are those which relate to both the user preferences and the video being viewed. Next in the list are the ones related to the video being viewed. Lastly, the videos related only to the users' preferences are suggested.

These features were described for the situation when there is a TV or wider screen and a tablet as a second screen, where each device can do what it does best, in a crossmedia scenario. But, for flexibility reasons, when we only have one device, both video and controls could be presented in the same device, like in previous versions of the application on the web [13,14].

3.7 Windy Sight Surfers Architecture

Windy Sight Surfers is based on a webservice architecture where the user requests the data from a backend (Fig.6). The mobile application's frontend is implemented in the Android platform as a Java application, and the TV frontend is implemented on HTML 5, supported by most recent browsers, and uses WebGL (as it has been described in section 3.6).

The Metadata Capture Module is implemented in the Android platform as a Service running in the background and it is responsible for the capture and recording of all the metadata described in section 3.1.

The communication between the mobile and "TV" applications is established through the JWebSocket bidirectional solution, which consists of a pure Java server that enables both the mobile (Java) and the TV (JavaScript) clients to communicate between them.

The backend is implemented using the PHP MVC (Model-View-Controller) framework CodeIgniter, that provides a very complete set of libraries to handle the HTTP requests and the database connection for the transactions and the requests for data that is done using ORM's (Object Relational Mapping) that allows a much more readable set of data. The used DBMS is a PostgreSQL as it is free and allows for smooth scalability of the system. It also provides a very complete set of geometric types and functions. With this, we can select the most pertinent routes right from the database, without the need for further computing.

To submit a video to Windy Sight Surfers, the user must be registered in the system. The Session Handler is the responsible for the validation and authentication of the users trying to log in the Windy Sight Surfers application, creating persistent sessions that last until the user logs out. The Submission Handler is responsible for handling each video and its associated metadata file submission to the server. The database is updated using the Nginx software.

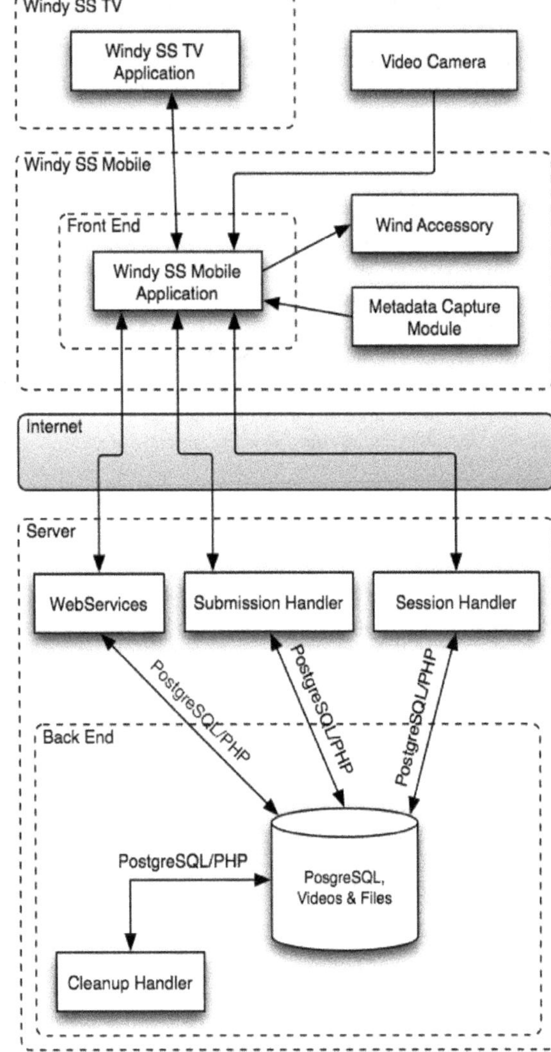

Figure 6. Windy Sight Surfers Architecture.

The Cleanup Handler is a tool that fires at regular periods of time (using the Unix Crontab) to clean incomplete submissions, after an extended period of time. It is necessary as the route submission is made in two phases, the video sent and the metadata sent are dependent on each other to complete the submission. Thus, the Cleanup Handler prevents the accumulation of incomplete and unnecessary data on server side.

4. USER EVALUATION

We conducted a user evaluation of Windy Sight Surfers to investigate whether the created experience sensing and context awareness features contribute to a more immersive environment to the user. We learned about its perceived Usefulness, users' Satisfaction and Ease of use [10], if users understood and appreciated the concept, and had a good user experience using the application. We looked for the existence of differences between the two feature categories (experience sensing and context awareness), and evaluated the immersiveness of Windy Sight Surfers.

4.1 Method

We performed a task-oriented evaluation based mainly on Observation, Questionnaires and semi-structured Interviews. After explaining the purpose of the evaluation and a short briefing about the concept behind Windy Sight Surfers, demographic questions were asked, followed by a task-oriented activity. Errors, hesitations and performance were observed and annotated. At the end of each of the sixteen tasks, users provided a 1-5 USE (Usefulness, Satisfaction, and Ease of use) rating, and could make comments and suggestions. At the end of the session, users were asked to rate the overall application.

To evaluate the existence of differences between the two feature categories (experience sensing and context awareness), users were asked to watch and interact with a video in four modes: 1) with all the created features activated; 2) with the experience sensing features disabled; 3) with the context awareness features disabled; and 4) with all the features disabled (so it was just bare video).The order in which the videos were reproduced was randomized for each of the users. Slater states that Presence is a human reaction to Immersion [17]. Therefore, by evaluating presence, one can tell about immersion capabilities of the system. To do so, users completed an adapted version of the seven-point scale format Immersive Tendencies Questionnaire (ITQ) before the experiment, and an adapted version of the Presence Questionnaire (PQ) after the experiment, with 28 questions each [18, 23].

We had 21 participant users (8 female, 13 male) between 18-57 years old. In terms of literacy, all users had at least finished high school, they were all familiar with the concept of accessing videos on the Internet, but only 7 had heard about 360º videos. The foreseen time for the completion of the 12 tasks was 35 minutes, which was met by all users, and they detected the new functionality underlying each new task they were asked to perform.

4.2 Results

Results are divided in six sections. First, the experience sensing features were evaluated (section 4.2.1), as well as the context awareness features (section 4.2.2). The following section evaluated the features to interact with TV & wider screens (section 4.2.3). Next, the conjunction of the experience sensing and context awareness features was evaluated (section 4.2.4). Results are commented for these tasks, highlighting the Mean (M) for Usefulness, Satisfaction, and Ease of use, and summarized in Table 1, where the standard deviation is also presented. Subsequently, the results of the Immersive Tendencies and the Presence Questionnaires are presented (section 4.2.5) and summarized in Table 2 and 3. The last section reports the global user appreciation of Windy Sight Surfers (section 4.2.6), summarized in the chart of Figure 7.

4.2.1 Experience Sensing Features

Users were asked to: move around a 360º video by moving the tablet around and by using the drag interface (T1: U:4.8; S:4.5; E:4.8); and view a 360º video with the wind accessory and identifying the wind direction (T2: 4.3; 4.6; 4.9). Users appreciated the tested features, especially the video navigation by moving the tablet around, which the reported to be a more natural approach when compared to the touch interface; and the wind accessory, which allowed a more realistic sense of speed in video watching.

4.2.2 Context Awareness Features

Users were asked to: view a video with the overlay, albeit with only the permanent information activated, and identify each category of information provided (T3: 4.3; 4.3; 4.6); view a predefined set of three videos about a specific theme with the overlay that has both the permanent and momentary information activated, identifying the permanent category related to each example of momentary information, being that all the momentary information available was always presented (T4: 2.9; 2.8;4.6); and (T5: 4.2; 4.0; 4.6), similar to T4, except that the momentary information presented took into account the user preferences. Some users stated that, when it did not take into consideration the users' preferences, some of the information being presented momentarily were not appreciated, problem that was solved with the mechanism tested in T5. Still on the context awareness features evaluation, users were asked to, while watching a 360° video, find and follow a link to a crossing trajectory (T6: 4.2; 4.2; 4.9); and find and follow a link to a similar experience (T7: 4.8; 4.8; 4.8); Users could easily find and follow the links.

4.2.3 Interaction with TV & Wider Screens Features

While evaluating the interaction with TV & wider screens features, users were asked to connect the mobile application to a TV set, select a 360° video and navigate it through both the "steering wheel" and the minimap displayed on the tablet (T8: 4.8; 4.8; 4.6). The minimap was especially appreciated, as it provided the user with a reference to the full 360° angle. Regarding the "steering wheel", users pointed out that its precision should be carefully adjusted, so that the application disregards false detections (it must successfully distinguish the angle changes that correspond to a turn by the user and the ones which simply represent ordinary hand movement), even when they were told that they could adjust the precision level to their own preferences. This suggests that the default configuration needs to be less sensible to movement, and that this precision level could be calculated according to a more sophisticated method, possibly learning from users behavior while using this feature.

4.2.4 Experience Sensing and Context Awareness in Conjunction

Users were asked to: watch and interact with a video in four modes: 1) with all of the created features (the experience sensing and the context awareness features) activated (T9: 4.9; 4.9; 4.9); 2) with the experience sensing features disabled (T10: 4.6; 4.3; 4.7); 3) with the context awareness features disabled (T11: 4.4; 4.5; 4.9); and 4) with all the features disabled (just bare video) (T12: 2.7; 2.8; 4.9). The order in which these four tests were performed was randomized for each of the users. All users preferred the environment reproduced on T9, with all features activated. But users tended to prefer experience sensing features in terms of satisfaction, and to prefer context awareness features in terms of usefulness.

4.2.5 Presence and Immersion

The Immersive Tendencies Questionnaire revealed a slightly above average score (table 2), whereas the Presence Questionnaire showed a high degree of self-reported presence in the application (table 3). As Presence is a human reaction to immersion, the ITQ score reveals the immersiveness of Windy SS. Moreover, as the ITQ relates to the users' immersive tendencies and the PQ relates to the experienced presence values (which relate to immersion), the improvement from the ITQ to the PQ reveals that Windy SS surpassed user's expectations.

Table 1. USE evaluation of Windy Sight Surfers (Scale: 1-5).

Features in Task:	Usefulness		Satisfaction		Ease of Use	
	M	σ	M	σ	M	σ
Experience Sensing Features						
T1	4.8	0.4	4.5	0.5	4.8	0.4
T2	4.3	0.6	4.6	0.7	4.9	0.3
Context Awareness Features						
T3	4.3	0.7	4.3	0.7	4.6	0.5
T4	2.9	1.1	2.8	1	4.6	0.5
T5	4.2	0.6	4	0.5	4.6	0.5
T6	4.2	0.9	4.2	0.8	4.9	0.3
T7	4.8	0.4	4.8	0.4	4.8	0.4
Interaction with TV & Wider Screens Features						
T8	4.8	0.4	4.8	0.4	4.6	0.5
Experience Sensing and Context Awareness in Conjunction						
T9	4.9	0.4	4.9	0.4	4.9	0.4
T10	4.6	0.5	4.3	0.6	4.7	0.5
T11	4.4	0.6	4.5	0.6	4.9	0.3
T12	2.7	0.7	2.8	0.7	4.9	0.3
Overall	**4.2**	**0.6**	**4.2**	**0.6**	**4.8**	**0.4**

Table 2. Results for Immersive Tendencies Questionnaire Scale: 1(low)-7(high degree).

Subscale	Mean	σ
Tendency to maintain focus on current activities	4.2	1.55
Tendency to become involved in activities	4.3	1.57
Tendency to view videos	5.1	1.37

Table 3. Results for Presence Questionnaire Scale: 1(low)-7(high degree).

Major factor category	Mean	σ
Control factors	6.1	0.86
Sensory factors	6.4	0.72
Distraction factors	5.3	0.93
Realism factors	5.6	0.77
Involvement/Control	6.2	0.86
Natural	6.2	0.76
Interface quality	6.1	0.86

4.2.6 *Global Appreciation*

As a final and global appreciation, users found Windy Sight Surfers very useful (4.6), satisfactory (4.7), fun (4.9), easy to use (4.8) and fairly easy to understand (4.3), and they would use it again (4.8). See Fig.7 for a more detailed and comparative view of these results. Final suggestions were aligned with the rest of the interview, and final comments were very positive and encouraging, along the lines: "Much fun.", "Very interesting.", "A nice ideia.", "Quite good".

Figure 7. Global Evaluation: (U)seful; (S)atisfactory; (F)un to use; (E)asy to Use; (EU) Easy to Understand; (UA) Use Again

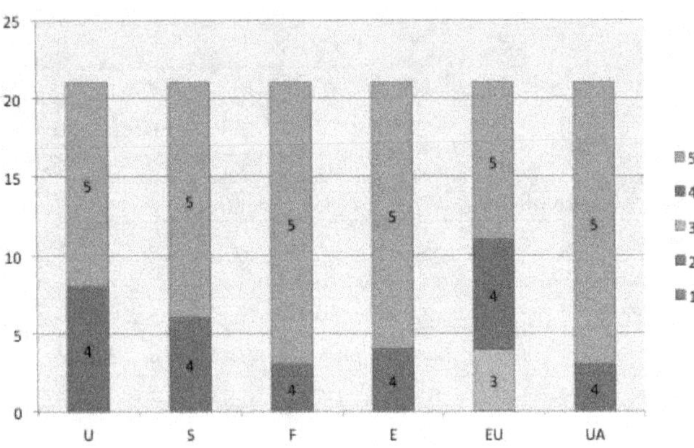

5. CONCLUSIONS AND PESPECTIVES

We presented the motivation and challenges, and described the design and user evaluation, of Windy Sight Surfers, an immersive mobile application that uses a wind accessory for the visualization and sensing of georeferenced 360º hypervideos.

Two main categories of features that could augment presence and immersion were designed: sensing experience (e.g. viewing and sensing the video around, the wind and speed); and context awareness (e.g. be conscious about current orientation in 360º, speed, wind and geo-location and trajectories). The user evaluation showed that Windy Sight Surfers increases the sense of presence and immersion, and that users appreciated having both categories of features, finding all of them very useful, satisfactory and easy to use. As for the different purposes served by these categories, users expressed that sensing experience features are slightly more satisfactory and fun to use, whereas context awareness features were considered more useful.

Next steps include refining and extending our current solutions, exploring further more immersive settings, like the CAVE and wide screens. The use of mobile devices as 2nd screen can be extended to a broader array of scenarios. More sensing modalities are to be explored. Namely, 3D high definition might be an interesting way to provide an enhanced sensation of movement, using low frequency sounds in a cyclic behavior as background sound for example. Consider georeferencing, ambient computing and augmented reality scenarios, e.g. as access points to videos shot in the same place at a different time, to compare them 'overlaid', or access videos with similar speed as the current speed experienced by the user (e.g. while traveling on a train). This concept can be extended for further filters (e.g. access videos in same time of day, or with similar weather conditions). Relying on reality aid in finding videos and feeling more immersed in the virtual video being experienced. Populating the system with a wide variety of videos is vital, as it enables to build up our application and do more testing with users, which will provide us the means to see in a more precise way what type of techniques are more useful.

6. ACKNOWLEDGMENTS

This work is partially supported by FCT through LASIGE Multiannual Funding and the ImTV research project (UTA-Est/MAI/0010/2009).

7. REFERENCES

[1] Bleumers, L., Broeck, W., Lievens, B., Pierson, J. 2012. Seeing the Bigger Picture: A User Perspective on 360° TV. In *Proc. of EuroiTV'12*. Berlin, Germany, July 4-6, pp.115-124.

[2] Burigat, S., Chittaro, L., 2011. Visualizing references to off-screen content on mobile devices: A comparison of Arrows, Wedge, and Overview + Detail. Interacting with Computers 23, pp.156-166.

[3] Cardin, S., Thalmann, D., and Vexo, F. 2007, Head Mounted Wind. In *Proceedings of Computer Animation and Social Agents*. Hasselt, Belgium, June 11-13. 101-108

[4] Courtois, C., D'heer, E. 2012. Second screen applications and tablet users: constellation, awareness, experience, and interest. In *Proc. of EuroiTV'12*. Berlin, Germany, July 4-6, pp.153-156.

[5] Douglas, Y., and Hargadon, A. 2000. The Pleasure Principle: Immersion, Engagement, Flow. ACM Hypertext'00. San Antonio, Texas, USA, pp.153-160.

[6] Google Street View: www.google.com/intl/en_us/help/maps/streetview/

[7] Immersive Media: www.immersivemedia.com

[8] Kojima, Y., Hashimoto, Y., Kajimoto, H. 2009. A novel wearable device to present localized sensation of wind. In *Proc. of ACE'09*. Rome, Italy, January 12-14, pp.61-65.

[9] Lehmann, A., Geiger, C., Wöldecke, B. and Stöcklein, J. 2009, Poster: Design and Evaluation of 3D Content with Wind Output. In *Proceedings of 3DUI'09*. Lafayette, Louisiana, USA, March, pp.14-15.

[10] Lund, A. M. 2001. Measuring usability with the USE questionnaire. *Usability and User Experience, 8(2)*.

[11] Mendes, M. 2010. RTiVISS | Real-Time Video Interactive Systems for Sustainability. *Proc. Artech'10*. Guimarães, Portugal, April 22-23, pp.29-38.

[12] Moon, T., Kim, G. J., and Vexo, F. 2004, Design and evaluation of a wind display for virtual reality. In *Proceedings of the ACM Symposium on Virtual Reality Software and Technology*. Hong Kong, China, November 10-12.

[13] Neng, L., and Chambel, T., 2010. Get Around 360º Hypervideo. In Proc.of MindTrek (2010). Tampere, Finland, October 22-23, pp.119-122.

[14] Noronha, G., Álvares, C., and Chambel, T., 2012. Sight Surfers: 360º Videos and Maps Navigation. In Proc.of GeoMM'12 ACM Multimedia'12. Nara, Japan, October 29-November 2, pp.19-22.

[15] Quake III Arena in a 360 Panoramic VR Installation with PanoramaScreen, game demo. www.youtube.com/watch?v=Cqh_F4iN8pg

[16] Seo, B., Hao, J., & Wang, G. 2011. Sensor-rich Video Exploration on a Map Interface. In Proc. of ACM Multimedia'11, Scottsdale, Arizona, USA, Nov 28 - Dec 1, pp.1013-1016.

[17] Slater, M., Lotto, B., Arnold, M.M., Sanchez-Vives, M.M. 2009. How we experience immersive virtual environments: the concept of presence and its measurement, Anuario de Psicología, 40(2), Fac. Psicologia, Univ. Barcelona, 193-210.

[18] Slater, M. 1999. Measuring Presence: A Response to the Witmer and Singer Presence Questionnaire, *Presence 8(5),* Oct. 560-565.

[19] Slater, M., & Wilbur, S. 1997. A framework for immersive virtual environments (five): Speculations on the role of presence in virtual environments. Presence,6, 603-616.

[20] Sullivan, D. 2009. Google Holodeck: StreetView In 360 Degrees, May 27. http://searchengineland.com

[21] Visch, T., Tan, S., and Molenaar, D. 2010. The emotional and cognitive effect of immersion in film viewing. Cognition & Emotion, 24: 8, pp.1439-1445.

[22] Wirth, W., Hartmann, T., Bocking, S., Vorderer, P.,Klimmt, Ch., Schramm, H., et al. 2007. A process model of the formation of spatial presence experiences. *Media Psychology*, 9, 493-525.

[23] Witmer, B., Singer, M.J. 1998. Measuring Presence in Virtual, Environments: A Presence Questionnaire. Presence, 7(3), Jun. (1998),MIT, 225–240.

[24] Yu, T., Moon, Y., Chow, K., Wong, H., & Li, Y., 2008. CityWalker: A Mobile GPS for Walking Travelers. In SAC'08. Fortaleza, Ceará, Brazil, March 16-20. ACM.

Time Shifting Patterns in Browsing and Search Behavior for Catch-up TV on the Web

Mika Rautiainen[1], Arto Heikkinen[1], Jouni Sarvanko[1],
Konstantinos Chorianopoulos[2], Vassilis Kostakos[1], Mika Ylianttila[3]

[1] Mediateam, University of Oulu
P.O.BOX 4500
FIN-90014 University of Oulu, Finland
+358 294 48 0000
firstname.lastname@ee.oulu.fi

[2] Ionian University
Plateia Tsirigoti 7, Corfu
49100, Greece
+30 26610 87707
choko@ionio.gr

[3]Center for Internet Excellence
P.O.BOX 1001
FI-90014 University of Oulu, Finland
+358 8 553 7651
mika.ylianttila@cie.fi

ABSTRACT
Catch-up TV services on the Web have facilitated time-shifted TV viewing. However, there is limited information about user search behavior with regard to recently time-shifted versus archival TV content. We deployed two distinct content-based web services to explore information retrieval of time-shifted TV content. The first web service is based on a browsing metaphor, while the second is based on free text content search metaphor. We analyzed more than 5000 user sessions from 12 months of logs and found that the programs accessed via browsing categorized program content summaries were typically less than one week old. In contrast, the programs accessed via free text search on subtitle content were typically more than a week old. Our findings provide a first assessment of user behavior in accessing time-shifted and archival TV content. Further research should develop the user experience for content-based TV access and explore the sharing patterns of archival TV content on social networks.

Categories and Subject Descriptors
H.5.1 [INFORMATION INTERFACES AND PRESENTATION]: Multimedia Information Systems, H.3.5 [INFORMATION STORAGE AND RETRIEVAL]: Online Information Services H.2.4 [DATABASE MANAGEMENT]: Multimedia databases

General Terms
Design, Experimentation, Human Factors

Keywords
User behavior, content-based retrieval, catch-up TV

1. INTRODUCTION
Digitalization and the Internet have created many opportunities for time-shifted TV viewing [4]. It is necessary to study what types of information seeking behavior span over long periods of time in the archive and for how long. The above issues are important to new digital service development for interactive TV. Looms [14] identified four distinct modes of TV viewing: 1) *Broadcast viewing* occurs in real-time to the broadcast 2) *Off-set* or *buffer viewing* occurs in almost real-time to the broadcast, 3) *Time-shifted viewing* occurs at least one hour after the broadcast, and 4) *Archival viewing* occurs over one week after the broadcast. Carey [3] noted that archival viewing has not been utilized due to user interface issues with solutions available at the time. Whereas traditional TV schedule is influenced by the expected popularity of the content, time-shifting TV services provide an opportunity to reach a wide audience for niche archival TV content [1]. Van den Bulck [16] suggests that it may be more productive to look at time shifting as another TV channel. Also Ferguson [7] and Kang [11] point out that new technologies such as video recorders, cable television and the Web have increased the channel repertoire of TV viewers.

In addition to Looms' modes of TV viewing, several information seeking and retrieval models can describe users' TV viewing behavior. The classical information retrieval model consists of a query with an explicit information need and process of matching the query to document representations to identify maximally relevant document set. Bates [2] proposed an alternative model called "berrypicking" where user needs and search strategies are constantly evolving during the seeking process. Cove and Walsh [5] described a serendipitous model as an unstructured random activity potentially leading to accidental discovery of interesting information. Marchionini [15] introduced exploratory search as a three-stage information behavior process consisting of lookup search (fact retrieval), learning search (exploration and comprehension of neighbourhood) and investigative search (analysis, synthesis and evaluation). According to the definition by Hjorland [8] browsing implies relevance assessment and decision about acquisition or selection of some of the information objects. We adopt Hjorlad's definition of relevance in our experiments: if a user clicks any program related information on the user interface, we interpret this action as an indication of relevance. In this paper we introduce new user services to facilitate flexible content-based searching and browsing strategies for finding TV programs and catch-up TV streams for time-shifted and archival viewing.

There are several studies that examine user behavior in content-based TV and video archives. Huurnink et al. [10] studied the search behavior of media professionals with audiovisual archives and suggested that increased support for fine-grained access to audiovisual material through e.g. content-based analysis might be beneficial for the archive use. In another study, Huurnink et al. [11] recommend audiovisual archives to invest in embedding content retrieval into their work-flow and that audiovisual archives prioritize video retrieval using transcripts. Lee et al. [12]

introduced Físchlár-News as one of the first automatic, content-based broadcast news analysis and archival systems that allowed users to search, browse, and play it in an easy-to-use manner with a conventional web browser. Informedia Digital Library is the landmark research program in video retrieval that was started in 1994. Christel [6] calls for longitudinal studies with broader user populations, broader selection of unstructured content genres and experimenting with browsing tasks in addition to retrieval.

2. NEW CONTENT-BASED SERVICES FOR BROADCAST TELEVISION

In order to collect a representative data-set, we developed an online TV platform that indexes and summarizes TV programs in near real-time using detected subtitle text from seven free-to-air TV channels. Our service architecture consists of three parts and an integrated browser component for viewing program content excerpts.

The *Service backend* processes content dynamically in DVB-C video data streams for seven national channels. Subtitle transcripts are detected and recognized from the video stream, and program descriptions are de-multiplexed from the MPEG transport stream.

The *TV Program Search* service provides free text queries on program subtitle content, program title, and broadcasting channel. Program subtitles and descriptions have been extracted from DVB broadcast. Title search finds programs by title, while channel search is used to filter out programs from other channels.

Figure 1 shows the results of conducting a content search using the query *"Fiat 500" auto.* The service allows free text searching of recent and relevant TV programs. Users can use the service to find specific information from the metadata of 180 000 broadcasted programs and use it to find web streams of relevant programs from the broadcaster sites.

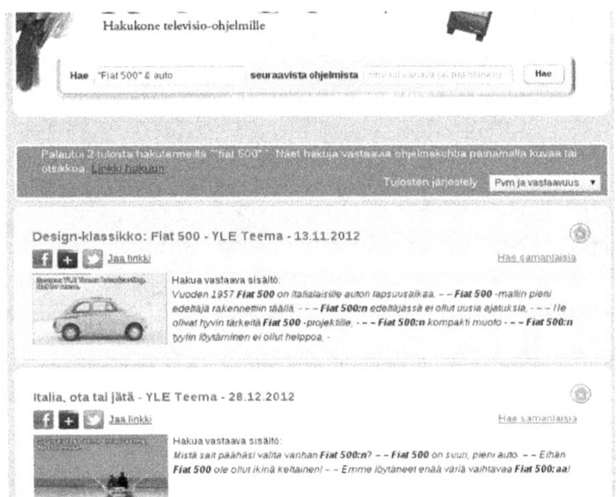

Figure 1. Program Search interface depicting program query with terms "Fiat 500" & auto

Catch-up TV Guide is a service that collects, categorizes and analyzes TV programs from seven linear Finnish TV channels. The service supports browsing and following programs on the Web, and programs can be watched online if the catch-up web stream is available on the broadcaster's site. The service relies on machine learning and data mining techniques to detect novelty keywords in program subtitle content, and allows users to view dynamic quotes from the program. The service allows users to browse Finnish TV broadcasting according to predefined program categories (*news, cooking, nature, living, documentaries, science* etc.) and time scales (*most recent, past week* and *past month*). Figure 2 shows the user interface of Catch-up TV Guide with a program and highlighted novelty words that allow users to view excerpts of the program content using Program Content Browser.

The *Program Content Browser* allows users to inspect program content information dynamically. It assists users in determining relevance of content without requiring playback of the TV program. The Program Content Browser is an integrated popup window that opens when users access program-related links from either Catch-up TV Guide or TV Program Search service. It displays basic information about the program and provides search and browsing mechanisms to inspect parts of program through dynamically extracted picture quotes. Figure 2 depicts the Program Content Browser with a quote from a program. The quote shows "grain fields", a concept that has been highlighted automatically as a novelty word in a TV program about national landscapes of Finland. Automatically highlighted novelty words help users to examine parts of content information before viewing the actual program stream at the broadcaster site.

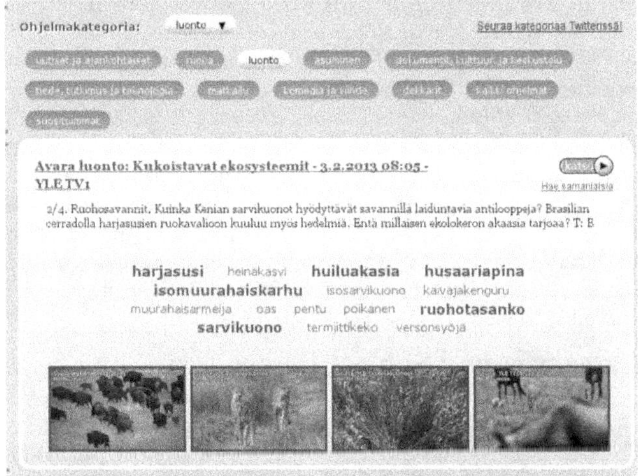

Figure 2. The Catch-up TV Guide user interface showing most recent program from the category Nature.

Figure 3. Program Content Browser shows an excerpt of "grain fields". It is part of the automatic program summary.

3. LONGITUDINAL EVALUATION OF A LIVE DEPLOYMENT

People searching for TV programs have traditionally used TV guides that show only upcoming programs. Today catch-up TV services and video-on-demand are liberating users from the present, i.e. linear broadcasting schedules. So will users change their behavior? Is recency still important in catch-up and archival viewing?

We deployed our platform on the Internet and conducted a 12-month evaluation of the system with real users. The users were members of the public and were not directly recruited by the researchers. Users were free to interact with the service using their own equipment at any time suitable to them. Our service did not stream any copyrighted media but instead offered users pictorial quotes from the program metadata.

We wanted to investigate what is the distribution of age of the programs that people find through the services. Our user logs span 12 months (02/2012-02/2013). We collected 5032 user sessions for analysis: 2500 sessions from the Catch-up TV Guide service and 2738 sessions from the Program Search service. Over 46000 activity events were logged, of which over 18000 were activities in the Program Content Browser. Next we present an analysis of user behavior in the Catch-up TV Guide and Program Search services respectively. For each session we recorded the behavior of users by logging activities at the service, i.e. pressing buttons and clicking URL access links that result in further examination of program content information with Program Content Browser.

3.1 Differences between browsing and searching strategies

The functionality for searching program information is different in the Catch-up TV Guide and Program Search services. Whereas the former provides recent, weekly and monthly summaries of programs in different categories for browsing, the latter allows free search of programs from 180 000 program database containing broadcasted program metadata of almost three years' time. Here we describe how these differences affect usage logs.

The premise of the strategies is that users are browsing most recent programs via Catch-up TV Guide service and using Program Search service for archival search. It is possible to use the Search service to retrieve most recent programs as well, but Table 2 shows that this is not the case.

Table 2 gives basic statistics of mean, median and percentiles of the respective age distributions (ADs).

Table 2. Age statistics of the accessed programs. Numbers are in days.

	Catch-up TV Guide	Program Search
mean	1.7546	196.371
stdev	2.05322	275.558
25th percentile	1	2
median	1	19
75th percentile	2	358
95th percentile	4	777

A significant sign of heavy tail effect can be seen in the Program Search AD where the differences between median, mean and standard deviation are large. The 95th percentile shows the heaviness of the tail: the oldest programs that are needed to cumulatively reach 95% of the accessed programs are 777 days i.e. 2.1 years old. In contrast, 95% of the accessed programs in Catch-up TV Guide service were up to four days old.

Figure 4 displays a cumulative age histogram of accessed programs in Catch-up TV Guide and Program Search service with programs up to 30 days of age. The age of a program is defined as the difference between access time in the service and original broadcast time. As can be seen from the figure, the majority of the programs accessed via the Catch-up TV Guide fall within the age of seven days, whereas only 42.8% of the programs accessed in Program Search service are less than seven days old. The age distribution of both histograms follows a power law, with the distribution of the Program Search having a heavier tail than the Catch-up TV Guide distribution.

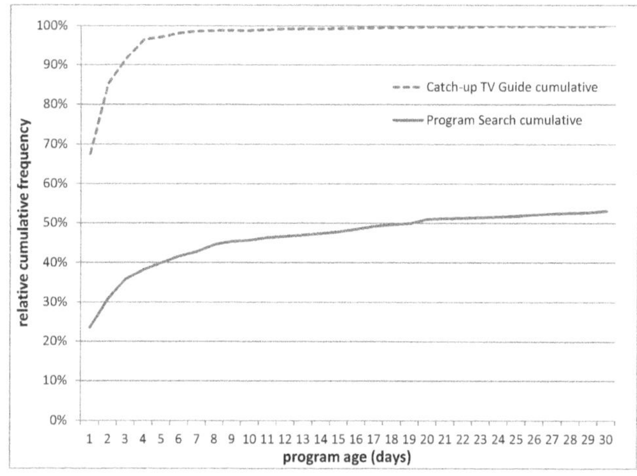

Figure 4. Cumulative age distribution of accessed programs in Catch-up TV Guide and Program Search (first 30 days)

Figure 5 illustrates the full cumulative frequency distribution of program ages in the Program Search service. The oldest accessed programs were 986 days (2.7 years) old.

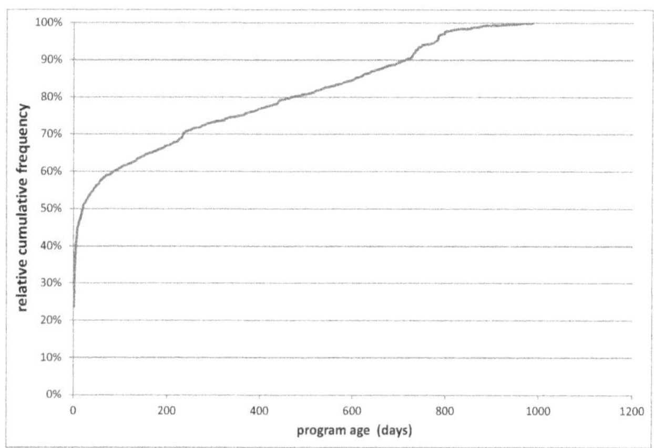

Figure 5. Cumulative age distribution of accessed programs in Program Search

4. DISCUSSION

The differences between the distributions of age of the accessed programs, namely the much heavier tail in Program Search than the Catch-up TV Guide, demonstrate that our users retrieved and accessed programs of any age with the Program Search service while used the Catch-up TV Guide to find the most recent programs within days. This is very likely due to the design of the services, but also indicates that people treat browsing TV programs with strong temporality in mind: 93.6% of the page views in the Catch-up TV Guide were using *most recent* time scale, the rest used *past week* and *past month* time scales. The 95th percentile of the age of accessed programs is only 4 days, suggesting that users were considering: "let's see what was shown recently on cooking programs." In contrast, a more goal oriented searching reveals a heavy tail effect of relevant information suggesting that user behavior is more age neutral ("Show me cooking programs that use lamb in cooking") even when the likelihood of finding active catch-up TV sources is low.

Previous work has explored time-shifted TV viewing in the short time-frame provided by local storage devices, but they have not regarded engagement with older TV content. In particular, Looms [14] found that the users of digital video recorders use them for pausing broadcast TV content or time-shifting a favorite program for later viewing in the same or the following day. There has been limited research in the scale of weeks and months old archives. Our results indicate that users access older TV content information if it is retrieved and relevant, so broadcasters and content providers should consider this when designing new services for end users. Our findings about the use of archival TV content are in agreement with Levy [13] who examined the use of home video recorders for time-shifting and found that many home recordings are not replayed within a diary week.

5. CONCLUSIONS

In this paper, we presented a fully operational online service for content analysis of Finnish TV broadcasting. We evaluated this framework by introducing new user services with content-based access for broadcast TV content on the web. We studied the significance of temporality in accessing program content information with thousands of user sessions.

We have introduced content-based service platform with new end user tools for accessing digital TV broadcast content on the web. With continuous indexing and 180 000 indexed programs, we believe our service framework provides a significant experimental platform for future end-user TV services development. We described two end user web services for searching and browsing TV broadcasts. The Catch-up TV Guide service supports time-shifted viewing and the Program Search is a content-based search tool for archived TV programs. We analyzed one year of usage data that revealed quantitative differences in user behavior between the two end user services. We showed that user queries retrieve relevant information from TV programs with heavy tailed age distribution indicating that users are not just limiting their search to recent programs, but are more broadly interested of any program content that fits to their search interests. Future work involves improving the content-based user experience and combining social media with content-based TV services.

Better utilization of personal preferences and context pose the need for technological frameworks capable of semantic understanding of media via content-based technologies. This is an important step in maximizing content relevance for the end users, improving user satisfaction with catch-up services, and more efficiently adopting new, non-mainstream content for broader audiences.

6. ACKNOWLEDGMENTS

We would like to thank Academy of Finland for supporting this work.

7. REFERENCES

[1] Anderson, C. (2004). The long tail. Business Books.

[2] Bates, M.J. (1989). The Design of Browsing and Berrypicking Techniques for the Online Search Interface. Online review. (5), 407-424.

[3] Carey, J. (2002, May). The evolution of TV viewing. In Proc. 4th Annu. TV Meets the Web Seminar.

[4] Cesar, P., & Chorianopoulos, K. (2009). The Evolution of TV Systems, Content, and Users Toward Interactivity. Foundations and Trends® in Human-Computer Interaction (Vol. 2, p. 95). doi:10.1561/1100000008

[5] Cove, J.F. and B.C. Walsh. (1988) Online text retrieval via browsing, Information Processing and Management, Vol. 24, No. 1, p. 31-37.

[6] Christel, M.G. (2006) Evaluation and User Studies with Respect to Video Summarization and Browsing http://repository.cmu.edu/cgi/viewcontent.cgi?article=1380&context=compsci

[7] Ferguson, D. A. (1992) "Channel repertoire in the presence of remote control devices, VCRs, and cable television," Journal of Broadcasting and Electronic Media, vol. 36, no. 1, pp. 83–91.

[8] Hjørland, B. (2011) Theoretical clarity is not "Manicheanism": A reply to Marcia Bates. In: Journal of Information Science. 37(5), p. 546-552

[9] Huurnink, B., Hollink, L., van den Heuvel, W. and de Rijke, M. (2010) Search Behavior of Media Professionals at an Audiovisual Archive: A Transaction Log Analysis. Journal of the American Society for Information Science and Technology, 61(6):1180–1197.

[10] Huurnink, B., Snoek C.G.M., de Rijke, M., and Smeulders, A. W. M. (2010) Today's and tomorrow's retrieval practice in the audiovisual archive. In Proceedings of the ACM International Conference on Image and Video Retrieval, CIVR '10, pages 18–25, New York, NY, USA.

[11] Kang, M.H. "Interactivity in television: Use and impact of an interactive program guide," Journal of Broadcasting and Electronic Media, vol. 46, no. 3, pp. 330–345, 2002.

[12] Lee, H., Smeaton, A.F., O'Connor, N.E. and Smyth, B. (2006) User evaluation of Físchlár-News: an automatic broadcast news delivery system. ACM Transactions on Information Systems (TOIS), 24 (2). pp. 145-189.

[13] Levy, M. R. (1983). The time-shifting use of home video recorders. Journal of Broadcasting & Electronic Media, 27(3), 263-268.

[14] Looms, P. (2005). Recent Developments with PVRs and the Free-To-Air Television Market in Europe. DR/EBU.

[15] Marchionini G. (2006). Exploratory Search, from Finding to Understanding. Communication of the ACM, 49(4): 41-46.

[16] Van den Bulck, J. (1999). VCR use and patterns of time shifting and selectivity. Journal of Broadcasting & Electronic Media, 43(3), 316-326.

Freehand Gestural Text Entry for Interactive TV

Gang Ren
Department of Computer Science
University of Bath
Bath, UK, BA2 7AY
garry.ren@gmail.com

Eamonn O'Neill
Department of Computer Science
University of Bath
Bath, UK, BA2 7AY
eamonn@cs.bath.ac.uk

ABSTRACT
Users increasingly expect more interactive experiences with TV. Combined with the recent development of freehand gestural interaction enabled by inexpensive sensors, interactive television has the potential to offer a highly usable and engaging experience. However, common interaction tasks such as text input are still challenging with such systems. In this paper, we investigate text entry using freehand gestures captured with a low-cost sensor system. Two virtual keyboard layouts and three selection techniques were designed and evaluated. Results show that a text entry method with dual circle layout and an expanding target selection technique offers ease of use and error tolerance, key features if we are to increase the use and enhance the experience of interactive TV in the living room.

Categories and Subject Descriptors
H.5.2 [**Information interfaces and presentation**]: User Interfaces – *Input devices and strategies, Interaction styles.*

General Terms
Human Factors; Design; Measurement.

Keywords
Text entry; Freehand gesture; Expanding Target.

1. INTRODUCTION
The increasing use of interactive TV in the living room brings novel opportunities and requirements for rich and engaging interactive experiences. There are many different ways to interact with an interactive TV. The most common input method is using a traditional remote, however, the remote normally offers only a limited set of buttons and does not lend itself to offering richer means of interaction. Other input methods used for computers or mobile devices can also be used with interactive TVs, such as keyboard, mouse, and touch-sensitive displays. However, these input devices are usually installed close to or even contiguous with the screen so they are not suitable for typical use scenarios with an interactive TV where the user is often at a distance from the screen. Mobile devices such as phones or tablets can also be used to interact with remote displays, but configuration is often needed to connect the personal devices so this may not be convenient in some scenarios.

Gestural input is increasingly popular, using hands-on input devices (e.g. Wii Remote) or freehand motion tracking by a camera (e.g. Microsoft Kinect). As gestural input moves beyond home gaming settings, freehand gestural interaction, which has no need for hands-on input devices and so enables easier and more convenient "walk up and use" [3], is likely to become more important in interactions with TV in everyday settings.

Currently, however, it is still difficult to perform some common tasks such as text entry with freehand gestural interaction. For example, when a person is trying to search for a program or a video clip on an interactive TV, her text entry task may be challenging due to several factors including, for example, the relatively low resolution of many remote gesture sensors and the distance to the TV screen.

Although research has been conducted on text input with various input devices and techniques, most techniques use handheld input devices and so cannot be used directly in freehand interaction. Therefore, we are motivated to investigate freehand gestural text input methods. Here, we report findings from the design and evaluation of some candidate virtual keyboard layouts and input techniques.

2. RELATED WORK
2.1 Gestural Interaction and Interactive TV
Gestural interaction has been investigated for a long time and many different gesture types have been designed and evaluated, and efforts made to summarize and classify different types of gesture [18, 37, 39]. Karam and Schraefel [18] classified gesture styles as deictic, manipulation, gesticulation, semaphores and sign language. Deictic and manipulative gestures are similar to pointing and manipulating in real life interaction, and they could be used without special learning or training. Gesticulation is the gesturing that accompanies everyday speech so it too requires no special training. In contrast, semaphoric gesture and sign language require a dictionary and even grammatical structures, thus training is necessary before using these gesture types with an interactive system.

Various input devices and gestures have been investigated with interactive TV. Bobeth et al. [5] tested freehand menu selection for interactive TV with 4 different designs, and found that freehand gestures could be an appropriate way for older adults to control a TV. A selection task was also investigated in [29], and participants preferred freehand gestural pointing to using a hand-held pointing device. Drawing different shapes in the air can also be used to select objects or menu items [2] with interactive TV, however, certain shapes are not easy to perform and remember, and have low recognition rates. User defined gestures for TV were also evaluated in [38]. The results showed that a pointing action was frequently used and a desktop interaction style, such as a push in mid-air to simulate clicking, was observed in many cases.

EuroITV'13, June 24-26, 2013, Como, Italy.

However, there has been some criticism of semaphoric and sign language style gestural interaction, including arguments that gesture is a step backwards to command-line interfaces [27], natural user interfaces are not natural [26], and that gestural interaction should be based on well designed metaphors rather than gesture design [16].

2.2 Text Entry and Interactive TV

The handheld remote control is by far the main input device for TVs, and many text input methods designed for TV remotes have been proposed and investigated [7, 14, 35]. Geleijnse et al. [9] also compared the physical Qwerty keyboard and remote control for text entry and suggested that the Qwerty keyboard is better.

Gestural text input methods have gained more research interest recently and many gestural text entry methods have been proposed and investigated. Most of these methods could also be used with interactive TV. For example, Jones et al [17] used accelerometer-based gesture enabled by a Wii remote and virtual keyboard for text entry. Users achieved 3.7 words per minute (wpm) in first time use and 5.4 wpm after 4 days' practice. A stroke-based text entry method was also designed with data gloves and fiducial markers [25]. Users reached 6.5 wpm without word completion after 2 weeks' practice. A Wii remote was used for text input with large displays in [34]. Three different layouts – circle, Qwerty and 3D cube – were evaluated and the Qwerty layout had the best performance (18.9 wpm), but decreased significantly with more errors as the user moved away from the display. Kristensson and Zhai [20] investigated shape writing recognition to perform word-based text input with a stylus keyboard, and saw high performance in informal trials. However, a test of text entry methods on mobile touch screens showed Qwerty was faster than handwriting and shape writing text entry, and handwriting was the slowest and least accurate text entry technique [6]. Other text entry methods originally designed for stylus and touch screen, such as FlowMenu [12] or Quikwriting [15, 28] could also be used with freehand gestural text input.

Besides holding a tracked device in the hand (e.g. Wii remote) or wearing a data glove or fidual markers, it is also possible to track freehand gesture with low-cost remote cameras, such as Microsoft Kinect[1] or ASUS Xtion[2]. This type of tracking device has the advantage of enabling freehand tracking in 3D space without requiring the user to hold any device in the hand or use fiducial markers. However, such tracking techniques with a single remote camera normally have low resolution and tracking accuracy. For example, the accuracy of the Microsoft Kinect depth sensor is about 3mm in the image plane and about 1cm in depth at a distance of 2 meters [30, 40]. In practice, the skeleton tracking based on the raw depth data can be even noisier.

Kristensson et al. [19] investigated freehand text entry using freeform alphabetic character recognition, and the evaluation shows a recognition accuracy of 92.7%–96.2%, however, no evaluation on text entry performance is available from their study. Freehand gesture was also used with speech recognition for text entry [13], and 5.18 wpm text input speed was achieved. Although freehand gestural text input with a virtual keyboard is widely used in commercial products (such as games designed for Microsoft Kinect), there is very little research available on this topic. A previous study showed that with the default Xbox 360 gesture based text input interface, the input speed was only 1.83 wpm [13].

Text input with a virtual keyboard is basically a sequence selection of small targets packed together on the keyboard. This is a challenging task for noisy motion tracking with a low-cost single camera system. For example, one typical issue for freehand gesture text input is the difficulty of gesture delimiter design [3]. Accot and Zhai proposed a cross technique which can be used for selection without clicking [1]. A similar method was used for freehand selection in [31], in which users reach towards the target to select without the need to stay inside the target. Furthermore, although an "expanding target" can be used to support selection for small targets, it can only magnify in visual space but not in motor space when targets are closely packed [24]. Although a predictor could be used to increase the motor space before the cursor enters the target area, the benefit is very limited [24].

3. DESIGN OF FREEHAND TEXT ENTRY

3.1 Design Considerations

Our aim is to implement text entry methods that facilitate "walk-up-and-use" – or in the case of interactive TV, more likely "sit-down-and-use" – interaction experiences [3], while retaining the simplicity and directness of freehand interaction. Most previous gestural text entry methods use handheld devices or fiducial markers for tracking motion, which can offer accurate tracking of hand, wrist and fingers. Freehand motion tracking enabled by an inexpensive remote camera, on the other hand, can track hand motion robustly but not the small motions of wrists or fingers, especially when users are at a distance from the display/sensor. Besides the lack of fine movement tracking, there are also some other challenges for freehand text input, such as noisy motion tracking, no physical button or surface to click or touch, and no physical support or tactile feedback for the hand.

Text entry methods based on freeform alphabetic character recognition have been investigated in considerable previous work [e.g. [19, 25, 41]. With such methods, however, users need to learn and remember a set of gestures. Such learning demands may not be suitable for scenarios with interactive TV where quick and easy interaction is important. Text entry methods based on word level prediction, such as shorthand writing [20, 43], Swype[3] or text input based on speech recognition [13] are also possible for freehand text entry. But with interactive TV applications, the requirements of non-dictionary word entry could be high (e.g. entering user name, password, email address, or url), thus character based text input may be better suited to interactive TV.

A virtual keyboard can provide easy recognition and learning [17, 34] and, therefore, may be more suitable for interactive TV text entry. Although character arrangement on a virtual keyboard can be optimized according to the context of use and alternative arrangements may improve the performance of expert users [4, 12, 15, 17, 33], the Qwerty layout still has some benefits [34, 42], is the basis of many improved text entry methods [8, 21, 42] and has the advantage of familiarity to many users. For "walk up and use" and entertainment scenarios with interactive TV, the familiarity of the Qwerty layout is very important to users, with less demand for extra learning and less visual scan time. Thus we designed our gestural text entry based mainly on the Qwerty keyboard layout and a character based text entry method.

[1] http://www.microsoft.com/en-us/kinectforwindows/

[2] http://www.asus.com/Multimedia/Xtion_PRO/

[3] http://www.swype.com/

One benefit of freehand gestural interaction is that the hand can move in 3D space, which means that the virtual keyboard can be in 3D. However, results from previous research [34] indicate that 3D layout text entry has low performance. And previous work on 2D and 3D option selection with freehand gesture [32] also suggests that freehand selection with a 3D layout is less accurate than with a 2D layout. Thus, we designed and evaluated a 2D keyboard layout in this study.

3.2 Keyboard Layout

3.2.1 Qwerty

As noted above, Qwerty is a very familiar keyboard layout and has been shown to perform well as a virtual keyboard with a mid-air handheld device [34]. It is therefore a reasonable candidate for freehand text entry. For our prototype design and evaluation, we used a Qwerty layout of 28 characters (26 English letters, space and backspace), similar to the keyboard used for touch-screens such as Windows Phone (Figure 1). More keys could also be added in different applications.

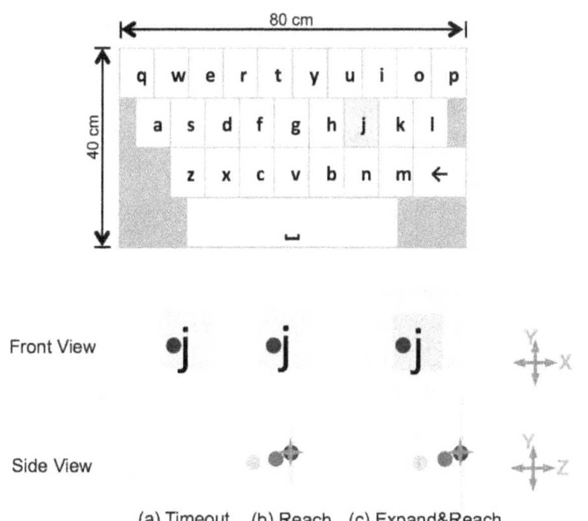

Figure 1. Up: Qwerty layout. Down: Selection techniques; the spherical grey cursor is controlled by the user's hand position.

3.2.2 Dual-circle

Besides the Qwerty layout, another virtual keyboard layout that has been investigated is circle. Although the circle layout is not as effective as Qwerty with a mid-air handheld device [34], it may bring benefits for freehand interaction. For example, rather than being tiled, characters are arranged to offer easy access to each character from the center of the circle. With accurate handheld devices such as in [34], all characters could be distributed in a single circle. However, for freehand motion tracked by low cost camera, the character size could be too small for reliable use with noisy tracking input in an interactive TV scenario.

To address this issue and to leverage the two-handed operation that freehand gestural interaction allows, we proposed a Dual-circle layout for text entry (Figures 2 and 3). The characters are evenly distributed in 2 circles next to each other. Thus each character can be bigger than if they were distributed in one similarly sized circle. To leverage users' familiarity with the Qwerty layout, we based the character distribution on Qwerty: the top and bottom of each circle is used for characters located in the top row and bottom row in Qwerty, and the middle row of the Qwerty layout is turned vertically and put sideways in each circle based on the corresponding hand and fingers when using a Qwerty keyboard. Gaps are used to separate the different character groups for a clearer mapping to the familiar Qwerty layout.

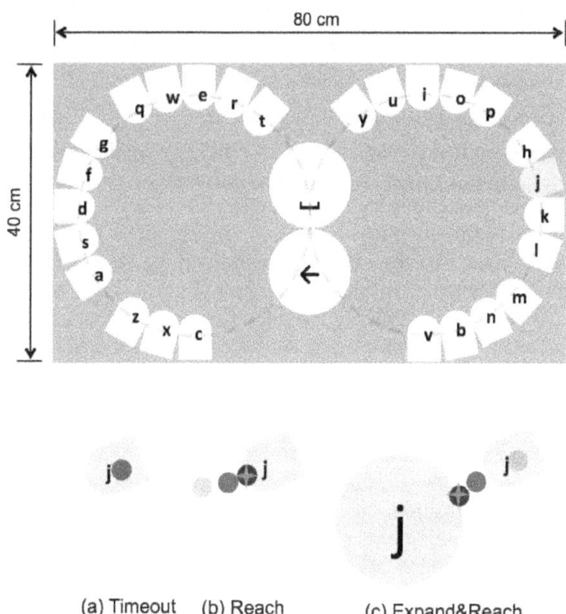

Figure 2. Up: Dual-circle layout. Down: Selection techniques; spherical grey cursor controlled by the user's hand position.

Space and backspace are represented by circles located in the middle of the keyboard for easy access with both hands. As users dislike selecting in the left-down direction with the right hand [31], the left-down portion of the right circle and the mirrored portion of the left circle are left blank.

3.3 Character Selection

As noted above, without a physical device in the hand and buttons to click, freehand gestural selection can be challenging. And although finger and wrist movement are used in some previous work, they are less suited to freehand interaction tracked by a remote inexpensive camera, so other techniques are required.

3.3.1 Timeout

One common method for freehand selection is pointing to the target and waiting for a timeout threshold. It is easy for novices to understand and perform and is accurate for large targets. The primary disadvantage is that the dwell time can slow overall selection time. For both our layouts, the timeout selection method can be used: the user points to a character by the X-Y position of her hand and waits for the timeout threshold, e.g. 1.2 s, to select the character (Figure 1.a, Figure 2.a).

3.3.2 Reach

As an alternative to timeout, hand motion can also be used for selection confirmation. The Reach technique has been used for target selection with freehand gestures [31], in which users select by moving their hands to reach into the target in 3D. For a virtual Qwerty keyboard placed vertically in front of the user's body position (in this case at 40 cm), the user can point to the desired character by X-Y movement of the hand and then reach forward in the Z-dimension to select the character (Figure 1.b). For the Dual-circle layout, the character can be selected by moving the hand's X-Y position to reach across the border of the desired character tab (Figure 2.b).

With the Reach technique, although 3D hand position is required in the Qwerty layout while only X-Y position is required in the Dual-circle layout, with both layouts the user's hands move freely in 3D space to reach the characters. In practice, with both Qwerty and Dual-circle layouts, the user tends to move the selecting hand forward and towards the target character simultaneously in one fluid movement.

3.3.3 Expand&Reach

In both layouts, the character size is relatively small so could be difficult and error-prone for freehand selection. It is also difficult to expand the target in motor space for tiled targets in 2D interfaces [24]. However, since the hand can move in 3D space, the combination of the extra dimension and the Reach selection technique brings new interaction opportunities. We designed additional Expand&Reach techniques for both keyboard layouts. With the Qwerty layout, when the user points to a character, an expanded target appears along the Z-dimension (e.g. 5 cm further away than the current hand position and 2 times bigger than original size). The user moves her hand forward to reach the expanded target in order to select (Figure 1.c). With the Dual-circle layout, when the user points to a character, an expanded character tab containing the target character appears in the center of the corresponding circle (Figure 3). The user moves her hand to reach the expanded target to select it (Figure 2.c).

There are several potential advantages of the Expand&Reach technique: (i) easier selection – the target expands in both visual and motor space; (ii) error tolerance – users need to move their hand to reach the expanded target to confirm selection, so if they notice a selection error before reaching the expanded target, they have a chance to (re)select the right character; (iii) requires only hand position tracking – no fine finger movement or posture tracking is needed.

4. EXPERIMENTAL EVALUATION

A controlled experimental evaluation was conducted to investigate the effects of the different keyboard layouts and selection techniques.

4.1 Independent variables

The independent variables were Layout (Qwerty, Dual-circle), Selection Techniques (Timeout, Reach, Expand&Reach), and Day (1 to 5).

4.2 Participants

6 participants (4 males, 2 females) were recruited from the local campus, mean age 26 (sd = 1.7), all right handed and with some experience of gestural interaction for gaming.

4.3 Procedure

The evaluation lasted for 5 days with 6 sessions every day. Each session tested a combination of Layouts and Selection techniques. In each session, 4 sentences were presented for the participant to reproduce, the first as practice followed by 3 test sentences. Six sets of sentences were randomly selected from MacKenzie and Soukoreff's phrase sets [22] and were assigned randomly to different sessions. User preferences and NASA task load index (TLX) data were collected on the first and last days.

Each character is white and highlights yellow when pointed at. With Timeout selection, the color gradually changes from yellow to green until timeout. Using Reach and Expand&Reach with Qwerty, the character gradually changes from yellow to green as the hand moves forward to reach it. The selected character appears immediately below the target sentence. A "typing" sound is played when a character is entered correctly. If the entry is incorrect, an error sound is played instead and the input is shown in red. All mistakes must be corrected for each sentence.

Both keyboards were 80 cm x 40 cm, with the top edge at the same height as the user's shoulders in motor space. Two spheres sized 2 cm were controlled with the hands. With the Qwerty layout, the hand in front of the other is enabled and rendered in black, the other is rendered grey and disabled to avoid accidental selection. With the Dual-circle layout, the blank center area is large enough to accommodate an idle hand without accidental selection, so no disable mechanism was used. For timeout selection technique with both layouts, the dwell time was 1.2 s. The 1.2 s dwell time was based on a pilot study which showed that less than 1.2 s produced more errors, and the observation that almost all commercial Kinect interfaces with timeout selection use more than 1.5 s.

4. 4 Experimental Setting

A Sanyo PDG-DWL2500 3D projector was used at 1280 x 720 resolution with a 203 x 115 cm screen centered at 130 cm height to simulate a large interactive TV display. A Microsoft Kinect camera was used with a refresh rate of 30 fps and the Kinect for Windows SDK V1.5 on Windows 7. The Kinect camera was placed 50 cm in front of the screen at a height of 70 cm. The user stood 250 cm from the screen (Figure 3).

Figure 3. Experimental setting.

5. RESULTS

5.1 Typing Speed

A repeated-measures ANOVA for Layout x Selection Technique x Day was used to analyze the text input speed. Main effects were found for Selection Technique ($F_{2,10}$=144.27, p<.001) and Day ($F_{4,20}$=21.49, p<.001). Layout had no significant effect ($F_{1,5}$=1.38, p=.29). Interaction effects were found for Layout x Selection Technique ($F_{2,10}$=11.69, p<.01) and Day x Selection Technique ($F_{8,40}$=6.05, p<.001).

Post hoc Bonferroni pairwise comparisons showed that Reach and Expand&Reach were both significantly faster than Timeout (p<.001), with no significant difference between them. Text input speed in the last 3 days was significantly faster than on the first day (p<.05), with no significant difference between the last 3 days. Mean text input speeds across all conditions are shown in Figure 4 and Table 1.

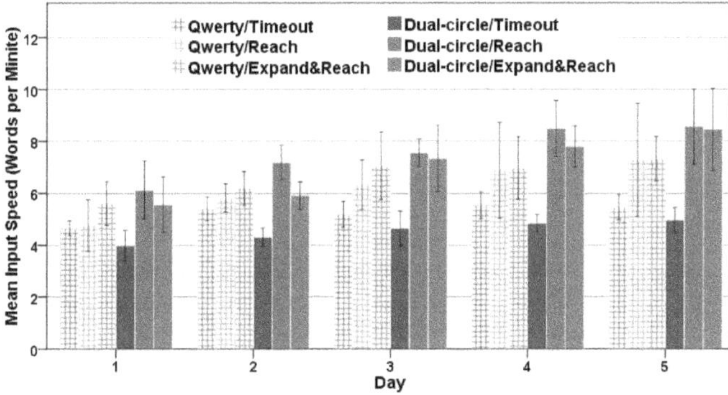

Figure 4. Mean text input speed. In this and later charts, error bars represent 95% confidence intervals.

Table 1. Mean speed (wpm) in day 1, day 5 and over 5 days.

	Qwerty			Dual-circle		
	Time-out	Reach	Expand &Reach	Time-out	Reach	Expand &Reach
Day 1	4.65	4.76	5.61	3.99	6.11	5.56
Day 5	5.46	7.29	7.31	4.96	8.57	8.46
5 days overall	5.25	6.22	6.63	4.55	7.58	7.01

5.2 Error Rate

A repeated-measures ANOVA for Layout x Selection Technique x Day was used to analyze error rate. Main effects were found for Layout ($F_{1,5}$=10.86, p<.05) and Selection Technique ($F_{2,10}$=11.09, p<.01). Day had no significant effect ($F_{4,20}$=1.94, p=.14). No interaction effect was found. Mean error rates are shown in Figure 5 and Table 2.

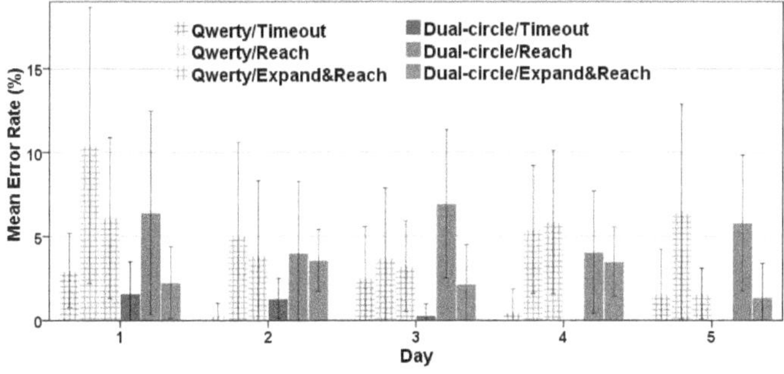

Figure 5. Mean error rate.

Post hoc Bonferroni pairwise comparisons showed Qwerty to be significantly more error-prone than Dual-Circle (p<.05). Timeout had significantly fewer errors than Reach (p<.05). There were no other significant differences between selection techniques.

Table 2. Mean error rate (%) in day 1, day 5 and over 5 days.

	Qwerty			Dual-circle		
	Time-out	Reach	Expand &Reach	Time-out	Reach	Expand &Reach
Day 1	3.00	10.44	6.11	1.63	6.42	2.27
Day 5	1.61	6.46	1.60	0.00	5.81	1.38
5 days overall	1.60	6.24	4.14	0.64	5.45	2.58

5.3 Hand Movement in 3D Space

We also recorded the hand movement distance per character in day 5. A repeated-measures ANOVA for Layout x Selection Technique was used to analyze hand movement distance per character. Main effects were found for Layout ($F_{1,5}$=39.42, p<.01) and Selection Technique ($F_{2,10}$=10.21, p<.01). No interaction effect was found. Post hoc Bonferroni pairwise comparisons showed that Qwerty layout required significantly more hand movement distance than Dual-Circle (p<.01). The Timeout selection technique had significantly less movement distance than Reach (p<.05). There were no other significant differences between selection techniques. Mean hand movement distance per character is shown in Figure 6(a).

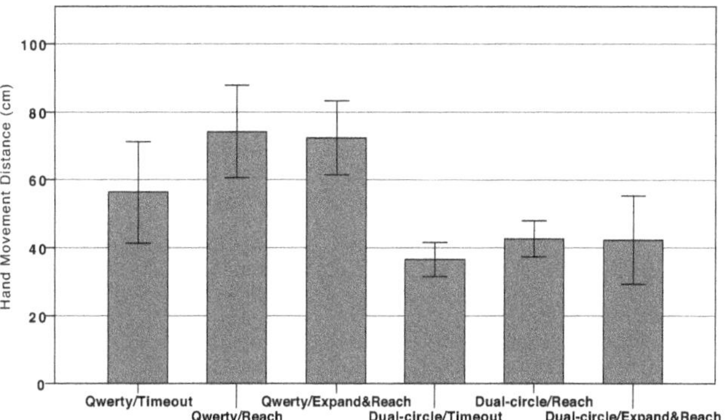

(a)

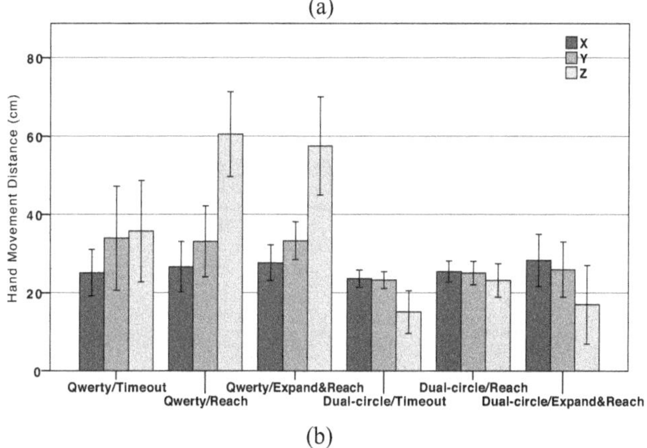

(b)

Figure 6. Mean hands movement distance per character. (a) Hand movement distance in 3D space (b) Hand movement distance in X, Y and Z axis.

We analyzed the hand movement distance in different axes (i.e. X, Y, Z) using one-way ANOVA for six text entry methods. We found that with Qwerty/Timeout and Dual-circle/Reach there was no significant effect of axis (p<.05). With Qwerty/Reach, Qwerty/Expand&Reach and Dual-circle/Timeout, main effects were found for Axis (p<.001). Post hoc Bonferroni pairwise comparisons showed that with Qwerty/Reach and Qwerty/Expand&Reach methods, hand movement in the Z axis was significantly more than in the X and Y axes (p<.05), while with Dual-circle/Timeout and Dual-circle/Expand&Reach methods, hand movement in the Z axis was significantly less than in the X and Y axes (p<.01). Mean hand movement distance per character in the X, Y and Z axes is shown in Figure 6(b).

5.4 Task Load

A repeated-measures ANOVA for Layout x Selection Technique x Day was used to analyze the NASA task load index (TLX).

Main effects were found for Layout ($F_{1,5}$=9.21, p<.05), Selection Technique ($F_{2,10}$=5.75, p<.05) and Day ($F_{1,5}$=7.61, p<.05). No interaction effects were found. Post hoc Bonferroni pairwise comparisons showed that the Qwerty layout had significantly higher task load than the Dual-circle layout (p<.05), the Reach selection technique had significantly higher task load than Expand&Reach (p<.05), and the task load on day 1 was significantly higher than on day 5 (p<.05). Figure 7(a) shows the overall workload. Figures 7(b) to (g) show users' mental demand, physical demand, temporal demand, performance, effort and frustration.

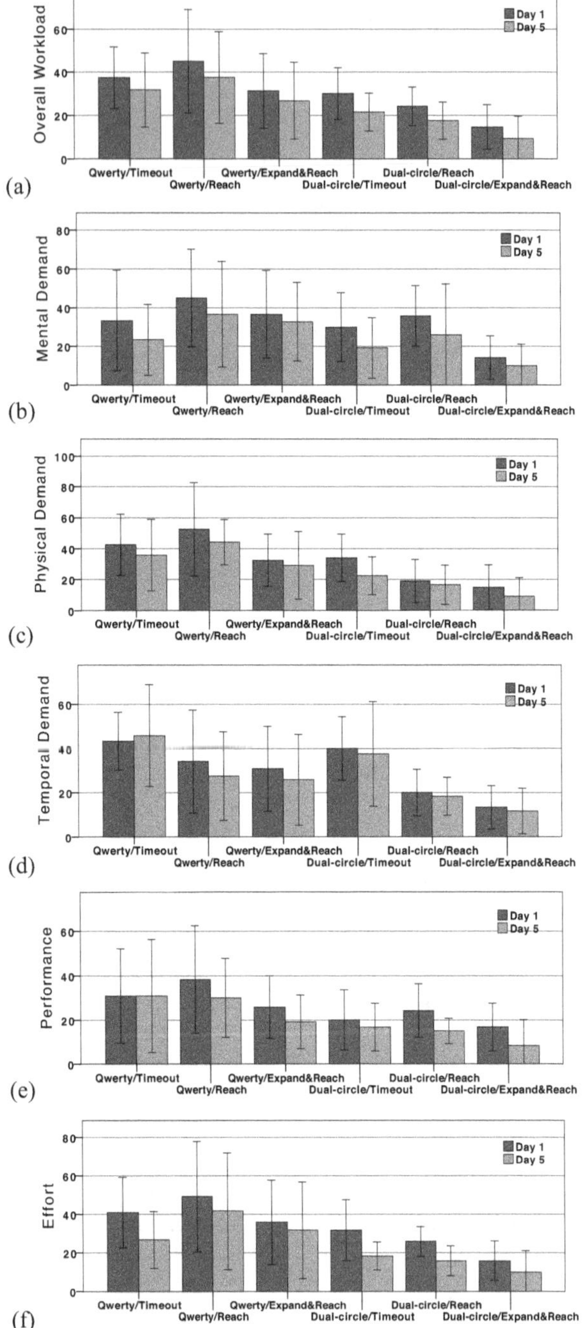

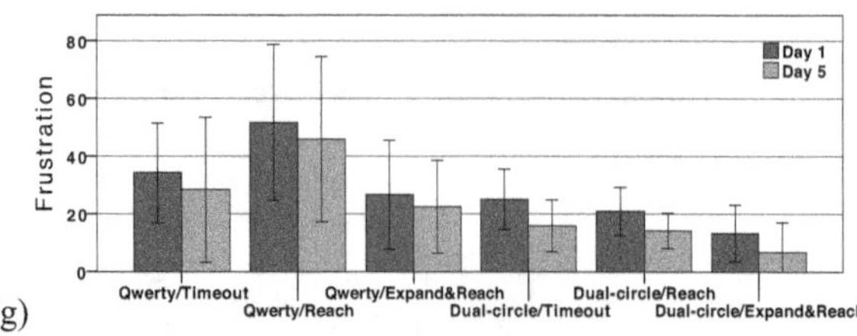

Figure 7. NASA task load index (TLX) results. (a) Overall workload (b) Mental demand (c) Physical demand (d) Temporal Demand (e) Performance (f) Effort (g) Frustration

5.5 User Preference

User preference data for each input method was collected on the first day and the last days. Users gave their preference (from 1 for strongly dislike to 10 for strongly like) after each text entry method, as shown in Figure 8. Overall, all participants preferred Dual-circle/Expand&Reach on both the first and last days of the study. The Dual-circle/Expand&Reach technique could enable people to input text comfortably in both "walk up and use" and the slightly longer term use (5 days) of our study. Its error tolerance also allows more casual hand movements without high concentration and physical effort.

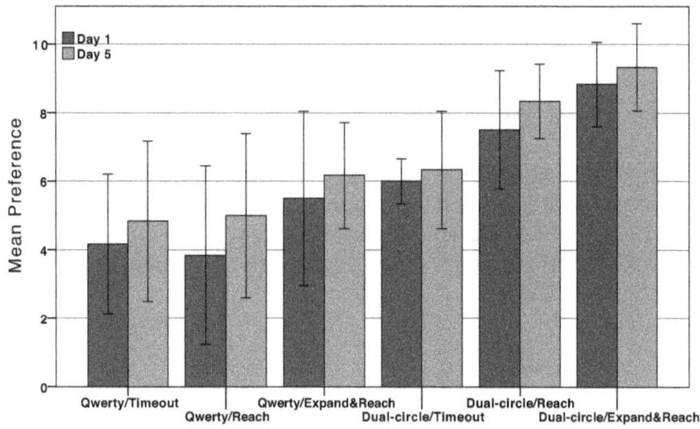

Figure 8. User preference

6. DISCUSSION

6.1 Text Entry Method

The dual-circle layout had better performance and lower task load than the Qwerty layout. Although users are more familiar with the Qwerty keyboard, its characters are packed in 2D, so character selection is more difficult than with the dual-circle layout. We noticed that when users select a character with one hand, they normally put down the other hand to avoid careless error selection. With the dual-circle layout, they can just relax the hand because there is enough blank space in the center to prevent careless error selection.

When using the Qwerty/Reach text entry method, users felt it was difficult to find the reach point in the beginning due to the absolute character position. High physical demand was also reported for Qwerty/Reach (Figure 7c). With Qwerty/Expand&Reach, on the other hand, as relative location was used, it was easier to select. However, both Qwerty/Reach and Qwerty/Expand&Reach required long movement distance when inputting text (Figure 6a), which was largely due to the hand movement for character selection along the Z dimension (Figure 6b). In contrast, as no hand movement along the Z dimension was required with the dual-circle layout, and users felt more easy and relaxed in their text entry tasks.

Participants also noticed their improvement for the Reach and Expand&Reach techniques with both keyboard layouts. Especially with the Dual-circle layout, typing speed increased continuously over 5 days (Figure 4). Timeout selection had few errors with both layouts, but the dwell time slowed text entry speed so it was not liked by users. Typing speed with the timeout method, with both layouts, did not improve in the course of the study as other text entry methods did.

Overall, Dual-circle/Expand&Reach was the best text input method. It is error tolerant thanks to the Expand&Reach selection technique so users make fewer careless wrong selections. The dual-circle layout also leaves enough blank space in the center to help people to relax their hands when not inputting text without worrying about triggering a selection when relaxing the hand. Thus Dual-circle/Expand&Reach is a practical text entry method for interactive TV. The Dual-circle/Reach method is more straightforward for selection but was found to be error-prone in this study. It could be a fast entry method given future improvements in tracking resolution. Qwerty/Expand&Reach could be used in scenarios in which familiarity with the standard Qwerty keyboard is desirable. The Qwerty/Reach method, however, may be a less promising method, given its high error rate and the high mental and physical effort noted by the participants.

6.2 Timeout Dwell Time

Typing speed with the timeout method, with both layouts, did not improve with practice. As noted above, the 1.2 s dwell time was based on a pilot study that showed a dwell time of less than 1.2 s produced more errors for beginners, and the observation that most commercial Kinect timeout based interaction uses more than 1.5 s. Our results suggest that the 1.2 s dwell time was suitable for first time use, however, it could be unnecessarily long for more experienced users, thus limiting performance. We also found that error rates using the timeout selection technique were low. With Dual-circle/Timeout, there were no errors at all in the last two days (Figure 5). Users also commented that the dwell time for the Dual-circle/Timeout method could be shorter after some practice.

For the dual-circle layout, it is possible to reduce the dwell time to any value. For example, the dwell time could be reduced even to zero, and in this extreme case, the Dual-circle/Timeout method will be equivalent to the Dual-circle/Reach method, leading to faster typing speed but more errors. With the Qwerty layout, however, the dwell time cannot be reduced so much. This is because users must move their hands over other characters to select the desired character in the Qwerty layout, thus a very small dwell time would trigger the undesired selection easily and lead to a huge increase in errors.

In a previous study using eye gaze for text entry, Majaranta et al. [23] showed that adjustable dwell time could improve text entry performance with a Qwerty layout keyboard, and this could also be the case for freehand text entry. However, in their experiment, participants used a 282 ms dwell time in the last session [23], which may be too short for a Qwerty layout with freehand gestural input. Further investigations of optimal dwell times for freehand interaction could form part of future work.

6.3 Tracking Sensors and Freehand Gestural Text Input

Our work aims to facilitate freehand gestural interaction tracked by a low-cost remote single camera with no requirement for a user to carry, wear or pick up an input device or fiducial markers. We used Microsoft Kinect in this study as it is a typical currently available low-cost remote gesture tracking sensor. Some more accurate tracking systems, such as Vicon cameras[4], can offer high accuracy but they require markers on the body and calibration before use, which we explicitly want to avoid. Other sensors, such as the soon to be available LeapMotion sensor[5], may provide accurate hand tracking without markers, but the tracking range is relatively short and thus may not suitable for interactive TV.

Even with remote single tracking sensors having higher resolution and become more affordable, there are still some common limitations of freehand interaction that cannot be solved by increasing tracking resolution: e.g. (i) having only one tracking angle brings occlusion of finger/wrist motion, (ii) previous work has shown that when the hands are moving freely in 3D, the user cannot point as accurately as with a 2D surface [36], and (iii) motor control with large muscle groups, e.g. the shoulder, is less accurate than with small muscle groups, e.g. fingers [10, 32]. Even with technical advances, remote camera tracking will share many of the current benefits and limitations and our main findings are broadly applicable across this class of devices. It is highly unlikely that improvements in tracking resolution would have a detrimental effect on Expand&Reach. Dual-circle/Reach, as suggested above, although error-prone in this study, could be a fast entry method given future improvements in tracking accuracy.

The performance of Dual-circle/Expand&Reach (5.56 wpm on the first day, 8.46 wpm on the last day and a mean of 7.01 wpm over 5 days) also compares well to similar gestural text input without word prediction, such as a mean of 5.4 wpm after 4 days' practice using accelerometer sensors in hand held devices [17], a mean of 6.5 wpm over 2 weeks with a data glove and fiducial trackers [25], and a mean of 5.18 wpm with a text entry method combing speech and gesture [13]. The performance of Dual-circle/ Expand&Reach is also much better than the 1.83 wpm text entry speed reported with the default Xbox gestural text input [13]. On the other hand, the performance of Dual-circle/Expand&Reach achieved in our study still cannot compete with some text entry methods enabled by more accurate tracking devices or input models, such as a Qwerty virtual keyboard and handheld devices (18.9 wpm) [34], or text entry using gaze (nearly 20 wpm) [23].

6.4 Design Suggestions

Interestingly, previous work using hand held devices for text input showed that the Qwerty layout was faster and had fewer errors compared to a circle layout when the users were about 2.5 m away from the display [34]. In our study, on the other hand, the results suggested exactly the opposite. Such differences are largely due to differences in user behavior with different input methods. Moving the bare hand and arm freely in the air is different from holding and moving a mouse on the desktop, tapping the thumb on a mobile phone, or multi-touch movements with the fingers on a touch sensitive surface. For example, in [34], the desired character is selected by pointing and pressing a button using a Wii

[4] http://www.vicon.com/products/cameras.html

[5] https://www.leapmotion.com/

remote, however, in freehand gestural interaction button pressing is not available and so must be replaced by other techniques, such as Timeout, Reach, and Expand&Reach.

From the findings of our study, some design guidelines emerge for freehand interaction.

(1) Although the hand can move freely in 3D space, and hand motion in the Z dimension could be used as a gesture delimiter or trigger [32], frequent movements in the Z dimension are not recommended for freehand gestural interaction. This is not only because the hand moves more slowly forward and backward [11], and actions in the Z dimension can be error-prone [32], but also because frequent movements in the Z dimension can increase the hand movement distance and corresponding physical demands, as shown in this study.

(2) The circle layout is useful for interface design with freehand gesture. In the circle layout, all items have one side facing to the center and, therefore, can be reached directly without moving the hands over other objects. And the large blank area in the center can allow users to relax their hands without worrying about triggering undesired actions, thus potentially reducing arm fatigue.

(3) Combing the expanding target and reach selection techniques is useful for freehand interaction, especially for hand motion tracking by inexpensive sensors with low resolution. When used with the reach technique, the target can expand not only in virtual space but also in motor space. Thus, using the expanding target and reach techniques together addresses the limitation that targets can expand only in virtual space [24], and it is also well suited to freehand interaction techniques due to its error tolerance.

7. CONCLUSIONS

Freehand gesture is a promising interaction technique for interactive TV. The work presented here reports an initial investigation of the design and evaluation of potential text entry methods. The designs proposed in this paper, such as the circle layout, expanding target and the reach selection technique, are not entirely new individually. But their combination and their use with freehand gesture have never previously been explored. This novel combination of layouts familiar to Qwerty users and selection techniques designed specifically for freehand gestural text input are well suited to the intended interactive TV context, and we have demonstrated the effectiveness of this combination. Future work includes integration of word-level functions such as word prediction and completion, as well as evaluation of emerging more advanced tracking sensors.

8. ACKNOWLEDGMENTS

This research was supported in part by a Google Research Award.

9. REFERENCES

[1] J. Accot and S. Zhai. More than dotting the i's --- foundations for crossing-based interfaces. In *Proceedings of the SIGCHI Conference on Human Factors in Computing Systems*, CHI '02, 73--80.

[2] R. Aoki, B. Chan, M. Ihara, T. Kobayashi, M. Kobayashi, and S. Kagami. A gesture recognition algorithm for vision-based unicursal gesture interfaces. In *Proceedings of the 10th European Conference on Interactive TV and Video*, EuroiTV '12, 53--56, 2012. ACM.

[3] H. Benko. Beyond flat surface computing: challenges of depth-aware and curved interfaces. In *Proceedings of Multimedia 2012*, MM '12, 935--944.

[4] X. Bi, B. A. Smith, and S. Zhai. Quasi-qwerty soft keyboard optimization. In *Proceedings of the SIGCHI Conference on Human Factors in Computing Systems*, CHI '10, 283--286, 2010. ACM.

[5] J. Bobeth, S. Schmehl, E. Kruijff, S. Deutsch, and M. Tscheligi. Evaluating performance and acceptance of older adults using freehand gestures for tv menu control. In *Proceedings of the 10th European Conference on Interactive TV and Video*, EuroiTV '12, 35--44, 2012. ACM.

[6] S. J. Castellucci and I. S. MacKenzie. Gathering text entry metrics on android devices. In *CHI '11 Extended Abstracts on Human Factors in Computing Systems*, CHI EA '11, 1507--1512, 2011. ACM.

[7] D. Chatterjee, A. Sinha, A. Pal, and A. Basu. An iterative methodolgy to improve TV onscreen keyboard layout design through evaluation of user studies. *Advances in Computing*, 2(5):81-91, 2012.

[8] M. D. Dunlop, N. Durga, S. Motaparti, P. Dona, and V. Medapuram. Qwerth: an optimized semi-ambiguous keyboard design. In *Proceedings of the 14th International Conference on Human-Computer Interaction with Mobile devices and Services*, MobileHCI '12, 23--28, 2012. ACM.

[9] G. Geleijnse, D. Aliakseyeu, and E. Sarroukh. Comparing text entry methods for interactive television applications. In *Proceedings of the 7th European Conference on Interactive TV and Video*, EuroiTV '09, 145--148, 2009. ACM.

[11] T. Grossman and R. Balakrishnan. Pointing at trivariate targets in 3d environments. In *Proceedings of the SIGCHI Conference on Human Factors in Computing Systems*, CHI '04, 447--454, 2004. ACM.

[12] F. Guimbretiére and T. Winograd. Flowmenu: combining command, text, and data entry. In *Proceedings of the 13rd Annual ACM Symposium on User Interface Software and Technology*, UIST '00, 213--216.

[13] L. Hoste, B. Dumas, and B. Signer. Speeg: a multimodal speech- and gesture-based text input solution. In *Proceedings of the International Working Conference on Advanced Visual Interfaces*, AVI '12, 156--163, 2012. ACM.

[14] A. Iatrino and S. Modeo. Text editing in digital terrestrial television: a comparison of three interfaces. In *Proceedings of the 4th European Conference on Interactive TV and Video*, EuroiTV '06.

[15] P. Isokoski and R. Raisamo. Quikwriting as a multi-device text entry method. In *Proceedings of the Third Nordic Conference on Human-Computer Interaction*, NordiCHI '04, 105--108, 2004. ACM.

[16] H.-C. Jetter, J. Gerken, and H. Reiterer. Why we need better modelworlds, not better gestures. In *Natural User Interfaces (Workshop CHI 2010)*, 1-4.

[17] E. Jones, J. Alexander, A. Andreou, P. Irani, and S. Subramanian. Gestext: accelerometer-based gestural text-entry systems. In *Proceedings of the SIGCHI Conference on Human Factors in Computing Systems*, CHI '10, 2173--2182.

[18] M. Karam and m. c. schraefel. A taxonomy of gestures in human computer interactions. Technical report, University of Southampton, 2005.

[19] P. O. Kristensson, T. Nicholson, and A. Quigley. Continuous recognition of one-handed and two-handed gestures using 3d

full-body motion tracking sensors. In *Proceedings of the 2012 ACM international conference on Intelligent User Interfaces*, IUI '12, 89--92, 2012. ACM.

[20] P. O. Kristensson and S. Zhai. Shark2: a large vocabulary shorthand writing system for pen-based computers. In *Proceedings of the 17th Annual ACM Symposium on User Interface Software and Technology*, UIST '04, 43--52.

[21] F. C. Y. Li, R. T. Guy, K. Yatani, and K. N. Truong. The 1line keyboard: a qwerty layout in a single line. In *Proceedings of the 24th Annual ACM Symposium on User Interface Software and Technology*, UIST '11, 461--470, 2011. ACM.

[22] I. S. MacKenzie and R. W. Soukoreff. Phrase sets for evaluating text entry techniques. In *CHI '03 Extended Abstracts on Human Factors in Computing Systems*, CHI EA '03, 754--755, 2003. ACM.

[23] P. Majaranta, U.-K. Ahola, and O. \v{S}pakov. Fast gaze typing with an adjustable dwell time. In *Proceedings of the SIGCHI Conference on Human Factors in Computing Systems*, CHI '09, 357--360, 2009. ACM.

[24] M. J. McGuffin and R. Balakrishnan. Fitts' law and expanding targets: experimental studies and designs for user interfaces. *ACM TOCHI*, 12(4):388--422, 2005.

[25] T. Ni, D. Bowman, and C. North. Airstroke: bringing unistroke text entry to freehand gesture interfaces. In *Proceedings of the SIGCHI Conference on Human Factors in Computing Systems*, CHI '11, 2473--2476.

[26] D. A. Norman. Natural user interfaces are not natural. *Interactions*, 17(3):6--10, May 2010.

[27] D. A. Norman and J. Nielsen. Gestural interfaces: a step backward in usability. *Interactions*, 17(5):46--49, September 2010.

[28] K. Perlin. Quikwriting: continuous stylus-based text entry. In *Proceedings of the 11th Annual ACM Symposium on User Interface Software and Technology*, UIST '98, 215--216, 1998. ACM.

[29] O. Polacek, M. Klima, A. J. Sporka, P. Zak, M. Hradis, P. Zemcik, and V. Prochazka. A comparative study on distant free-hand pointing. In *Proceedings of the 10th European Conference on Interactive TV and Video*, EuroiTV '12, 139--142, 2012. ACM.

[30] Primesense. The primesense 3D awareness sensor. .

[31] G. Ren and E. O'Neill. 3D marking menu selection with freehand gestures. In *IEEE Symposium on 3D User Interfaces*, 3DUI '12, 61-68.

[32] G. Ren and E. O'Neill. 3D selection with freehand gesture. *Computers & Graphics*, 2013.

[33] J. Rick. Performance optimizations of virtual keyboards for stroke-based text entry on a touch-based tabletop. In *Proceedings of the 23nd Annual ACM Symposium on User Interface Software and Technology*, UIST '10, 77--86, 2010. ACM.

[34] G. Shoemaker, L. Findlater, J. Q. Dawson, and K. S. Booth. Mid-air text input techniques for very large wall displays. In *Proceedings of Graphics Interface 2009*, GI '09, 231--238.

[35] A. J. Sporka, O. Polacek, and P. Slavik. Comparison of two text entry methods on interactive tv. In *Proceedings of the 10th European Conference on Interactive TV and Video*, EuroiTV '12, 49--52, 2012. ACM.

[36] R. J. Teather and W. Stuerzlinger. Assessing the effects of orientation and device on (constrained) 3d movement techniques. In *IEEE Symposium on 3D User Interfaces*, 3DUI '08, 43--50.

[37] R.-D. Vatavu. Interfaces that should feel right: natural interaction with multimedia information. In M. Grgic, K. Delac, and M. Ghanbari, editors, *Recent advances in multimedia signal processing and communications*, volume 231, page 145-170. Springer Berlin / Heidelberg, 2009.

[38] R.-D. Vatavu. User-defined gestures for free-hand tv control. In *Proceedings of the 10th European Conference on Interactive TV and Video*, EuroiTV '12, 45--48, 2012. ACM.

[39] A. Wexelblat. Research challenges in gesture: open issues and unsolved problems. In I. Wachsmuth and M. Fröhlich, editors, *Gesture and sign language in human-computer interaction*, volume 1371, page 1-11. Springer Berlin / Heidelberg, 1998.

[40] A. D. Wilson. Using a depth camera as a touch sensor. In *Proceedings of ACM International Conference on Interactive Tabletops and Surfaces*, ITS 2010, 69--72.

[41] J. O. Wobbrock, B. A. Myers, and J. A. Kembel. Edgewrite: a stylus-based text entry method designed for high accuracy and stability of motion. In *Proceedings of the 16th Annual ACM Symposium on User Interface Software and Technology*, UIST '03, 61--70, 2003. ACM.

[42] S. Zhai and P. O. Kristensson. Interlaced qwerty: accommodating ease of visual search and input flexibility in shape writing. In *Proceedings of the SIGCHI Conference on Human Factors in Computing Systems*, CHI '08, 593--596, 2008. ACM.

[43] S. Zhai and P.-O. Kristensson. Shorthand writing on stylus keyboard. In *Proceedings of the SIGCHI Conference on Human Factors in Computing Systems*, CHI '03, 97--104.

Live-Streaming Changes the (Video) Game

Thomas P. B. Smith
Culture Lab
Newcastle University
Newcastle Upon Tyne
thomas.smith@ncl.ac.uk

Marianna Obrist
Culture Lab
Newcastle University
Newcastle Upon Tyne
marianna.obrist@ncl.ac.uk

Peter Wright
Culture Lab
Newcastle University
Newcastle Upon Tyne
p.c.wright@ncl.ac.uk

ABSTRACT

Video games are inherently an active medium, without interaction a video game is benign. Yet there is a growing community of video game spectating that exists on the Internet, at events across the world and, in part, as traditional television broadcasts. In this paper we look at the different communities that have grown around video game spectating, the incentives of all stakeholders and the technologies involved. An interesting part of this phenomenon is its relation to the malleability of activity and passivity; video games are traditionally active but spectatorship brings an element of passivity, whereas television is traditionally passive but interactive television brings an element of activity. We explore this phenomenon based on selected examples and stimulate a discussion around how such understanding from the video game field could be interesting for interactive television.

Categories and Subject Descriptors

K.8.0 [**Personal Computing**]: General: *Games,* H.5.1 Multimedia Information Systems: *interactive television.*

General Terms

Human Factors, Documentation

Keywords

Live-streaming, live streaming, video games, spectatorship, spectator, interactive television, broadcaster, TwitchTV, Let's Play community, passive usage, active usage.

1. LIVE-STREAMING CHANGES THE GAME

Live-streaming is not a new phenomenon and has existed on the Internet for the last 20 years but recently, with advances in Internet bandwidth and new web services, live-streaming has become democratized, allowing anyone to stream anything they want. Members of the live-streaming community have become more focused in recent years as the technology and users' grasp of it has matured, with specific shows publishing daily or weekly live-streams. Many different people and organizations live-stream for different reasons, for example, a wildlife organization may live-stream a birds nest from a hidden camera [22] or an artist may live-stream an interactive horror show [23].

EuroITV'13, June 24–26, 2013, Como, Italy.

Regardless of incentive, all of these acts have one thing in common; they use a live-streaming platform to broadcast their content to users, which Bruns [2], for instance, discussed with respect to conventional television broadcasting as DIY broadcasting. He reflects on the shift from traditional television to new models of broadcasting, especially discussing new streaming services, which better fit the interests of users and are more likely to be taken up due to better broadband options.

A popular category on live-streaming services is video games. Growing popularity in video game live-streaming communities has turned video games into not only a player experience but also a viewer experience. From highly competitive international championships to players recording themselves playing horror games for laughs, the community has grown so large there is a live-stream web service just for video games (see TwitchTV [19]). Within this paper, we are exploring the practices and incentives of performers and viewers in the video game live-stream community, aiming to gain insights that can be of interest to the interactive television community. Three of the most popular video game live-streaming communities are e-sports, speedrunning and "Let's Play". We conducted an exploratory analysis of these communities in general and the "Let's Play" community specifically. We focus on "Let's Play" in detail because it allows people broader experiences of playing and spectating a game compared to the other two communities.

1.1 Approach

The basis for the exploratory study presented in this paper is the first authors' participation in and observation of video game live-streaming, using Cheung and Huang's [3] framework of spectator experiences as an analytical framework.

The first author is an active viewer in the live-stream community, using websites such as TwitchTV and YouTube Live to view performers for over a year. The first author is also an avid gamer, playing many of the games he watches on live-streaming platforms either prior to or after viewing broadcasts. The fact that he is a gamer is beneficial to the study as familiarity with the nuances of the community such as the vocabulary used and knowledge of the platforms is invaluable for understanding these communities. The second author has worked on game research projects but is not a gamer or live-stream viewer. It was important to be impartial in our analysis (not to let the first author's tacit knowledge act as a bias) and as such the authors conducted a series of discussions to unpick the first author's understanding and observations of the live-streaming community and objective study of live-streaming platforms.

Analysis of the "Let's Play" community was guided by Cheung and Huang's [3] work on the e-sports spectator community. The authors provide a series of spectator personas that exist in the community and a theoretical framework of facets of the spectator experience. By applying this existing knowledge (particularly the personas) to the "Let's Play" community, we were able to unpick the first author's knowledge and observations of "Let's Play" and form an understanding of how the community functions. This guided our discussion of the phenomenon with respect to its malleability of activity and passivity, which we used to reflect on interactive television and the community's effort to support and encourage a more interactive TV experience and surrounding technologies (e.g., second screen, mobile TV).

To aid the exploratory nature of this study using Cheung and Huang's [3] framework, we were guided by the following questions: (1) For what reasons do "Let's Play" players stream themselves? (2) How do viewers watch these broadcasts? (3) What insights can we consider for interactive television?

2. VIDEO GAME LIVE-STREAMING COMMUNITIES

We have found that the video game live-streaming community is not unified in content or practice but technology. Not all live-streamers are live-streaming the same games, for the same reasons or to the same audience. There are different forms of video game live-streaming; major e-sports organizations stream competitive matches while others informally play through entire games and others still engage in the act of trying to complete a game as quickly as possible.

Although there are possibly many more use cases for video game live-streaming, these three forms make up the three most popular communities we have seen on the live-streaming services and as such are the communities we focus on. The most popular community is the e-sports community, which stream competitive matches between players and usually a separate commentator. The "Let's Play" community is less popular but its popularity is rising on YouTube and live-streaming websites. "Let's Play" broadcasts are more informal than e-sports matches and much of the entertainment comes from the player as they play through a game. Finally, the speedrunning community is the least popular but is also rising in popularity. Speedrunning is the idea of trying to complete a game as fast as possible, and so like e-sports, viewers watch to see these competitive feats.

We will now discuss the three communities before going onto discuss different live-streaming platforms. We then use these understandings to discuss the "Let's Play" community in-depth.

2.1 E-sports

E-sports is a term to describe the phenomenon of competitively playing video games. E-sports began in the era of arcade games where players would challenge each other to compete for the top place on an arcade machine hi-score board. This practice was formalized with the Space Invaders Tournament [5] in 1980 held by Atari, the developer of Space Invaders, which attracted around 10,000 participants. The e-sports community is now very large with professional leagues and tournaments mirroring those of traditional sports.

Cheung and Huang [3] analyzed the spectator's incentives to watch e-sports rather than play the games themselves. They make three important contributions to the discussion: they (1) detail nine personas of spectators, (2) provide a dissection of the spectator ecosystem and (3) outline what makes watching e-sports entertaining. The authors point out that a person might be a mix of multiple personas; however the nine main personas they identified for e-sports are summarized below [3]:

1. The Bystander – There are two forms of Bystander, the *uninformed* and *uninvested.* The *uninformed* lacks knowledge of the meaning of in-game acts. The *uninvested* have a basic understanding of the game but they are spectating by accident, they have "stumbled" into being a spectator.
2. The Curious – This persona is interested in filling knowledge gaps they have with the game.
3. The Inspired – This persona is keen to play the game after watching a match, perhaps attempting strategies they saw or purely because they are interested in the game.
4. The Pupil – The Pupil is very similar to *The Curious* persona but they are focused on turning what they have learnt into practice and develop their skills.
5. The Unsatisfied – This persona sees spectating as a weaker form of playing but watches because it is still more interesting than not consuming the content at all.
6. The Entertained – The Entertained is more or less the opposite of *The Unsatisfied*; they find watching entertaining in itself, not as a weaker form of playing.
7. The Assistant – The Assistant impacts the game by helping a player, for example this could be in the form of pointing out information the player may have missed.
8. The Commentator – This persona is a spectator who is also a performer to a larger group of spectators, much like a traditional sports broadcast.
9. The Crowd – The Crowd persona details the delight in spectating as a group.

The authors further detail the above findings in rich descriptions of the different personas [3]. Although the usage of the terminology 'personas' could be seen confusing, as the descriptions rather seem to highlight different roles people could have over time interacting with a game, we found the empirically grounded spectator perspectives very relevant for our exploratory purposes. Thus, we used the personas as a framework to discuss the video game live-streaming communities, in particularly the 'Let's Play' community.

2.2 Speedrunning

Speedrunning is the challenge of attempting to complete a game as fast as possible; this began as simply as acquiring sufficient skill to play a particular game as quickly as possible but now encompasses exploiting bugs in a game to skip as much game content as possible.

Speedrunning has been around for a long time before live-streaming was a possibility, but the advent of live-streaming has allowed the "sport" to turn into a real spectator sport. There is an inherent difference between watching a pre-record of someone performing a speedrun after the fact and watching a player as they attempt to perform a speedrun. As Jared Rea (community manager of TwitchTV) has said "it humanizes inhuman abilities" [12]. In terms of live-streams, speedrunning is usually framed as a way of imparting tips and strategies to other speedrunners because finding and exploiting bugs is a community effort.

2.3 "Let's Play"

The "Let's Play" (which will also be referred to as LP) community is different to the speedrunning and e-sports communities in that the experience of playing a game is not framed in a larger competitive context (for example an international league in the e-sports community) or focusing on a certain player skill (completing games as fast as possible in the speedrunning community), giving rise to a very diverse community in terms of the games played and players themselves. LP broadcasts are usually informal and enjoyment as a viewer comes from what the player/performer adds to the experience. The success of a LP broadcaster is not based on their skill in the game (as it is with e-sports and speedrunning) but how entertaining they are as they play the game. One important difference between the LP and e-sports community is that LP viewers are watching LP performers exclusively through live-streaming and video sharing services which is in contrast to e-sports spectators who use these services in conjunction with watching tournaments in person and broadcasts in bars and pubs.

The term "Let's Play" originates from the posts on the Something Awful [18] forum in which users made annotated screenshot stories of their play through video games. Even without live-streaming technology, this past-time was viable given its focus on what the player/performer adds to the experience, unlike the other two communities discussed. When screen-capture became viable, video-based LPs began to surface on video sharing services such as YouTube. LPs are now broadcasted as pre-recorded videos on video sharing services as well as live-streamed on services such as TwitchTV.

An LP is generally an episodic account of a player's journey through a particular game or creative play in a non-linear game, following the same "diary structure" as the image-based Something Awful forum threads. The skill level of the player is not very important; it is the entertainment the player creates around the experience that draws interest. To be entertaining, every LP contains commentary from the player which usually light-hearted or comedic. The player/performer's voice is very important; framing the action of the game in a certain light and acting as a driver of the content as well as reactionary to the content, it is what sets an LP apart from a walkthrough, which usually focuses on telling the viewer how to complete the game. Therefore an LP can be thought of a narrative commentary on an experience as it is being experienced.

3. LIVE-STREAMING PLATFORMS

Below we detail different examples for live-streaming platforms used for video games and how their user experience design assists the broadcasters and viewers in ways a traditional television broadcast of a video game would not. We dedicate most attention to TwitchTV [19], which is popular among the 'Let's Play' community, a community type we are going to discuss in more detail afterwards. However, in order to show the particularities of different platforms we also describe OnLive [17] and YouTube Live [20] as alternative practice examples.

3.1 TwitchTV

TwitchTV is the dominant live-streaming service for all of the communities. Twitch used to be the "Gaming" section of JustinTV but after it grew much larger than expected, JustinTV spun it out into its own website. This independence allowed the service designers to focus on designing specifically for video games. Users choose channels by browsing the featured channels or view channels by game or broadcaster. Figure 1 shows how the TwitchTV interface looks while watching a channel. This channel is a "Let's Play" broadcast in which we can see the broadcaster's screen-capture and webcam feed in the bottom-left corner. The title tells us that the broadcaster is showing off a particular feat that he has created. To the right is the live chat window is which viewers are discussing what the broadcaster is showing and asking him questions. Below we detail the relevant aspects of TwitchTV, which will be discussed later.

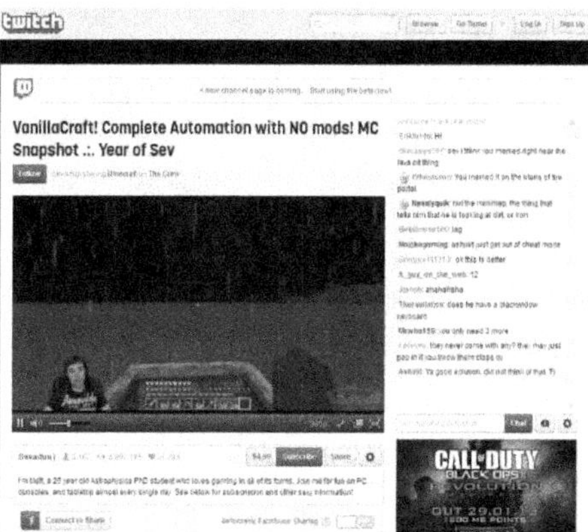

Figure 1. TwitchTV interface when watching a stream.

3.1.1 *Live Chatting*

The TwitchTV platform allows viewers to live chat while players are broadcasting, discussing what is happening with other viewers as well as conversing with the broadcaster directly. The form these conversations take is very dependent on the community; some discussions are about tactics and strategies whereas others are suggestions and comical discussions. For example, the most popular Minecraft channel espouses itself as a highly interactive channel, asking viewers to tell the broadcaster what to do/make and discuss Minecraft as he plays [24].

3.1.2 *Featured Channels*

Featured channels are channels which are shown on the homepage of the website. These channels are owned by large organizations, that are in the live-stream community rather than amateur channels that have been promoted by TwitchTV. There are many forms of featured channels but they usually have to be daily channels of some sort. One example is the *Major League Gaming* channel, which broadcasts e-sporting events such as tournaments or league games. Members of the video gaming press such as *Gamespot* and *Destructoid* also have featured channels, which broadcast different shows every day. Another example are e-sports learning channels in which famous e-sporting personalities teaches strategies to amateur players.

3.1.3 *Game Integration*

TwitchTV have integrated their service directly into games and the first such game they have integrated themselves into is *Call of Duty: Black Ops 2*, the fastest selling game. Users on the PC, Xbox and PS3 are now able to enable streaming straight to TwitchTV.

3.1.4 *Monetization*

In order to continue service, TwitchTV makes money by placing advertisements over broadcast content. When viewing a broadcast, every few minutes an advertisement will interrupt the broadcast and run for 30 seconds. They have recently announced

Twitch Turbo which allows users to pay a monthly fee to remove advertisements as well as other cosmetic benefits such as a "turbo" logo by their name in chat windows and a larger set of emoticons to use in chat.

Monetization is also possible for the broadcasters and may be an incentive for some to broadcast. Broadcasters can share a cut of the advertisements placed on their page and also set up their own monthly charge to remove ads from their broadcasts.

3.2 OnLive

OnLive is not a live-streaming service like TwitchTV but is a service in the emerging area of "cloud gaming". Cloud gaming services allow users to log in from PC, Mac, Android and iOS devices to play any game that they have unlocked in their account. Technically, the games are running on a server at one of OnLive's buildings and streaming each video frame over the Internet to the user's screen. When a user uses an input device, the input data is streamed over the Internet to the OnLive server, allowing players to play graphically intensive games on systems that would not be able to support the games natively.

Figure 2. The OnLive live-stream spectator interface, each grid square is different stream that can be viewed full-screen.

The fact the games are streamed from a server under OnLive's control is an advantage for live-streaming, OnLive make every game currently being played available to anyone to view. Figure 2 shows how the live-streaming interface looks; each tile is showing a game that is being played and as a viewer we can choose any tile to watch, at this point it fills the screen. This form of live-streaming is very different to TwitchTV from the perspective of the player; a TwitchTV broadcaster has gone out of their way to stream themselves playing whereas an OnLive player is live-streamed anytime they are playing. This difference in incentive has affected the way OnLive implements the live-streaming service, viewers can only give a "thumbs up" or "thumbs down" which is presented to the player unobtrusively in the top corner of the screen, rather than the richer participation viewers have on TwitchTV.

3.3 YouTube Live

Apart from the previous example, YouTube is the largest video sharing website and is primarily used by uploading videos after they have been recorded. Recently, YouTube have implemented a live-streaming service called YouTube Live. This service acts much like other live-streaming services technically; allowing users to live-stream any recording device, be that a screen-capture or video camera. Although YouTube is the most popular video sharing website for "Let's Play", it is not very popular for live-

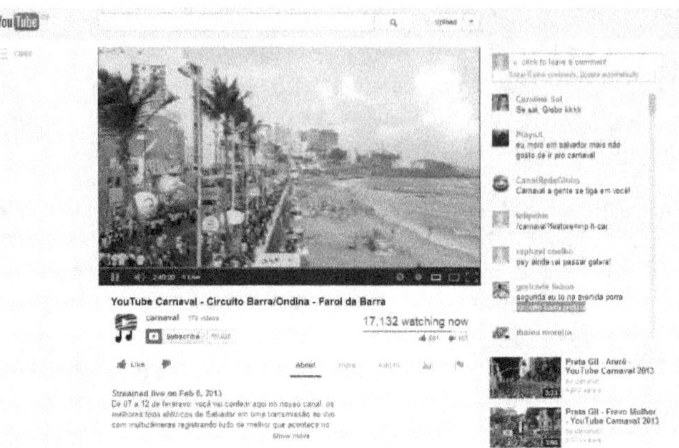

Figure 3. Youtube Live interface.

streams, which can be explained by its lack of video game-specific features.

The interface allows users to browse categories such as "News & Politics" and "Gaming" with each video thumbnail showing how many people are currently watching and if the channel is currently live. Upon choosing a channel, the webpage in Figure 3 is displayed, looking very much like the standard YouTube page. The main difference is that the comments do not flow underneath the video but beside it in a live-chat format. This makes commenting a more important part of the experience than usual as comments are always in view. We see this design mirrored in other live-streaming websites such as TwitchTV.

4. THE "LET'S PLAY" COMMUNITY

As previously mentioned, we are focusing on the Let's Play community for further analysis. There are a number of popular LP performers who use TwitchTV to live-stream as well as YouTube to share pre-recorded LPs. A popular LP performer is Felix Kjellberg who goes by the YouTube username *PewDiePie* [21]. Felix's gained popularity through playing horror games and entertaining viewers through his scared reactions to situations in the game *Amnesia*, he has since made LP videos his profession. We have chosen Felix as an illustrative example in our exploratory analysis due to the author's familiarity with his work and his position as the most popular LP performer on YouTube with 6.5 million subscribers and 1.4 billion video views. Although Felix is primarily popular on YouTube through uploaded videos he does perform live-streams and the differences between his uploaded videos and live-streams (in terms of audience participation and recording practices) are not significant enough to separate them.

In the following, we are using Cheung and Huang's [3] research findings as an analyzing framework. With regard to the nine identified personas (as described above) we are only using six for discussing the LP community. Three of the personas (*The Bystander, The Pupil* and *The Crowd*) were considered as less relevant when exploring the Let's Play community because *The Bystander* and *The Crowd* applies to viewing contexts that don't exist for LP performances and *The Pupil* applies to the viewer wanting to improve their skills in a game, which is not an apparent incentive in our study of the LP community. However, the choice taken within this exploratory analysis does not imply that they should be disregarded in future work.

4.1 Player as Commentator

Cheung and Huang's [3] list of personas includes *The Commentator*, this persona is derived from the e-sports commentators that are spectators of an e-sports match but with the special privilege of controlling the "camera angle" the spectators see and commenting on proceedings. In an LP broadcast, there isn't a commentator spectator; *The Commentator* is a part of the LP player/performer. Being the performer and *The Commentator* at the same time gives the performer much more control over how the narrative of the experience develops, this allows an LP to be much more curated than an e-sports broadcast where the player and *The Commentator* can have a different amount of control over the experience and different incentives on how the match will unfold. For example, a commentator may want a very exciting and suspenseful match whereas the player mainly wants to win and progress to the next round, even if that means playing in quite a boring way. This problem does not occur for the LP performer as they are curating the experience in the way they see fit (other than the constraints of the game).

Another interesting difference that arises from *The Commentator* also being the performer is that there is no way for the performer to manipulate how information is released to the player. Cheung and Huang discuss the idea of information asymmetry, describing the effect that the release of information has on spectator entertainment. There are three forms of information asymmetry; where the player holds information the view does not, where the viewer holds information the player does not and where neither of them holds a particular piece of information. In e-sports, good commentators deliberately hide information from the spectator until the most opportune moment as to surprise the spectator. In LPs, as the performer also plays the role of *The Commentator*, they have less control over this form of information asymmetry because the viewer is seeing the same content as the performer. Yet, an LP elevates the "Player Unknown, Viewer Unknown" form of information asymmetry as revealing the unknown is experienced by both the performer and the viewer at the same time.

It is possible to have the "Player Unknown, Viewer Known" form of information asymmetry in an LP but only on a small scale. When a viewer knows what is coming up in a game because they have played it before, they can look forward to when the performer is reaching that point.

4.2 Performer Incentives

In his Nonick 2012 talk [15], Felix discusses how he saw another LP performer playing the horror game Amnesia, which made him want to play too (see *The Inspired* persona). When he began to play he was scared and talking to an imaginary audience, gave him the confidence to carry on playing. Although this is a specific case of why a performer started, it gives us insight into why performing as a player to an audience is more desirable than playing alone. The act of performing to someone changes the way a player plays; in the simple scenario of someone walking into a room where someone is playing a game, the player becomes more of a performer if they know someone is watching: narrating what is happening, what parts of the game are good and what parts are bad, etc.

Cheung and Huang [3] discuss the spectator ecosystem as a way of describing the experience of viewers and players as inter-related; players promise not to disappoint fans and viewers judge players according to sportsmanship. In e-sports, the promise not to disappoint fans comes from the incentive to win games as all e-sports are inherently competitive. This incentive does not exist for the LP performer as there is nothing to "win", but the incentive to not disappoint fans comes more from the performance as a separate layer to the play. An LP performer can be bad at a game (in terms of not being skilled in the way the game requires) but this does not mean they will be judged by viewers, their performance is what is judged by viewers. One way to explain this is that both e-sports and LP broadcasts are meta-games on top of the original game; e-sports players are playing a video game to progress in the "game" of the tournament, whereas LP performers are playing a video game to construct interesting and entertaining situations for the viewers. It is common a performer to play around with one object in a game while commenting for a long time, whereas doing so while playing a game alone isn't.

4.3 Viewer Incentives

Felix began LPing in 2011 and, as previously mentioned, his initial reasons for recording himself playing games while narrating was so that he didn't feel alone while playing horror games that he thought he would be scared by. He imagined a future audience that he was conversing with. This audience began to form a community who started to converse back in the form of YouTube comments and the TwitchTV chat window. Some comments include information about the game by experienced players telling Felix what to do next or interesting parts of the game that he missed. Felix would reply to the community in following videos, mentioning things people had asked him to do in the game and then changing the way he played based on this. This reciprocal nature of the performer's persona as the one driving the content as well as reacting to commenters to alter the content helps build a community around their LP broadcasts.

Here we see *The Assistant* persona in the spectatorship. *The Assistant* finds interest in advising the player on what to do; telling Felix what to do next, where to go next or describing interesting parts he has missed; *The Assistant* has a voice on the experience, altering how the game is played. We see assistants more on TwitchTV broadcasts where the viewers can directly influence the game as it is being played. It is common to hear the performers answering questions from viewers and reacting to their suggestions. *The Assistant* also has an interesting role in information asymmetry, as discussed earlier; a spectator can derive entertainment from knowing what is coming up in a game and directing the performer accordingly.

The Entertained persona prefers watching than playing which is very prevalent in the LP community. This can be explained by the extra content the performers provide through their commentary, it adds to the entertainment viewers would get from playing the game. For example, Felix creates characters and voices for elements of the games. Felix discussed in his Nonick [15] talk how he tries to create a "universe" around the games, which helps create the community of viewers.

The Curious are interested in filling knowledge gaps. In the e-sports community these are knowledge gaps in the game, but in the LP community this can represent knowledge gaps in the game or the LP player's universe. For example, after watching more broadcasts of a game, a viewer may be interested in learning more about the game, but also they may be interested in learning more about Felix and why he always makes jokes about certain objects in the game. Watching broadcasts of his other games will fill these knowledge gaps and make *The Curious* feel more a part of the community.

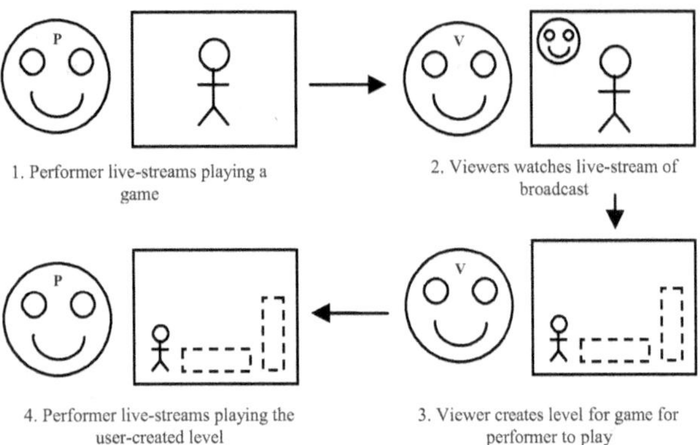

Figure 4. Performer (P) and Viewer (V) co-laboring in play.

There are two more personas that fit the LP viewer community: *The Inspired* and *The Unsatisfied.* These two personas are somewhat opposites in their incentives to watch but both derive enjoyment from watching. *The Inspired* is eager to play a game after seeing it, as the LP performers play a myriad of different types of game for the enjoyment of the viewers and themselves, there is a high chance the viewer will see a game they have not heard of before or in a genre they enjoy. Compared to an e-sports broadcast, the informal nature of an LP broadcast shows *The Inspired* the game is for fun rather than competition and is therefore more inviting to try out. *The Unsatisfied* watches a broadcast as a weaker substitute for playing. As many "Let's Play" games are single-player games or obscure games (chosen for their originality), *The Unsatisfied* may not have the time to invest or the equipment required to play the games, but experiencing the story of the game (along with the performer's added content) as a viewer is more interesting than not experiencing it at all.

4.4 Performer-Viewer Incentives

We have discussed the performer's and viewer's incentives separately so far but there is a third category of incentive which includes both the performer and the viewer. Cheung and Huang [3] discuss the idea of *Co-laboring in Spectatorship* – the fact that spectators work together to create an enjoyable experience. This is described as how *The Commentator* and observer-cameraman (camera operator) shape the information asymmetry for the spectators and how spectators share informal information between themselves to heighten the experience.

In the LP community, this emphasis is much more on *Co-laboring in play* rather than *Co-laboring in Spectatorship.* As previously mentioned, viewers can chat while a performer is playing, telling the performer to do certain things in the game or asking them questions. This act is for the purposes of entertainment; the viewer is heightening their enjoyment, and hopefully others' enjoyment, by interacting with the performer as *The Assistant.*

LP performers play any game they decide will be entertaining (in contrast to e-sports) and there are many games, which allow user-generated content such as user-generated levels or items. This allows viewers to create content for their favorite LP performer, which the performer will then play in the future (see Figure 4). We see this with Felix when he plays the game *Happy Wheels* [14]; some viewers make "PewDiePie Quiz" levels or other levels themed with characters from Felix's LP universe and we then see LPs by Felix where he is playing these levels in *Happy Wheels.* This is a reciprocal process in which a viewer has watched a performer's videos, found that they play a certain game that allows user-generated content, created content for that performer and the performer then creates a video including the viewer's content, this strengthens the LP performer's universe and community. This process is interesting in that it shows the importance of creating a cohesive universe around the different games an LP performer plays and how viewers can be a part of the creation.

Now that we have discussed the LP community using Cheung and Huangs' personas and framework, we will discuss the different platform approaches to designing live-streaming experiences for both the broadcaster and viewer as well as how these insights can be of interest and value for designing interactive television services and experiences.

5. DISCUSSION

Within this paper we aimed to explore the questions around players' motivation for streaming themselves (becoming a broadcaster) and generate an understanding why other people (viewer) would watch the streamed content. Each live-streaming service described above (TwitchTV, OnLive, YouTube Live) approaches the design of the broadcaster experience and viewer experience differently based on the incentives of the stakeholders.

5.1 Live-Streaming Platform Comparison

In the case of YouTube Live, the service has grown out of an existing video sharing service, which is used mainly for uploading pre-recorded videos. As a result, it is relegated somewhat to a "add on" feature rather than an integrated part of the service provided. In contrast to TwitchTV, the ability to discover content, especially video game content in YouTube Live, is not as flexible or contextual and as a result there is less of a community around certain categories or games. This is a problem TwitchTV doesn't seem to have, by designing their content discovery and viewer participation around featured channels and game hubs, users are much more part of a community, with regular viewers recognizable by broadcasters in the chat, and thus giving evidence of repeat viewing.

OnLive's implementation is vastly different to the previous two services, as it seems like a side-effect of the technical architecture of their service. Interestingly, the "always on" streaming when playing a game on the service is influencing the way other services harness content. For example, TwitchTV recently announced a partnership with Activision which builds TwitchTV live-streaming directly into Call of Duty: Black Ops 2 for PC, PS3 and Xbox 360, vastly reducing the number of technical obstacles required to live-stream. In the same way as OnLive, it makes it very simple and thus easily accessible for a broader audience.

All live-streaming platforms have been designed for different ways of interaction, diverse reasons and provide the communities (such as Let's Play, e-sports, speedrunning) different experiences. Based on our discussion of the 'Let's Play' community we also explored the roles of broadcaster and viewer based on Cheung and Huang's framework. Thus we were not only able to reflect on the practices in the streaming communities, but also relate them back to a set of existing types of viewers (personas) revealed from the gaming context. We highlight what and why spectating as well as acting as broadcaster is entertaining and reciprocal some times, such as in the act of co-creating content.

5.2 Value for Interactive Television

As stated in the beginning, the most interesting part of the live-streaming – broadcaster and spectator – phenomenon is its relation to the malleability of activity and passivity; video games are usually enjoyed actively and television is enjoyed passively, but through studying the video game live-streaming community, we can see that these are not concrete concepts and viewers sometimes play active or passive roles. The question now is what we can learn from these insights from the gaming world for interactive television? Several themes can be directly linked to interactive TV activities and efforts, such as engaging viewers in social interaction while watching TV. Below we outline some insights for interactive television, which should be seen as inspiration for discussion as well as stimulation for further explorations of this phenomenon by the iTV community.

5.2.1 Up-Down Reciprocity

Social TV applications (e.g. [5][6]) have been studied increasingly within the iTV community. For instance, Geerts and De Grooff [7] suggested a set of sociability heuristics covering a wide range of social and shared usages around the TV. Still, insights from the video game live-streaming community on live chatting could provide additional food for thought on these heuristics, in particular, with respect to extending the concept of social TV towards the broadcaster, thus breaking down the power relationship between viewer and broadcaster.

5.2.2 Flat and Democratic Practices

User-generated content platforms such as YouTube have stimulated several discussions within the iTV community [8]. Producing and distributing media content is not the exclusive domain of broadcasters and media professionals anymore. End-users are participating in peer-to-peer networks by producing, distributing and consuming content and communities – such as from the gaming context – show that the content is increasingly influenced and co-created by the viewer. The changing and dynamic relationship between player and broadcaster or performer enriches the viewing experience and makes, as Cheung and Huang have shown spectating more entertaining. Although interactive television pushed content creation by viewers, it did not become successful as much as it is in the gaming context. An interesting element seen in our example on 'Let's Play' is the flat and democratic practices, which needs more emphasize in the TV related context in order to be successful. However, more analysis of a variety of cases is need for supporting this assumption.

5.2.3 Learning Through Others

Interactive television has since its beginnings been seen as a potential learning and education platform and medium, especially due to its pervasiveness and embeddedness in an entertainment context. Chorianopoulos and Lekakos [10] stress the point that while commercial broadcasters might financially not be interested in developing educational content, they could integrate existing content from thematic documentary channels or children's channels in interactive television applications, such as add interactive browsing or a game. That highlights another anchor point to the insights gained from the video game community, where the viewers' incentive amongst others lays in learning playing strategies by viewing the live-streamed game play. Enhanced through additional live chats and the ability to even influence the players' next game (giving suggestions on what the player should try to do), the viewer her/himself is receiving a very active and engaging role as part of the audience, and can further inspire design strategies around learning through interactive television services (e.g., [10][8]).

5.2.4 Behavior-based Ads

The exploration of the TwitchTV platform also revealed insights on advertisement strategies, which might be of interest to consider for broadcasters in the interactive television context. For instance, the insights gained by Vennou et al. [11] on evaluating new advertising (ads) formats for digital interactive television could be re-thought and reflected based on the described live-streaming examples, in particular looking beyond formats itself and foster the focus on the viewers' behavior as an active element for positioning ads.

5.2.5 Cross-Platform Support

Finally, the example introduced for the Let's Play community 'Felix' shows clearly a link to the proliferation of multi-platform services (active on TwitchTV and YouTube Live). Beyond cross-platform experiences as explored by Obrist et al. [14] or second screen applications [1], interactive television service providers need to appreciate viewers (or broadcasters) desire to be active on different platforms.

6. FUTURE WORK

We are aware that more analysis beyond our selected examples is needed to get a better understanding on how such live-streaming platforms are changing the game around 'active' and 'passive' roles and how this can be beneficial for inspiring the design around interactive television services and platforms. There are also questions related to the cultural context of video games live-streaming and spectatorship, which need more attention as some genres of games and the ways they are played can be localized, meaning personas that apply to some spectators may not apply to others from a different cultural background.

7. CONCLUSIONS

What we provided in this paper is an overview and reflection on selected topics emerged from the exploration of video game live-streaming platforms and community practices and their relation to interactive TV. Through our exploratory analysis of live-streaming communities, particularly 'Let's Play', we have found that there is malleability between the roles of the active and passive in users' experiences. We cannot treat them as separate and conflicting ideas as live-stream viewers can co-labor in play in the same way a television viewing experience can float between passivity and activity in an interactive television context.

We have highlighted the potential opportunities for inspiring and improving the design spaces for interactive television services and experiences in our discussion and hope to provide stimulation for future work, in particular engaging in a more structured analysis of these two fields by learning from each other's through looking outside of the box.

8. ACKNOWLEDGMENTS

This work is funded by the Arts and Humanities Research Council through the Creative Exchange project, and supported by the EU Marie Curie IEF Action (FP7-PEOPLE-2010-IEF).

9. REFERENCES

[1] Basapur, S., Mandalia, H., Chaysinh, S., Lee, Y., Venkitaraman, N., and Metcalf, C. 2012. FANFEEDS: evaluation of socially generated information feed on second

screen as a TV show companion. *In Proceedings of the 10th European conference on Interactive tv and video* (EuroiTV '12). ACM, New York, NY, USA, 87-96.

[2] Bruns, A. 2009. The user-led disruption: self-(re)broadcasting at Justin.tv and elsewhere. In *Proceedings of the seventh european conference on European interactive television conference* (EuroITV '09). ACM, New York, NY, USA, 87-90.

[3] Cheung, G., and Huang, J. 2011. Starcraft from the stands: understanding the game spectator. In *CHI*, pages 763-772.

[4] Chorianopoulos, K., and Lekakos, G. 2007. Learn and play with interactive TV. *Comput. Entertain.* 5, 2, pages.

[5] Harboe, G., Massey, N., Metcalf, C., Wheatley, D., and Romano, G. 2008. The uses of social television. *Comput. Entertain.* 6, 1, Article 8 (May 2008), 15 pages.

[6] Geerts, D., Cesar, P., and Bulterman, D. 2008. The implications of program genres for the design of social television systems. *In Proceedings of the 1st international conference on Designing interactive user experiences for TV and video* (UXTV '08). ACM, New York, NY, USA, 71-80.

[7] Geerts, D., and De Grooff, D. 2009. Supporting the social uses of television: sociability heuristics for social tv. *In Proceedings of the SIGCHI Conference on Human Factors in Computing Systems* (CHI '09). ACM, New York, NY, USA, 595-604.

[8] Martins, D.S., Oliveira, L.S., and Pimentel, M.G.C. 2010. Designing the user experience in iTV-based interactive learning objects. *In Proceedings of the 28th ACM International Conference on Design of Communication* (SIGDOC '10). ACM, New York, NY, USA, 243-250.

[9] Obrist, M., Miletich, M., Holocher, T., Beck, E., Kepplinger, S., Muzak, P., Bernhaupt, R., and Tscheligi, M. 2009a. Local communities and IPTV: Lessons learned in an early design and development phase. *Comput. Entertain.* 7, 3, Article 44 (September 2009), 21 pages.

[10] Obrist, M., Moser, C., Alliez, D., Holocher. T, and Tscheligi, M. 2009b. Connecting TV & PC: an in-situ field evaluation of an unified electronic program guide concept. *In Proceedings of the seventh european conference on European interactive television conference* (EuroITV '09). ACM, New York, NY, USA, 91-100.

[11] Vennou, P., Mantzari, E., and Lekakos, G. 2011. Evaluating program-embedded advertisement format in interactive digital TV. *In Proceedings of the 9th international interactive conference on Interactive television* (EuroITV '11). ACM, New York, NY, USA, 145-154.

[12] Webster, A. 2013. Don't die: livestreaming turns video game speedruns into a spectator sport. The Verge: http://www.theverge.com/2013/1/21/3900406/video-game-speedruns-as-live-spectator-sport. Retrieved 02/2013.

[13] Web: Electronic Games. Issue 02. Vol 01/02/1982. March. Page 35. http://www.archive.org/stream/electronic-games-magazine-1982-03/Electronic_Games_Issue_02_Vol_01_02_1982_Mar#page/n35/mode/1up. Retrieved 02/2013.

[14] Web: Fancy Force. *Happy Wheels.* http://www.totaljerkface.com/happy_wheels.php. Retrieved 02/2013.

[15] Web: Nonick 2012. *Felix Kjellberg Keynote.* http://www.youtube.com/watch?v=5-iEQD5s97E&playnext=1&list=PL8AD0F546C4CF57E8&feature=results_video. Retrieved 02/2013.

[16] Web: OnGameNet. http://ko.twitch.tv/ongamenet. Retrieved 02/2013.

[17] Web: OnLive. http://www.onlive.com. Retrieved 02/2013.

[18] Web: Something Awful. *Let's Play! - SomethingAwful Forum.* http://forums.somethingawful.com/forumdisplay.php?forumid=191. Retrieved 02/2013.

[19] Web: TwitchTV. http://www.twitch.tv. Retrieved 02/2013.

[20] Web: YouTube Live. http://www.youtube.com/live. Retrieved 02/2013.

[21] Web: YouTube. *PewDiePie Channel.* http://www.youtube.com/pewdiepie. Retrieved 02/2013.

[22] Web: UStream. http://www.ustream.tv/SouthwestFloridaEagleCam. Retrieved 02/2013.

[23] Web: UStream. http://www.ustream.tv/aliceseyes. Retrieved 02/2013.

[24] Web: TwitchTV. http://www.twitch.tv/cyanideepic. Retrieved 04/2013.

TV-related Content Online – A Brief History of the Use of Webplatforms

Pauliina Tuomi
Tampere University of Technology
Pohjoisranta 11 A
P.O.Box 300, 28101 Pori, Finland
+358503609803
pauliina.tuomi@tut.fi

ABSTRACT

Television viewing has fragmented to several platforms and the use of web elicits new uses of TV & audience behaviors. Audiences are combining different mediums and content in order to gather a coherent media experience. In this paper the role of the online websites as a platform for TV-related content is analyzed trough a case study. The Finnish TV channel MTV3's website for British TV program Emmerdale is analyzed trough the linkage and purpose of the webplatform to the audience. It is done trough the webplatform data that covers the Emmerdale's site from the year 2002, 2006, 2010 and the current situation in 2013. By this the development of TV-related content online and the use of webplatform in order to engage the audiences outside the TV broadcast is presented. Also the role of the social media acting as a webplatform is acknowledged in order to tackle the current situation as well as the future position.

Categories and Subject Descriptors

H.5.2 (**Information Interfaces and Presentation**): User Interfaces – evaluation/methodology, user-centered design, graphical user interfaces (GUI)

General Terms

Documentation, Experimentation, Human Factors

Keywords

Participatory television, webplatform, web 2.0, social media

1. INTRODUCTION

Television has always been a social medium. It has also gone trough several interactive phases and tryouts for example via SMS-technology. Nowadays, popular TV programs are multi- and cross- media projects that are not restricted to the television. [6]; [7]; [9] However, already since 1990s interactivity has been one of the most significant features of the future media culture. The growth of media culture's playability is triggered by interactivity; cultural products are no longer just products that are watched, read or listened – they enable active participation as well. [5]

"TV meets the internet" is a global expression that characterizes digital and interactive TV (e.g. [3]) A user can be active in many ways, on various platforms, without breaking the link to a TV program **format**. Audience involvement has widened to include "passive" ways of using a broadcast platform, "active" uses of Web platforms, and in-between phenomena such as voting via mobiles and participating in net meetings [8].

Today's interaction, or rather participation is mainly based on social media services and networks like Facebook & Twitter. Before the coming of social media, television content was shared online trough old fashion websites and later as a part of web 2.0[1] multiplatform formats that combine TV and Web platforms in order to create a wider TV experience. Multiplatform itself as a term, in the computer industries and occasionally in media research, tends to focus on digital technology as providing a technological and design framework for incorporating existing media (e.g., text-based or audiovisual), platforms (e.g., chats, blogs, discussion groups) or software systems (e.g., Linux or Windows, gif or jpeg) [9] Today both of these ways to connect television and internet exist.

2. BACKGROUND OF THE STUDY

2.1 Participatory television before the age of social media

The phase of participatory TV is mainly the era of the coming of Web 2.0. The time when internet changed to be somewhat more social than before and it started to accept and be based of more and more user generated content. This had an impact on television viewing experience as well. Web 2.0 acted as second screen for television. Internet became an archive for all the TV format's other related material and viewers were able to browse trough asynchronous discussions + chats as well, blog texts and dedicated web pages offered by the TV channel. There were a lot of features and activities that served individuals particularly but also lots of user-generated content that enabled asynchronous communication between the viewers. Because of the media convergence there were lots of different intermedial products for sale, which means that the contents were, and still are, available in many mediums. [2] A committed TV format follower can easily dive into the world of different news, extra-material and take part in gallups and shadow votes. [7]

EuroITV'13, June 24-26, 2013, Como, Italy.

[1] The term Web 2.0 was coined by Tim O'Reilly, founder and CEO of O'Reilly Media, Inc The term became better known across the industry after the O'Reilly Media Web 2.0 conference in 2004. In the Web 2.0 model, users actively participate and contribute to a website.

2.2 Research methods and material

This study concentrates on the recent years of television history trough the features online platforms have offered on the coming of web 2.0 and before the break trough of the social media. This era is interesting and especially trough comparison. How has the websites survived after the social media and has their role changed? What are the purposes and functions of TV content related webplatforms and how has the material and engaging hooks changed during the last ten years? There are studies done concerning the convergence between television and internet, but not in focus of comparison (comparing the lice cycle of web platform in particular) and longer timespan. The participatory online features are addressed and analyzed trough The Finnish TV channel MTV3's website for British TV program Emmerdale. [2] Emmerdale is a long-running British soap opera; it has been aired since the year 1972. It is a TV show that offers completely linear material so it does not include any participatory or influencing features to the actual TV content. This is one of reason why Emmerdale was chosen to be the case in this study. The main interest lies purely on the online potential of the TV show and the other features such as SMS voting would make the studied field too broad. In addition these often around different TV- and media spectacles used features are already studied. [6]; [7]; [9]

The Mtv3 websites dedicated to Emmerdale are retrieved from the Internet Archive's *Wayback Machine*, which is a service that enables users to see archived versions of web pages across time. [10] By using the Wayback machine: http://archive.org/web/web.php the archived websites of Mtv3 [11] and Emmerdale from the years 2002, 2006 and 2010 were available for the research and comparison. The timespan, approximately four years, is chosen to address the possible bigger pictures taking place during the development of the site. The four-year span is frequent enough to show the possible differences but also enough not to repeat the similar features year by year. The sites retrieved via Wayback machine are presented in more detail in the Results-section.

The analysis concentrates on the participatory online features and on the dimensions (spatial, temporal, wider knowledge etc.) they have brought to the TV watching experience. This analysis will basically base on defining the each year's online features e.g. extra material, blogs, discussion forums and purposes e.g. additional information, social activity, and advertisement. The results will also indicate what is the current role of social media when compared to the usage of websites earlier. In this study, multidisciplinary qualitative content analysis is used. It is based on a catalog of criteria that relies on preview literary research work followed by an analysis of existing work in this case a website dedicated to Emmerdale TV-series. The research was executed by observing the current state of the linkage of linear TV and Web content.

2.3 The criteria: online supply and purposes

The participatory nature of convergence between TV and internet before the coming of social media was basically based on combining TV and web-content asynchronously together. First, before the web 2.0, there were simple web pages that offered some background information about the TV format, but these were platforms that were updated by the dedicated people, audiences were not able to affect these sites. After that the webpages turned into web platforms where the users control their own data [4].

Core features of a Web platform are services, participative architecture (e.g., uploading videos, comment functionality, votes), availability of social features (e.g., blogs, discussion forums, like / dislike functionality, polls) and may also include video clips [8]. There are different types of linkage between TV and internet content, which can be defined on a thematic or technical level. Especially the purpose of the linkage between Web and TV content tells us what is the additional value to the audiences. See Table 1.

Table 1. The uses and levels of web 2.0 in addition to television

Purpose	Content
1) Additional Information	Background, characters, spoilers, background images, trivia, pictures & video clips
2) Advertisement	Advertising the TV content or its elements / objects or to sell side products related to TV format
3) Interaction with / Participation in content	Chats and discussion forums, blogs, feedback, polls, contests
4) Transmission of content	Video-on-demand stream, online TV

There are multiple ways how webplatforms were engaging the audiences also outside the television broadcasts. Next the nature of participatory era will be presented trough an example.

3. RESULTS: EMMERDALE ONLINE

3.1 The year 2002 – Rich content

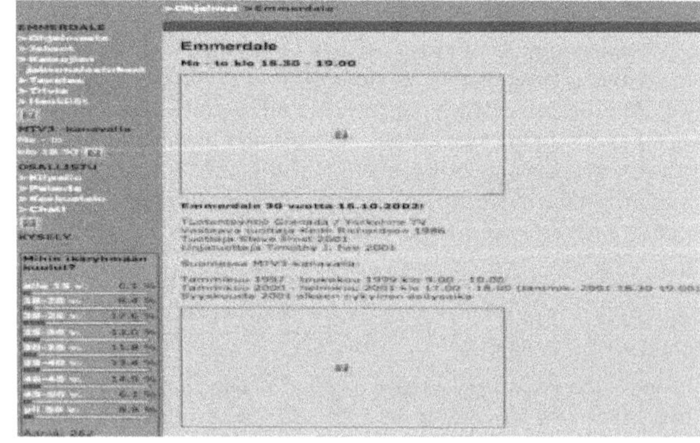

Figure 1 A screenshot from the webplatform of MTV3's Emmerdale from the 2002.

[2] MTV3 is a commercial free-to-air provider of entertainment and information, with its programming founded on news and current affairs, top sports, Finnish entertainment and drama, and international series and movies. The site for Emmerdale is found at: http://www.mtv3.fi/emmerdale/

Levels: 1 & 3

The site in 2002 is visually boring, but the content is surprisingly rich. There are selection of background, episodes, trivia & characters. There is a lot of participatory and social activity offered: contests, discussions and chats, blog and feedback. There is a current poll concerning the age of the Emmerdale viewers on the left. Also the site even offers video clips concerning the storyline concerning the Finnish fans of Emmerdale and even receipts of traditional Emmerdale food that could be served at Woolpack, the main pub in the show. Emmerdale or channel related material is not advertised to be sold. Neither there is an opportunity to watch episodes via online TV-application.

3.2 The year 2006 – Marketing place

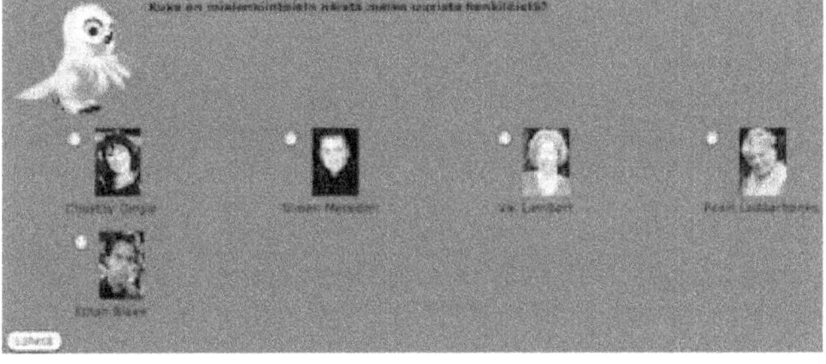

Figure 2 A screenshot from the webplatform of MTV3's Emmerdale from the 2006.

Levels: 1, 2 & 3

In the year 2006 we can see that the website has evolved quite a lot visually, but the actual content offered has not changed that much. The site still offers background information of the characters, episode presentations, spoiler information for those who want to peak the future and an extra photo material as well + the weekly polls that are now addressed by photos. Also the possibility to buy TV channel related objects is now available.

The main emphasizes seems to be on the content concerning the plot, episodes and future twists. There is for example a wide section for spoilers on advertised on the first page. In addition, Mtv3 is advertising its other services like teletext and video clips of next spring's new series. Still, online TV-application is not available.

3.3 The year 2010 – Online TV has landed

Figure 3 A screenshot from the webplatform of MTV3's Emmerdale from the 2010.

Levels: 1, 2, 3 & 4

As we can see, the webplatform is constructed in a way where the upper banner is holding the links to additional information and extra-material. The right side of the site is offering for example polls and shadow votes. From the up left the offering is the following: 1) front page, 2) episodes, 3) characters, 4) trivia & background, 5) the village, 6) background images, 7) Kaarina's (maintainer) blog, 8) discussion, 9) Katsomo – online TV application and 10) feedback.

The site offers additional information such as episode and character information and trivia & background. It also offers a place to viewers create content in discussions and blog-sections + there is a possibility to participate in weekly poll questions and give feedback. The week's poll is about: What is your opinion on Emily's behavior towards Ashley and several answer possibilities are offered. Katsomo, an online TV application, is launched and it offers the transmission of content since viewers are able to watch episodes outside the actual broadcast times. Also the aspects of advertisement and selling of program related merchandise exist. See Figure 4.

Figure 4 MTV3 store is selling an excellent collection of Emmerdale's special episodes on DVD.

3.4 The current situation in 2013 – Social media enters the stage

Figure 5 A screenshot from the webplatform of MTV3's Emmerdale from the 2013.

Levels: 1, 2, 3 & 4

Emmerdale site in the year 2013 seems to offer the same content than in previous years. Sections found on the site include: characters, coming up, get to know the village, online TV 'Katsomo' and feedback. It is clear that the main focus is on the online TV and as a new feature: social media – Mtv3's Facebook page link. The main material – additional information, social activity and episodes & spoilers are diminished and the material is not as rich as it still was in 2002 & 2006. The communication and user generated content seem to have switched to social media.

4. CONCLUSIONS

Television viewing has fragmented to several platforms and the possibilities to connect with TV have expanded greatly during the last 10 years. TV is in major transition where it is evolving from just scheduled broadcast to being always-on, participatory and social. [1] The TV broadcasters and channels struggle with the competition from online viewing, which is why they are eager to strike a balance between TV shows, online episodes and other extra material.

Results indicate that in the year 2002 there were surprisingly rich material on the website. Four years later, in 2006, the visual outlook of the site has improved e.g. the polls are now done trough photos. Also the possibility to buy MTV3 material was added. In 2010 the website got even more visually organized and the rich content the same: lots of extra-material, social services and background material. The Mtv3's Net-TV 'Katsomo' is introduced to people to watch the Emmerdale episodes whenever wanted. The current site in 2013 offers the link to Mtv3's Facebook and the space for the Emmerdale-site is shrunken. Quite similar material still exists, but the updating and most important material & news seems to be delivered preferably via FB-page.

Social media is definitely one of the easiest ways of linking TV and Web. Social media acts the same way web 2.0 platforms have done during the years. If the participatory TV was more asynchronous communication (discussion forums, blog comments), social media brings the real-time based interaction to the scene. Both can be used obviously for social purposes and to advertise online shops affiliated with TV programs and their contents. They can offer additional information and content transmitted from TV, for instance, in the form of video clips. Furthermore, they are places for audiences to influence and to create content for the actual TV broadcasts. Social media in particular gives the viewers the possibility to influence the course and/or characteristics of the TV content (e.g., by a voting mechanism) and to offer viewers the possibility to participate in the show (e.g., by concurrent forum and live TV discussion; the outcome of the e.g. FB-site is introduced to the discussion on TV). [8] Although, not in linear TV shows like Emmerdale. However, various social applications, for example, likes/dislikes, discussions, small-scale voting and polls on social networks such as Facebook, are remnants of the web 2.0 era.

In the future social media will be used for example to contact the audiences and in order to create communication between TV broadcasters, channels and the viewers. Webplatforms will probably remain, but their role may diminish due to the social media services. However, the webplatforms and -sites will remain as archives that hold up information that is too big or problematic to e.g. Facebook or Twitter to handle. Also devices they are supposed to be used on will have an impact. To conclude, there is space for lots of different TV related content online in general - whether it is done by the broadcasters or the viewers themselves.

5. REFERENCES

[1] Hayes, G. 2009. Social IPTV: Interactive and Personal. SPAA 2009. http://www.slideshare.net/hayesg31/social-iptv-interactive-and-personal?from=ss_embed

[2] Hautakangas, M. 2008. Yleisöä kaikki, tuottajia kaikki: toimijuuden neuvotteluja Suomen Big Brotherissa. In Nikunen, K (Ed.) Fanikirja: tutkimuksia nykykulttuurin fani-ilmiöistä. Jyväskylä University, Jyväskylä.

[3] Jensen, J. and Toscan, C. (1999). Interactive Television : TV of the Future or the Future of TV? Aarhus: Aarhus University Press.

[4] O'Reilly, T. (2005). What is Web 2.0? – Design patterns and business models for the next generation software. http://oreilly.com/Web2/archive/what-is-Web-20.html

[5] Parikka, Jussi (2004). Interaktiivisuuden kolme kritiikkiä. *Lähikuva* 2004:2–3, 83–97.

[6] Simmons, N. (2009) "Me TV: Towards changing TV viewimg practices? Euroitv09, June 3-5, 2009, Leuven, Belgium. ACM 978-1-60558-340-2/09(06

[7] Tuomi, P. 2010. The role of the Traditional TV in the Age of Intermedial Media Spectacles. In Proceedings of the 8th international conference on Interactive TV and Video 2010 (EuroiTV '10). ACM, New York, NY, USA, 5- 14.

[8] Tuomi, P & Bachmayer, S. 2011. The Convergence of TV and Web (2.0) in Austria and Finland. In Proceedings of the 9th EuroiTV. Lisbon, Portugal — June 29 - July 01, 2011. ACM New York, NY, USA 2011. Pp. 55-64

[9] Ytreberg, E. 2009. Extended liveness and eventfulness in multiplatform reality formats. New Media & Society, 11(4): 467–485.

[10] The Internet archives, Waybackmachine: http://archive.org/web/web.php

[11] Mtv3 – Emmerdale: http://www.mtv3.fi/emmerdale

There's a World outside Your TV: Exploring Interactions beyond the Physical TV Screen

Radu-Daniel Vatavu
University Stefan cel Mare of Suceava
13 Universitatii, 720229 Suceava
vatavu@eed.usv.ro

ABSTRACT

We explore interactions in the space surrounding the TV set, and use this space as a canvas to display additional TV content in the form of projected screens and customized controls and widgets. We present implementation details for a prototype that creates a hybrid, physical-projected, augmented TV space with off-the-shelf, low-cost technical equipment. The results of a participatory design study are reported to inform development of interaction techniques for multimedia content that spans the physical-projected TV space. We report a set of commands for twelve frequently-used TV tasks and compile guidelines for designing interactions for augmented TV spaces. With this work, we plan to make TV interface designers aware of the many opportunities offered by the space around the physical TV screen once turned interactive. It is our hope that this very first investigation of the interactive potential of the area surrounding the TV set will encourage new explorations of augmented TV spaces, and will foster the design of new applications and new entertainment experiences for our interactive, smart TV spaces of the future.

Categories and Subject Descriptors

H.5.1 [Multimedia Information Systems]: Artificial, augmented, and virtual realities; H.5.2 [Information interfaces and presentation]: User Interfaces. Input devices and strategies (e.g., mouse, touchscreen), Interaction styles (e.g., commands, menus, forms, direct manipulation).

Keywords

Interactive TV, multiple TV screens, augmented TV spaces, video projection, around TV, Wii, participatory design study, guidelines.

1. INTRODUCTION

Today's smart TVs are becoming highly interactive by leveraging computer-like operating systems and applications that deliver their users enriched functionality beyond the passive TV watching paradigm. Similar to a standard computer, the TV is used for web browsing, social networking, and sharing data with personal devices, in the attempt to provide users with the ability to create, edit, share, and control content [13]. However, the features of today's smart TVs struggle to meet their users' ever increasing demands for more content, information, and control, and, were lacking or incomplete, are substituted by users with applications on personal devices transposed into secondary screens [14,15].

EuroITV'13, June 24-26, 2013, Como, Italy.

Such auxiliary displays complement the primary TV screen that is either limited in functionality and control or simply cannot display more content without occluding and, therefore, affecting the experience of watching the primary TV transmission. For example, browsing the web on the TV screen cannot be done without ruining the full-screen experience of the broadcast; and watching two channels simultaneously is not possible without constraining either one to the reduced form factor of a picture-in-picture display. Such limitations come from the fact that the TV is essentially a single-screen computer, which constrains content to share the real-estate of the same display [43].

To overcome such limitations, we explore the space around the TV set that we turn into an interactive canvas and use it to display content that would otherwise occlude the main broadcast. To this end, we developed the AROUND-TV prototype that creates an extended augmented-TV space by employing off-the-shelf, low-cost equipments, such as a video projector and a remote controller with pointing functionality. The TV screen is extended to its surrounding area by means of video projections that disclose new, available space to support novel TV interactions: for example, the content displayed by the TV screen can be "dragged out" in the area surrounding the TV set, leaving the physical screen available to display new content; a projected screen can be dragged inside the boundaries of the physical TV set and thus become the primary transmission. Figure 1 illustrates an artist's vision of the concept. Towards this goal, we use previous research results from augmented reality that employ video projections to digitally enrich home and work spaces [16,43,45]. Also, we rely on the results of other researchers that experimented interactions beyond the standard form factors of today's computing devices, such as personal computers [27], mobile phones [10], and interactive tabletops [22,46]. However, interactions that span the physical area of the TV screen have not been investigated up to now, despite their potential to enrich the user experience and extend today's standards for home entertainment design [11,39].

This paper contributes to the little researched area of augmented home entertainment environments, and delivers the first investigation of the interactive potential of the area around the physical TV screen. Our contributions are as follows: (1) we discuss design guidelines for augmented-TV spaces; (2) present implementation details of a prototype that creates the hybrid, physical-projected augmented-TV space; and (3) we propose interaction techniques for multimedia content that spans the physical-projected TV space, as informed by the results of a participatory design study. With this work, we hope to encourage TV interface designers and practitioners to explore the many opportunities offered by the (previously non-interactive) space around the TV set and, therefore, foster development of new interactive TV applications that would live "outside the box" and enrich our future home entertainment experience.

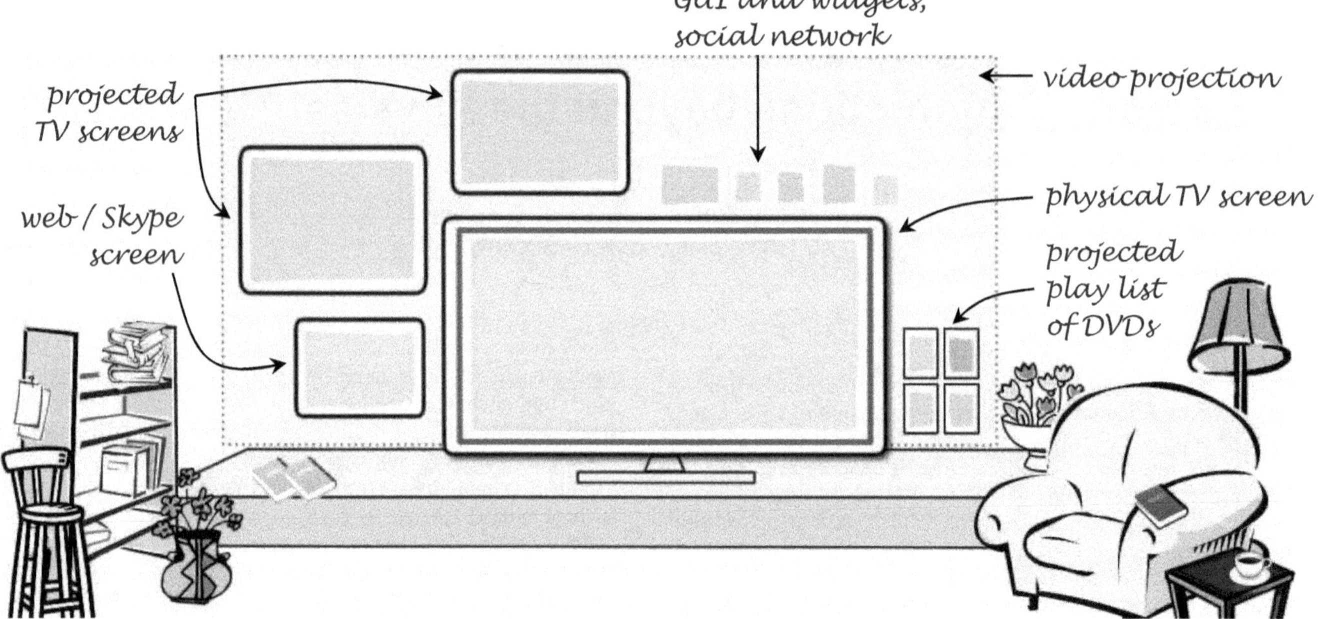

Figure 1. Artist's vision of the AROUND-TV concept. The physical area of the TV screen is extended in the surrounding ambient by employing a video projector that displays supplementary content in the form of additional TV screens, widgets and controls, and application data from various sources (e.g., web and social networks). The hybrid physical-projected TV space enables novel interactions that go beyond the physical form factors of today's TV screens (e.g., dragging content outside the physical area of the TV screen).

2. RELATED WORK

We relate in this section to previous work that investigates the use of secondary screens during television watching and discuss multiple-display TVs; review augmented reality systems that employ video projections to digitally enrich the home entertainment experience; discuss new interaction techniques for novel TV environments that employ gestures; and relate to previous work on designing interactions that go beyond the standard form factors of today's computing devices.

2.1 Secondary screens

Today's users' increasing demands for more content, feedback, and sophisticated control during television watching are not matched by single-screen TV sets, which makes users recur to employing additional devices. The recent availability and affordability of personal smart devices has led to new usage patterns in conjunction with television watching, with mobile devices turning into secondary screens [4,14,15]. In fact, a recent mobile market study revealed the high prevalence of such cross-device multi-tasking practice, "*with 42 percent of tablet owners using them* [mobile devices] *daily while watching television*" [35]. In a comprehensive effort, Cesar, Bulterman, and Jansen reviewed previous works on secondary screens and identified four main motivations for their usage in an interactive television environment, which are control, enrich, share, and transfer of television content [14]. Courtois and D'heer revealed multiple patterns for consuming media from various sources including television, personal devices, and print media [15]; they found that tablet devices are mostly used to support content search and social networking during television watching. Other applications of secondary screens include accessing electronic program guides [17], consuming secondary content synchronized with the main transmission [2,4], participation and voting [25], web surfing and social networking [15], and providing more sophisticate input modalities in order to compensate the shortcomings of today's TV remote controls [7].

2.2 Augmented reality home entertainment

Researchers have experimented before augmentation of work and home environments by video projecting digital content over the real space in order to enrich the user experience for various application scenarios. For example, Cotting and Gross explored freeform displays projected onto tabletops, and investigated applications for collaborative design, delivering presentations, and conducting activities in large-scale work environments [16] Vermeulen et al. used video projectors to display status visualizations for an ambient system in order to provide appropriate feedback to its users in the form of animated physical connections between sensors and actuators [45]. Other works experimented with personal video projectors that users can employ to augment the real space with their own personal content. One example is Cao, Forlines, and Balakrishnan, which investigated interaction techniques for hand-held projectors to support co-located collaboration; they report applications in gaming, formal group meetings, and casual communications [12]. Going further, other researchers explored on-body projections by using ceiling-mounted [20] or wearable video projectors [21,33]. Motorized platforms were implemented to project three-dimensional realistic content at any point of a room in the form of steerable displays [50].

Although most augmented reality designs for home entertainment focus on the opportunities that the technology has for gaming applications [26,41,47], some works have addressed augmenting the interactive TV. One example is the Ambilight technology[1] used by Phillips Smart TVs, which is a lighting effects system that generates colored light around the TV screen that is consistent with the color scheme of the content played on TV. amBX[2] is another example of lighting technology that delivers sensory surround entertainment experiences for computer games and TV

[1] http://www.research.philips.com/technologies/ambilight/index.html

[2] amBX, http://www.ambx.com/showcase/ambx-home

entertainment. And, the recent IllumiRoom prototype [26] enhances the gaming experience on Microsoft Xbox consoles by projecting visualizations of game elements and extended game views on the area surrounding the TV screen. However, neither Ambilight nor IllumiRoom are interactive, which limits these installations to only provide output. The Point & Click prototype [43] augments the real space with multiple interactive screens displayed on the walls and furniture of the living room, and allows users to control each screen independently with WIMP-like interaction metaphors; however, the system does not support intertwined usage of digital and physical screens.

2.3 Interactions beyond the form factors of today's computing devices

The standard form factors of today's computing devices (i.e., laptops, tablets, and smart phones) have often proved limited in terms of the functionality delivered to the user. Therefore, researchers have started to explore interactions that take place in the surrounding area of these devices, such as around [10,27,38], above [6,22,32], behind [5], or under the device [46]. Such techniques are meant to improve interactions with devices for which the small input area is impractical to use (e.g., the small touch-screen area of some mobile devices [10]), and for meeting application requirements that are not fit for the form factor of the device they are running on, such as three-dimensional data projected onto the two-dimensional surface of a tabletop [22].

Target devices that have been enhanced with interactions that span their surrounding area include PCs and laptops [27,53], mobile phones [10,38], and interactive tabletops [22,32,46]. For example, Bonfire [27] is a hybrid laptop-tabletop system that allows users to interact with content projected to the left and right sides of the laptop. The system employs two micro-projectors and two cameras mounted on the laptop together with mirrors that reflect the projected images to the two sides of the keyboard. SideSight [10] is a system meant to enhance interactions for devices with small touch input area by leveraging optical sensors positioned along each side of a mobile phone that sense fingers moving around the device. Hilliges et al. implemented in-the-air interactions above the surface of a tabletop [22], and Wigdor et al. investigated techniques for interacting under-the-table for touch-screen tabletop surfaces [46]. In a more unifying vision, Wilson and Benko designed LightSpace [49], which is a room with multiple depth cameras and video projectors that allows users to interact by touch with un-instrumented surfaces and to perform mid-air interactions between and around the projected displays.

2.4 Novel interfaces for the interactive TV

The new entertainment experience offered with the use of today's advanced processing and visualization technologies is in demand of appropriate interaction techniques that would provide users with the perception of powerful, natural, and fluid control. Following these requirements, gestural interfaces have been explored for home entertainment applications [3,8,43,44]. Such interfaces usually fall into two categories, according to the capture technology they rely on: embedding video cameras into the ambient or inside the TV to capture and recognize free-hand gestures [8,18,44], and using smart phones and remote controls augmented with acceleration sensors that capture motion [3,28,43]. Gestural interfaces can be customized so that users can define their own gesture commands for the various functions in the application, without being constrained by the rigid design of the remote control. However, such interfaces are only powerful where they are properly designed to meet their users' perceptions of natural, undemanding, and intuitive interaction [36].

2.5 Summary

Augmented home environments can offer enriched experiences for users in demand of more content and control. This observation has made researchers explore design options that range from the use of secondary devices [4,14,15], to multiple screens [43], and large augmented spaces [26,49]. For such new environments, researchers have designed natural interaction techniques that employ gestures [44,49], and explored the surrounding areas of today's computing devices to deliver increased interactivity for these devices [10,22,27]. However, there is no previous attempt to augment the space around the physical TV screen to enrich the experience of watching television. In this work we fill this gap by presenting the AROUND-TV prototype, a system that allows users to watch and interact with television content that spans beyond the physical limits of the TV screen. In doing so, we hope to encourage TV interface designers to start exploring the opportunities offered by interacting "outside the TV box".

3. THE AROUND-TV PROTOTYPE

This section describes implementation details of the AROUND-TV prototype, which reuses off-the-shelf, low-cost equipments that allow easy replication and installation of the system. We describe implemented functionality in accordance with several design principles that we set for such new augmented-TV environments.

3.1 Design principles

We start by identifying a set of design requirements for the new augmented-TV spaces; towards this goal, we follow previous works that offer guidelines to support the design of the home entertainment environment. These previous works report practical guidelines for visualizing information on remote TV screens and for providing appropriate forms of feedback [24]; guidelines for the creative design of TV interfaces, such as promoting new input modalities to deliver natural and intuitive interactions, designing shareable interfaces, and incorporating the dimension of "fun" into the interface [42]; principles for multi-screen environments, such as sophisticated control, configuration of individual displays, and design of appropriate assistance and feedback mechanisms [43]; current trends and practices of the industry of smart TVs and home entertainment (e.g., the design of the LG Magic Remote[3] and the Philips pointing devices for smart TVs[4]); as well as the elements of the edit–share–control taxonomy taking over the traditional produce–deliver–consume paradigm [13]. By following this literature on designing for the interactive TV, we were able to compile a set of design principles for the new augmented-TV environments, which can be grouped into six categories that address three inter-connected elements: the user, the home environment, and personal devices:

(1) Context-dependent augmentation of home entertainment. This principle addresses the fixed and rigid form factors of today's TV screens that offer limited options for installation (e.g., once mounted on the wall, moving the TV set to another location requires considerable effort and/or professional assistance) and no option for customization (e.g., one cannot increase the size of their TV screen without buying a new one). Addressing such aspects obviously cannot be done by acting on the form factor of

---

[3] LG Magic Remote, http://www.lg.com/global/magicremote/

[4] Philips. Pointing devices for smart TVs, http://www.uwand.com/smarttv/pionting-devices.html

the screen itself, but instead by virtually extending the television content outside the boundaries of the physical screen. We call this process "augmentation of the TV-centered home entertainment" that takes place at two levels: (a) the physical space around the TV screen is augmented by video projecting content in its surrounding area with the technologies of augmented reality [41,43]; and (b) the user experience of home entertainment is augmented by a system delivering more content and potentially more options for implementing lean-forward vs. lean-backward consumption of television content. This augmentation process needs to be context-dependent, where by context we understand: (a) the constraints of the physical space (e.g., the available free space for video projection); and (b) the personal preferences of each user in terms of displayed content, which make up a personal mixed-reality space [37].

(2) Multi-tasking home entertainment spaces. Augmented TV environments should be able to deliver content from multiple sources and providers (e.g., cable TV providers, local media, web content) in order to meet the demands of today's users in terms of more content, information, and socialization options. With such features lacking or incomplete in today's smart TVs, users recur to auxiliary solutions such as employing secondary screens [35]

(3) Sophisticated control. The individual components of an augmented-TV environment (e.g., TV screens, web content delivery screens, widgets and controls) should be customizable in terms of location, size, and content. For example, projected TV screens should offer flexibility not available with physical screens, such as adjustments of position, size, and aspect ratio of their display areas (e.g., a common scenario could be larger screens displaying more prioritary content and smaller screens showing low-priority transmissions; 16:9 and 4:3 aspect ratios could co-exist within the same TV space; Z-depth could be exploited when managing multiple screens, as a function of the importance of their displayed content). The augmented TV space should be customizable according to the user's preferences.

(4) Natural forms of interaction and gradual transition to new input modalities. This principle addresses the new technologies recently adopted by the TV industry for interfacing the interactive TV, such as voice and gesture commands. We argue that such natural input modalities are necessary for the new augmented-TV spaces that superimpose on the real environment. This actual superimposition of digital content (that needs to be controlled) over the real ambient (that is addressed naturally by making use of speech and gestures, e.g. "give me the red book that is over there [pointing]") represents the driving factor that will foster the adoption of these novel interfaces. However, although speech and gestures are promising means of interaction and have been researched extensively (actually starting with the 80's put-that-there system of Bolt [9]), today's practice does not necessarily meet the users' expectations of natural interaction [30,36]. As a consequence, a form of transition is needed to gradually shift from the old-fashioned TV remote control [7] to natural interfaces. Such options are available today in the form of remote controls with motion sensing and pointing functionalities, but they lack proper interaction techniques. Augmented-TV spaces should therefore allow gradual transition to natural input modalities, in the ultimate goal to attain the same naturalness of addressability for digital content that real environments posses today.

(5) Seamless integration with personal devices. The recent "digital revolution" [35] fostered by the wide availability of personal devices and high consumption of internet content, has created strong connections between users and their personal devices, with a consequence being the high frequency of 42% of tablet owners using their tablets daily while watching television. Augmented-TV spaces should offer such integration possibility, not necessarily in the form of secondary screen usage [14,35], but rather by making the personal device appear as a "box" of personal content, ready to be transferred and consumed into the augmented environment.

(6) Scalability to more viewers. This principle addresses the limited shareability of today's remote controls that implement "one viewer at a time" interaction. The larger scope offered by an augmented-TV space in terms of projected area should be complemented with corresponding multi-user control. One immediate application example is gaming for multiple players, but opportunities for social TV watching [19] and support for co-located collaboration [31,52] can also be envisaged.

3.2 Implementation

We implemented the AROUND-TV prototype using a desktop PC (2.4GHz CPU, running Windows), a standard video projector (Dell 3400MP), and a TV set (Sony BRAVIA, 40 inch/102 cm). We connected the video projector to one of the video outputs of the computer, and set it to function as the primary display. The TV screen was connected to the PC and set as the second display to which the Windows desktop was extended. During system installation, we tried to match the resolutions delivered by the two displays as best as possible in order to create smooth effects for the content transitioning between them. For example, the video projector delivered a maximum resolution of 1400×1050 pixels and, although the TV set was able to provide full-HD quality images, we reduced the resolution of the TV to a lower 1600×900 value, which best matched that of the projector. During tests, the difference between the two resolutions was negligible. The PC ran a software application that used YouTube videos and live-streaming TV to simulate television broadcasts. The Wii controller[5] was used to implement a TV remote control with pointing functionality. The controller was chosen due to the many input modalities it offers, its low-cost, and wide adoption in the research and practitioners communities [23,29,43]. The remote reports button presses (12 buttons), acceleration on three axes, and tracks the position of up to four infrared light sources at a resolution of 1024×768 pixels and 100 Hz. (see Figure 2c) The Wii remote controller was connected to the PC via Bluetooth and we read its reported data with the WiimoteLib library[6].

The installation was set up in a room of $3.5 \times 4 = 14$ m^2, with the TV and video projector installed at the maximum available distance in order to maximize the area of the projection (which was 2×1.5 m^2). The installation, as well as the standard equipments that we used, were purposely adopted to simulate a real-world scenario, replicable in many homes. As a practical note, short-throw projectors are available and deliver much larger projections from much shorter distances than the equipments we used for our prototyping.

Prior to actual usage, the system must go through a calibration procedure, in which the area of the physical TV screen is defined, as a minimum setting. This action is performed only once during installation and consists in drawing a rectangle on top of the video projection by pointing with the remote control and holding the "A" button pressed, similar to mouse drawing (see Figure 2c).

---

[5] http://wii.com

[6] http://wiimotelib.codeplex.com/

(a)

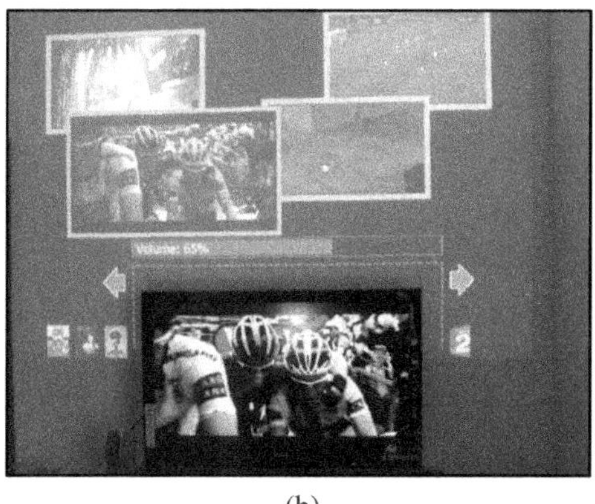

(b)

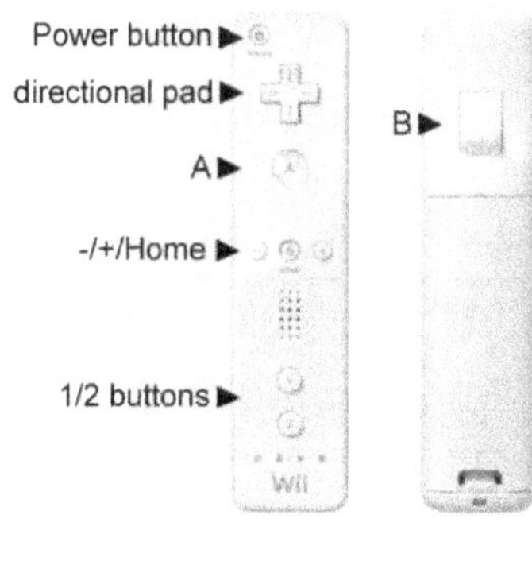

(c)

Figure 2. Snapshots of the AROUND-TV prototype: (a) multiple projected TV screens playing in the area surrounding the TV set; (b) mixed display of TV screens, widgets, and controls. The dotted line visible in (a) and (b) is the result of the calibration step that informs the system of the location of the physical TV screen. The Wii controller (c) was used as input device, as it resembles a TV remote and features many input modalities.

The location and size of the rectangle are stored and associated to the physical location and size of the TV screen. This calibration phase is needed in order for the system to "know" where the physical TV is located so that interactions that exchange content between the physical TV and the projected area can detect user actions accordingly. Other calibration steps were implemented to set up the area of the DVD player and specify regions inside the video projection that the user wants disabled from being interactive; however, such calibration steps are optional.

The AROUND-TV prototype implements some of the design guidelines we set in the previous section, leaving the others as future work; however, the current implementation successfully demonstrates the augmented-TV concept by showcasing:

(1) Creation of new TV screens in the projected space surrounding the area of the physical TV set. The screens are controlled independently in terms of position, size, and displayed content.
(2) Content transfer between physical and projected TVs: the content displayed by one screen is transferred to another. This implementation allows the TV transmission to overpass its boundaries and "escape" physical constraints.
(3) Display of WIMP controls for the physical TV set, such as shortcut icons for preferred channels, volume and channel control widgets.
(4) Display of auxiliary content related to the TV transmission, such as movie subtitles and scrolling news, without occluding the main content.

Figures 2a and 2b illustrate snapshots of the running prototype.

4. INTERACTION TECHNIQUES

By following the principle of designing natural interfaces for augmented-TV spaces, we focus on interaction techniques that take advantage of people's ability to use simple pointing gestures (i.e., in the tradition of "put-that-there" [9]) and, at the same time, reuse already acquired experience from working with windows on PC operating systems. In this attempt, we rely on a simple analogy between windows from PC operating systems and TV screens in the augmented-TV space. By doing so, we hope to provide a smooth transition from WIMP-like interaction metaphors (which seem to be deeply integrated into the users' models of interaction with a computing system [51]) towards new gestural interfaces [3,44]. For this purpose, we employ the Nintendo Wii Remote, which is essentially a remote controller with pointing and motion sensing functionalities. We rely on previous work that evaluated the remote for pointing tasks and gesture recognition [34,40], and explored it for prototyping novel interactions [29] and new TV gestural interfaces [3,43].

4.1 User elicitation experiment

In order to inform the design of commands for working with TV screens projected into the ambient, we employed a participatory design study in which commands were elicited from users. Such methodology has been successfully used before to compile high-agreement gesture sets for interactive tabletops [51], controlling devices in a smart home [28], and informing free-hand gesture design for controlling the TV set [44]. In the following, we employ the methodology [51] to elicit commands for projected TV screens that can be controlled with a hand-held motion sensing device with pointing functionality (the Wii remote).

Participants

Twenty volunteers (twelve females) participated in the study (mean age 27.4 years, $SD = 7.6$). Participants were of different technical backgrounds, such as nontechnical (7 participants), technical at student level (12), and technical at research level (1). Seven participants had played games on the Nintendo Wii console before, but witnessed little experience with the controller. All participants were right-handed.

Referents

Twelve commands were selected for the study (see Table 1). By following the terminology of Wobbrock et al. [51], commands are denoted in the following as referents. The set of referents covers: (1) screen manipulation tasks, such as creation of a new screen and positioning and resize of existing screens; and (2) content-related tasks, such as channel navigation, volume control, and accessing extra options through menus. As in a previous elicitation study for TV commands [44], we opted for a minimal set of referents, although many others could have obviously been included. One reason for working with this minimal set of referents is that people actually use few buttons on their remote controls [7]. This observation made short-sized versions of remotes a viable option, such as the case of Tek Pal Simple Remote[7] or Flipper[8] that feature six buttons only (i.e., power,

[7] TekPal Remote, http://www.bigbuttonremotes.com/remotes-tekpal.htm

[8] Flipper Remote, http://www.flipperremote.com/

channel +/-, volume +/-, mute; all included in our set of referents). A second reason is that the number of gesture commands people can remember for effective use is limited, while other commands can be grouped into menus [3]; consequently, we considered for our set commands for working with menus.

Table 1. Referents used for the elicitation experiment.

REFERENT		NOTES
1.	CREATE	Create a new TV screen
2.	SELECT	Select a specific TV screen
3.	CLOSE	Close the TV screen
4.	MOVE	Move the screen to a new location
5.	RESIZE	Resize the bounding box of the TV screen
6.	NEXT	Go to next channel
7.	PREVIOUS	Go to previous channel
8.	VOLUME +	Volume control
9.	VOLUME -	
10.	MUTE	
11.	OPEN MENU	Pop up contextual menu
12.	HIDE MENU	Hide contextual menu

Procedure

The study consisted in twelve trials for each participant. Each trial was presented with a short text description (e.g., "move the screen to a new location") and a video demonstration depicting the effect of the referent. After participants confirmed they understood the effect, they were asked to suggest a suitable command that would trigger the effect they had witnessed. Participants were allowed to use any of the input modalities offered by the remote controller when proposing commands; these included buttons, combination of buttons, variations on button press (such as long or quick press, double press, etc.), pointing and motion gestures, and combinations between all of these. Participants were also encouraged to think aloud while searching for the best command to match the task. No visual feedback was provided during the gesture elicitation phase. The order of referents was randomized across participants.

4.2 Results

We report results from 20 (participants) × 12 (referents) = 240 proposals for TV commands. Following the steps of the elicitation methodology [51], we grouped similar commands and calculated agreement rates for each referent:

$$A_r = \sum_{P_i \subseteq P_r} \left(\frac{|P_i|}{|P_r|}\right)^2$$

where P_r represents the set of all gestures proposed for referent r and P_i represent groups containing similar commands. The agreement rates A_r range from $|P_r|^{-1}$ denoting no agreement at all to 1 reflecting perfect agreement. During the procedure of grouping similar commands, we had to adopt a guiding rule on the use of buttons. As the remote presents twelve distinct buttons, participants disposed of a large variety of choices, had they decided to use a button for a given command. Therefore, we considered the action of a button press on the remote identical irrespective of the specific button that was used, as long as the conceptual model was the same. For example, we considered agreement being reached for the SELECT SCREEN referent if two participants had pressed either "A", "B", or any other button on the remote while pointing to the target screen (as the model of the interaction is the same: point and click). At the same time, we treated differently variations on button press, as they relate to different mental models for the task. For example, pressing a button twice or holding it pressed for a long time are considered different from a single button press.

The mean agreement was .53 (SD = .14). Individual agreement rates for each task are shown in Figure 3. The referents with the highest agreement were MOVE SCREEN, VOLUME DOWN, and VOLUME UP (.66), followed closely by NEXT and PREVIOUS (.63). The lowest agreement value was found for HIDE MENU and MUTE (with values below .30). 75% of the referents had agreement rates above .50, which shows a high level of user consensus[9].

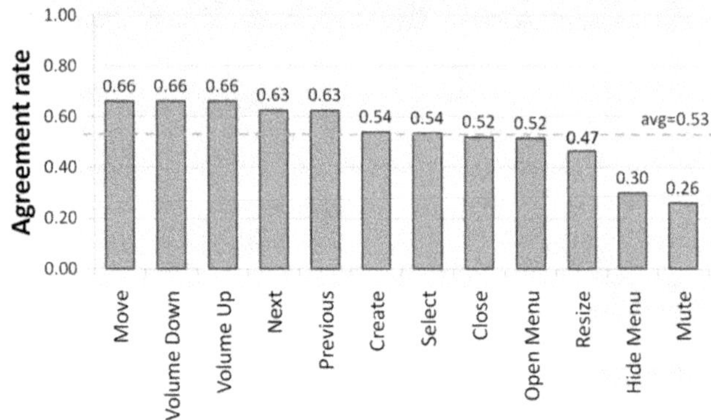

Figure 3. Agreement rates for user-defined commands, shown in descending order. Mean agreement was .53 for the entire set.

4.3 A set of commands for interacting with TV screens in the augmented-TV space

The agreement results show good levels of consensus between participants, which recommends compilation of a set of representative commands for the proposed referents. The motivation of this command set is to inform the design of augmented-TV interactions, as well as to serve as inspiration for designers and practitioners developing TV interfaces that would include similar referents. We adopted the majority-vote rule when selecting representative commands for each referent, according to which the command proposed by most participants wins the referent [44,51]. Figure 4 illustrates the set.

Screen selection was mostly performed by pointing at the target screen followed by a single button press (the "A" button on the remote, conveniently placed to be easily reached with the thumb finger). A new screen was created by participants by drawing a rectangle shape while holding a button pressed (again, the "A" button of the remote was the most occurring suggestion). Closing a screen was done with the help of a button (the "Power" button of the remote was indicated by half of the participants). MOVE and RESIZE were carried out with drag operations while holding a button pressed ("A"). For RESIZE, half participants dragged the corners of the screen, while the other half suggested the use of the directional keys on the remote. VOLUME UP and VOLUME DOWN operations were performed using the "+" and "-" buttons with high confidence (they were suggested by 16 out of 20 participants).

[9] To illustrate the agreement rate calculation with an example, an agreement of .50 can be obtained with two groups of 10 participants, $(10/20)^2 + (10/20)^2 = .50$, or with two groups of 12 and 8 participants respectively: $(12/20)^2 + (8/20)^2 = .52$, both revealing a high level of user consensus when proposing commands.

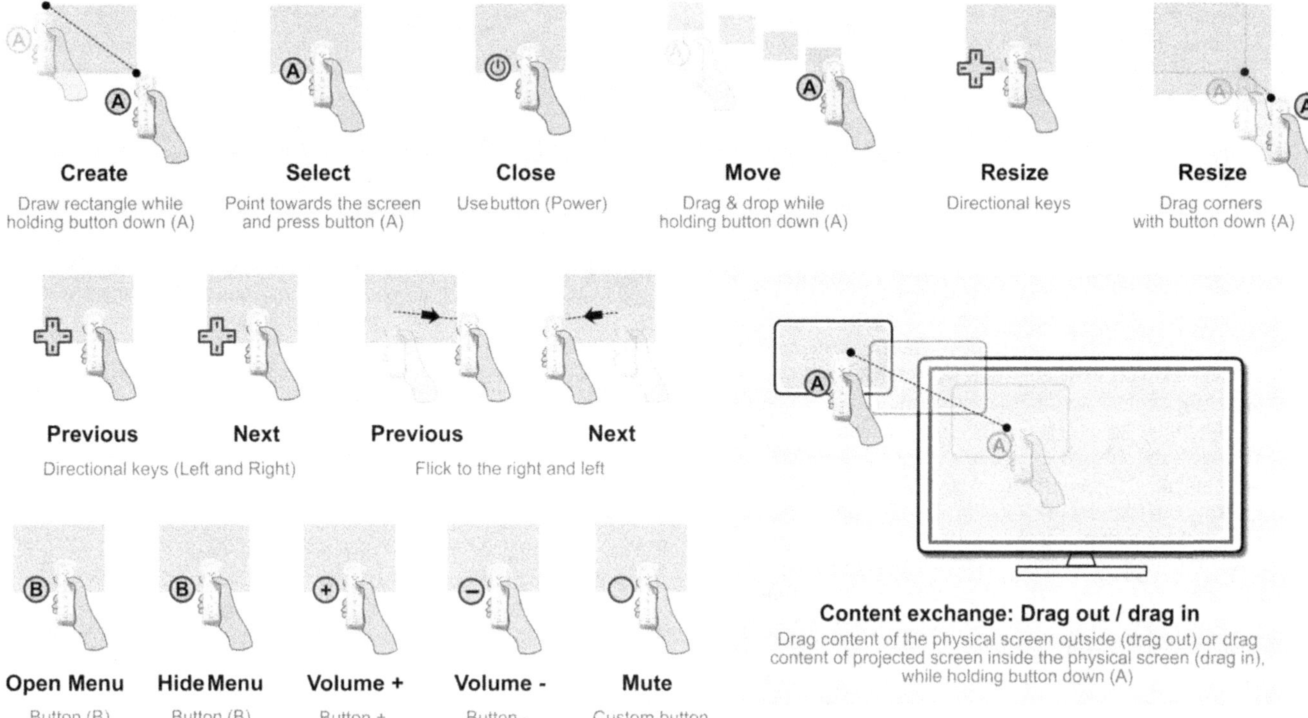

Figure 4. A set of commands for interacting within augmented-TV environments compiled from the results of a participatory design study: working with individual screens (top and left) and interacting between physical and projected displays (bottom-right).

The same conceptual model was found for PREVIOUS and NEXT, for which the majority of participants used the "right" and "left" directional buttons of the device (15/20), while the rest performed motions towards left and right. Low agreement was found for VOLUME MUTE (.26), and some participants felt the need of a special "mute" button. Most participants agreed on using buttons for opening (14/20) and hiding a contextual menu (10/20).

4.4 Interactions between physical and projected TV screens

The commands discussed so far only address individual screens; however, the hybrid nature of the augmented-TV environment that includes both projected displays and a physical screen offers the opportunity to design interactions that span between the two. The current implementation of the AROUND-TV prototype includes a content exchange operation, in which the content of the physical TV screen can be "dragged out" to create a new projected screen showing the same content. Reversely, the content of a projected TV screen can be "dragged in" inside the physical premises of the primary TV screen, which takes over that channel's transmission (see Figure 4, bottom-right, for a visual representation of the DRAG-OUT operation). Both operations implement the drag & drop interaction metaphor, which was observed during the elicitation study when moving individual screens, but extend it over the hybrid space of physical-projected displays. DRAG-OUT is another way to create a projected display that clones the transmission broadcasted on the physical TV screen. DRAG-OUT ultimately acts as a "save" operation of the current channel, leaving the main TV screen available to display other content. DRAG-IN is another way to maximize the content of a projected display to the whole area of the primary physical TV screen. The following Discussion section contains more pointers on how to further develop such hybrid interactions that span across physical and projected displays.

5. DISCUSSION

We discuss in this section additional findings revealed by the participatory design study, which are reflective of user behavior when interacting in augmented-TV environments. We translate these findings into practical guidelines in the hope they might prove useful to other practitioners working in similar augmented-TV spaces. We also present new opportunities to further develop interactions around the TV set and point to future work.

5.1 Guidelines for designing interactions for augmented-TV spaces

The participatory design study revealed a few interesting findings that we summarize in this section, as they reflect general user behavior that could potentially inform the design of new, but similar commands:

(1) **Symmetry for dichotomous tasks**. One of the findings that can be turned into a practical guideline is participants proposing similar commands for dichotomous operations: VOLUME UP and VOLUME DOWN, PREVIOUS and NEXT, and OPEN and HIDE MENU (e.g., VOLUME UP and VOLUME DOWN were performed using the "+" and "-" buttons of the remote). This preference has also been observed in other elicitation studies [44,51], and we confirm it for the interactive TV scenario. New tasks that present opposed meanings could benefit of similar symmetrical commands (e.g., play and pause, cut and paste, etc.).

(2) **Influence of the Windows paradigm.** Similar to Wobbrock et al. [51] that explored surface gestures, we found our participants thinking in terms of windows and widgets present in windows operating systems. For example, participants suggested drawing screen corners for performing RESIZE, used drag & drop operations for moving TV screens, and some imagined an "X" icon in the top-right corner of the

screens for CLOSE. As mentioned in the paper, reusing acquired experience with WIMP metaphors may be beneficial for multiple-screen augmented-TV spaces, for which the analogy window – TV screen seems reasonable and natural. New tasks may benefit of designs that replicate similar windows-like interaction metaphors.

(3) **Preference for button use over motion gestures**. When participants had to choose commands implemented with either buttons or motion gestures, buttons were preferred: 13 out of the 15 commands selected for our set use buttons (87%, see Figure 4), and 10 commands (67%) use buttons exclusively. Abstract tasks such as MUTE, OPEN, and HIDE MENU were all assigned buttons. The first intention of the participants was to search for a button that would match the command, which was motivated by the fact that buttons are much simpler to use (as well as to remember, as buttons are "visible" and labeled on the remote, while motion gestures are "invisible" by their nature and only exist in the user's memory). However, participants also mentioned that not too many buttons should exist on the remote, a remark which was revealed previously by ethnographic studies on the use of remote controls [7]. When employed, motions were always accompanied by buttons (e.g., point to a screen, then confirm selection with a button; move screen while holding a button pressed). Only in two cases (PREVIOUS and NEXT), motion commands made it into the set (suggested by 25% of the participants). Such preference for buttons is reflective of the participants' extensive experience with remote controls during television watching, and sustains our design principle of gradually moving towards natural interaction by momentarily employing hybrid solutions, such as remote controls with pointing functionality.

(4) **Working beyond the projected area.** In some cases, participants proposed commands that went beyond the visible area of the video projection. For example, two participants suggested pulling the screen outside the visible area in order to close it. Similar observations were reported by Wobbrock et al. [51] for surface gestures. This finding supports our exploration of interactions around the physical area of the TV screen (as some participants seem to form models of extended spaces around interactive areas that are visible); also, the finding suggests designing interactions that do not confine to the projection of a single wall, but rather span over the entire living-room (more in the next section).

(5) **"Fun" commands.** We were able to witness proposals for commands that were "fun" to perform. Examples include bringing the remote control close to the mouth to mute the volume, or "throwing" the screen out of the visible area of the video projection to remove/close it. Such findings are in agreement with ethnographic observations from the literature (e.g., children suggesting "funny" remotes with "cool shape" [7]) and with guidelines to promote fun for interface design [42]. The first command (bringing the remote close to the mouth) is an example of body-referenced gesture, also observed by other researchers in similar elicitation studies (e.g., shake the phone next to the ear to invoke setting a reminder [28]; lifting shoulders to invoke system help [44]). Body-referenced movements can therefore stand as valuable source of inspiration for designing new interactions with intuitive meanings inside a given community or culture.

5.2 Future work

There are many ways to further develop interactions around the physical TV screen, and we use this section to pinpoint a few: (a) a physics engine could be implemented to assign pseudo-physicality to the projected displays through the implementation of gravity, collision detection and frictions between screens, which are reflective of real-world dynamics [1,48]; (b) augmented-TV spaces could display social status and social connections (e.g., show friends that are watching the same channels); (c) provide sophisticated visual feedback for assisting users dealing with TV displays projected by "invisible" computers with hidden embedded complexity, in the line of work of Vermeulen et al. [45] (e.g., should the cable connecting the DVD player to the TV not work, an animated broken line could be projected between the two devices, making users aware of the fault and how to fix it; should the TV wireless connection be disabled, an animation of the "internet cloud" could be displayed near the TV screen); (d) other areas could be turned into interactive spaces, with the most notable example being the space "behind" the physical TV screen, where digital TV content could be "stored away" and retrieved when needed.

There are also technical aspects that need more investigation, such as projecting on spaces of different colors or materials or on walls that contain furniture. Investigations are needed to assure a proper contrast between the video projected images and the color of the canvas (i.e., walls that are painted in darker colors), which could turn into settings to be selected during system calibration. Future work will also need to address ethnographic studies of families using the technology over longer periods of time.

We highlight a potential business case in which scrolling news or advertisements are displayed next to the TV screen, without occluding the main transmission. The advertiser could gain more "air time", as the advertisements and the main show do not have to compete for the real-estate of the same screen.

We point to these interesting opportunities for future work in our hope to encourage TV interface designers to explore such augmented-TV spaces, develop new applications, and design new interactive experiences for our future TV environments.

6. CONCLUSION

We conducted the first exploration of the interactive area surrounding the physical TV set in order to create augmented television environments. We introduced design guidelines, presented implementation details of an easily-replicable prototype, and designed interaction techniques informed by a participatory design study. We hope that this work will turn into a valuable source of inspiration for the designers of our future home entertainment experience, and will help leading the way towards novel, highly interactive, and smart TV spaces implementing "out of the box" TV concepts.

7. REFERENCES

[1] Agarawala, A., Balakrishnan, R. 2006. Keepin' it real: pushing the desktop metaphor with physics, piles and the pen. In *Proceedings of the SIGCHI Conference on Human Factors in Computing Systems* (CHI '06). ACM, New York, NY, USA, 1283-1292

[2] Alfaro, S., Bove, V.M. 2011. Surround vision: a hand-held screen for accessing peripheral content around a main screen. In *Proceedings of the 4th international conference on Interactive Digital Storytelling* (ICIDS'11). Springer-Verlag, Berlin, Heidelberg, 342-345

[3] Bailly, G., Vo, D.B., Lecolinet, E., Guiard, Y. 2011. Gesture-aware remote controls: guidelines and interaction technique. In *Proceedings of the 13th international conference on*

multimodal interfaces (ICMI '11). ACM, New York, NY, USA, 263-270

[4] Basapur, S., Mandalia, H., Chaysinh, S., Lee, Y., Venkitaraman, N., Metcalf, C. 2012. FANFEEDS: evaluation of socially generated information feed on second screen as a TV show companion. In *Proceedings of the 10th European conference on Interactive TV and video* (EuroiTV '12). ACM, New York, NY, USA, 87-96

[5] Baudisch, P., Chu, G. 2009. Back-of-device interaction allows creating very small touch devices. In *Proceedings of the SIGCHI Conference on Human Factors in Computing Systems* (CHI '09). ACM, New York, NY, USA, 1923-1932

[6] Benko, H., Saponas, T.S., Morris, D., Tan, D. 2009. Enhancing input on and above the interactive surface with muscle sensing. In *Proceedings of the ACM International Conference on Interactive Tabletops and Surfaces* (ITS '09). ACM, New York, NY, USA, 93-100

[7] Bernhaupt, R., Obrist, M., Weiss, A., Beck, E., Tscheligi, M. 2008. Trends in the living room and beyond: results from ethnographic studies using creative and playful probing. Comput. Entertain. 6, 1, Article 5 (May 2008), 23 pages.

[8] Bobeth, J., Schmehl, S., Kruijff, E., Deutsch, S., Tscheligi, M. 2012. Evaluating performance and acceptance of older adults using freehand gestures for TV menu control. In *Proceedings of the 10th European conference on Interactive TV and video* (EuroiTV '12). ACM, New York, USA, 35-44

[9] Bolt, R.A. 1980. "Put-that-there": Voice and gesture at the graphics interface. In *Proceedings of the 7th annual conference on Computer graphics and interactive techniques* (SIGGRAPH '80). ACM, New York, NY, USA, 262-270

[10] Butler, A., Izadi, S., Hodges, S. 2008. SideSight: multi-"touch" interaction around small devices. In *Proceedings of the 21st annual ACM symposium on User interface software and technology* (UIST '08). ACM, New York, USA, 201-204

[11] Canizares, A.G. 2006. Fun Rooms: Home Theaters, Music Studios, Game Rooms, and More. Harper Design

[12] Cao, X., Forlines, C., Balakrishnan, R. 2007. Multi-user interaction using handheld projectors. In *Proceedings of the 20th annual ACM symposium on User interface software and technology* (UIST '07). ACM, New York, NY, USA, 43-52

[13] Cesar, P., Chorianopoulos, K. 2009. The Evolution of TV Systems, Content, and Users Toward Interactivity. *Foundations and Trends in Human.-Computer Interaction* 2 (4), 373-95

[14] Cesar, P, Bulterman, D.C., Jansen, A.J. 2008. Usages of the Secondary Screen in an Interactive Television Environment: Control, Enrich, Share, and Transfer Television Content. In *Proceedings of the 6th European conference on Changing Television Environments* (EuroITV '08). Springer-Verlag, Berlin, Heidelberg, 168-177

[15] Courtois, C., D'heer, E. 2012. Second screen applications and tablet users: constellation, awareness, experience, and interest. In *Proceedings of the 10th European conference on Interactive TV and video* (EuroITV '12). ACM, New York, NY, USA, 153-156

[16] Cotting, D., Gross, M. 2006. Interactive environment-aware display bubbles. In *Proceedings of the 19th annual ACM symposium on User interface software and technology* (UIST '06). ACM, New York, NY, USA, 245-254.

[17] Cruickshank, L., Tsekleves, E., Whitham, R., Hill, A. 2007. Making interactive TV easier to use: interface design for a second screen approach. *The Design Journal*, 10 (3)

[18] Freeman, W.T., Weissman, C.D. 1995. Television Control by Hand Gestures. In: *Proceedings of the IEEE Int. Workshop on Automatic Face and Gesture Recognition*, Zurich

[19] Harboe, G., Metcalf, C.J., Bentley, F., Tullio, J., Massey, N., Romano, G. 2008. Ambient social tv: drawing people into a shared experience. In *Proceedings of the SIGCHI Conference on Human Factors in Computing Systems* (CHI '08). ACM, New York, NY, USA, 1-10.

[20] Harrison, C., Ramamurthy, S., Hudson, S.E. 2012. On-body interaction: armed and dangerous. In *Proceedings of the Sixth International Conference on Tangible, Embedded and Embodied Interaction* (TEI '12). ACM, NY, USA, 69-76

[21] Harrison, C., Benko, H., Wilson, A.D. 2011. OmniTouch: wearable multitouch interaction everywhere. In *Proceedings of the 24th annual ACM symposium on User interface software and technology* (UIST '11). ACM, New York, NY, USA, 441-450

[22] Hilliges, O., Izadi, S., Wilson, A.D., Hodges, S., Garcia-Mendoza, A., Butz, A. 2009. Interactions in the air: adding further depth to interactive tabletops. In *Proceedings of the 22nd annual ACM symposium on User interface software and technology* (UIST '09). ACM, New York, USA, 139-148

[23] Hoffman, M., Varcholik, P., LaViola, J.J. Jr. 2010. Breaking the status quo: Improving 3D gesture recognition with spatially convenient input devices. In *Proceedings of the 2010 IEEE Virtual Reality Conference* (VR '10). IEEE Computer Society, Washington, DC, USA, 59-66

[24] Ibrahim, A., Johansson, P. 2002. Multimodal dialogue systems for interactive TV applications. In *Proceedings of the 4th IEEE International Conference on Multimodal Interfaces* (ICMI'02), Pittsburgh, USA, 117-122

[25] Jensen, J.F. 2005. Interactive television: new genres, new format, new content. In *Proceedings of the second Australasian conference on Interactive Entertainment* (IE '05). Creativity & Cognition Studios Press, Sydney, 89-96

[26] Jones, B., Benko, H., Ofek, E., Wilson, A.D. 2013. IllumiRoom: Peripheral Projected Illusions for Interactive Experiences. In *Proceedings of the SIGCHI Conference on Human Factors in Computing Systems* (CHI'13)

[27] Kane, S.K., Avrahami, D., Wobbrock, J.O., Harrison, B., Rea, A.D., Philipose, M., LaMarca, A. 2009. Bonfire: a nomadic system for hybrid laptop-tabletop interaction. In *Proceedings of the 22nd annual ACM symposium on User interface software and technology* (UIST '09). ACM, New York, NY, USA, 129-138

[28] Kühnel, C., Westermann, T., Hemmert, F., Kratz, S., Müller, A., Möller, S. 2011. I'm home: Defining and evaluating a gesture set for smart-home control. *International Journal of Human-Computer Studies*, 69 (11), 693-704

[29] Lee, J.C. 2008. Hacking the Nintendo Wii Remote. *IEEE Pervasive Computing* 7(3), July 2008, 39-45

[30] Malizia, A., Bellucci, A. 2012. The artificiality of natural user interfaces. *Communications of the ACM*, 55(3), 36-38

[31] MacKenzie, R., Hawkey, K., Booth, K.S., Liu, Z., Perswain, P., Dhillon, S.S. 2012. LACOME: a multi-user collaboration

system for shared large displays. In *Proceedings of the ACM 2012 conference on Computer Supported Cooperative Work Companion* (CSCW '12). ACM, New York, USA, 267-268

[32] Marquardt, N., Jota, R., Greenberg, S., Jorge, J.A. 2011. The continuous interaction space: interaction techniques unifying touch and gesture on and above a digital surface. In *Proc. of the 13th IFIP TC 13 international conference on Human-computer interaction* (INTERACT'11). Springer, 461-476

[33] Mistry, P., Maes, P., Chang, L. 2009. WUW - Wear Ur World: a wearable gestural interface. In *CHI '09 Extended Abstracts on Human Factors in Computing Systems* (CHI EA '09). ACM, New York, NY, USA, 4111-4116

[34] Natapov, D., Castellucci, S.J., MacKenzie, I.S. 2009. ISO 9241-9 evaluation of video game controllers. In *Proceedings of Graphics Interface* (GI '09). Canadian Information Processing Society, Toronto, Ont., Canada, Canada, 223-230

[35] Nielsen Group. 2012. The digital revolution: A look through the marketer's lens (April 2012) http://na.ad-tech.com/sf/wp-content/uploads/DigitalConsumer.pdf

[36] Norman, D.A. 2010. Natural user interfaces are not natural. *Interactions* 17(3), May 2010, 6-10

[37] Park, K.L., Park, J.K., Kim, S.D. 2008. An Effective Model and Management Scheme of Personal Space for Ubiquitous Computing Applications. *IEEE Transactions on Systems, Man, and Cybernetics. Part A: Systems and Humans*, 38(6), November 2008, 1295-1311

[38] Qin, Q., Rohs, M., Kratz, S. 2011. Dynamic ambient lighting for mobile devices. In *Proceedings of the 24th annual ACM symposium adjunct on User interface software and technology* (UIST '11 Adjunct). ACM, NY, USA, 51-52

[39] Rushing, K. 2006. Home Theater Design: Planning and Decorating. Media-Savvy Interiors. Quarry Books

[40] Schlömer, T., Poppinga, B., Henze, N., Boll, S. 2008. Gesture recognition with a Wii controller. In *Proceedings of the 2nd international conference on Tangible and embedded interaction* (TEI '08). ACM, New York, NY, USA, 11-14.

[41] Thomas, B.H. 2012. A survey of visual, mixed, and augmented reality gaming. *Computers in Entertainment* 10 (3), December 2012, 33 pages

[42] Vatavu, R.D. 2010. Creativity in Interactive TV: Personalize, Share, and Invent Interfaces. In A. Marcus, A. Cereijo Roibas, R. Sala (Eds.), *Mobile TV: Customizing Content and Experience*, Springer Human-Computer Interaction Series, Springer London, pp. 121-139

[43] Vatavu, R.D. 2012. Point & Click Mediated Interactions for Large Home Entertainment Displays. *Multimedia Tools and Applications*. 59 (1). Springer Netherlands, 113-128

[44] Vatavu, R.D. 2012. User-defined gestures for free-hand TV control. In *Proceedings of the 10th European conference on Interactive TV and video* (EuroITV '12). ACM, New York, NY, USA, 45-48

[45] Vermeulen, J., Slenders, J., Luyten, K., Coninx, K. 2009. I Bet You Look Good on the Wall: Making the Invisible Computer Visible. In *Proc. of the 3rd European Conference on Ambient Intelligence* (AmI'09). Springer, 196-205

[46] Wigdor, D., Leigh, D., Forlines, C., Shipman, S., Barnwell, J., Balakrishnan, R., Shen, C. 2006. Under the table interaction. In *Proceedings of the 19th annual ACM symposium on User interface software and technology* (UIST '06). ACM, New York, NY, USA, 259-268

[47] Wilson, A.D. 2005. PlayAnywhere: a compact interactive tabletop projection-vision system. In *Proceedings of the 18th annual ACM symposium on User interface software and technology* (UIST '05). ACM, New York, NY, USA, 83-92

[48] Wilson, A.D., Izadi, S., Hilliges, O., Garcia-Mendoza, A., Kirk, D. 2008. Bringing physics to the surface. In *Proceedings of the 21st annual ACM symposium on User interface software and technology* (UIST '08). ACM, New York, NY, USA, 67-76

[49] Wilson, A.D., Benko, H. 2010. Combining multiple depth cameras and projectors for interactions on, above and between surfaces. In *Proceedings of the 23nd annual ACM symposium on User interface software and technology* (UIST '10). ACM, New York, NY, USA, 273-282

[50] Wilson, A., Benko, H., Izadi, S., Hilliges, O. 2012. Steerable augmented reality with the beamatron. In Proceedings of the 25th annual ACM symposium on User interface software and technology (UIST '12). ACM, New York, USA, 413-422

[51] Wobbrock, J.O., Morris, M.R., Wilson, A.D. 2009. User-defined gestures for surface computing. In *Proceedings of the SIGCHI Conference on Human Factors in Computing Systems* (CHI '09). ACM, New York, NY, USA, 1083-1092.

[52] You, W., Fels, S., Lea, R. 2008. Studying vision-based multiple-user interaction with in-home large displays. In *Proceedings of the 3rd ACM international workshop on Human-centered computing* (HCC '08). ACM, New York, NY, USA, 19-26

[53] Ziola, R., Kellar, M., Inkpen, K. 2007. DeskJockey: exploiting passive surfaces to display peripheral information. In *Proceedings of the 11th IFIP TC 13 international conference on Human-computer interaction* (INTERACT'07). Springer Berlin, Heidelberg, 447-460

Panoramic Video: Design Challenges and Implications for Content Interaction

Goranka Zoric, Louise Barkhuus, Arvid Engström
Interactive Institute
Box 1197
SE-164 26 Kista, Sweden
{ goranka.zoric, louise.barkhuus, arvide }@tii.se

Elin Önnevall
Mobile Life @ Stockholm University
c/o SICS, Box 1263
SE-164 29 Kista, Sweden
elin@mobilelifecentre.org

ABSTRACT

In this paper we explore viewing and interaction in an emerging type of interactive TV, where viewers are presented with *panoramic ultrahigh-definition video* combined with *extensive interactive control* over view selection. Instead of delivering only what will be consumed, emerging TV services offer high-resolution panoramic video to the viewers, enabling them to more freely explore the broadcast content by selecting regions of interest and navigating within the larger panoramic image. However, as we open up the television space both in field of view and in terms of the freedom given to viewers, new interactional challenges emerge. We have done user studies on two systems for interacting with panoramic high-resolution video, one based on the tablet interaction and other on the gesture interaction. Our findings revealed a number of design challenges concerning properties specific to panoramic video. Based on findings from the user studies and the identified design challenges, we have compiled a set of the design recommendations on how to support interactive viewing of panoramic content.

Categories and Subject Descriptors

H.4 [Information Systems Applications]: Miscellaneous; H.5 [Information interfaces and presentation (HCI)]: User interfaces

General Terms: Design, Human Factors.

Keywords

Panorama, video, interactive TV, mobile TV, live sports, broadcasting, tablet, second screen, gesture interaction

1. INTRODUCTION

Within the area of interactive television there is a new emerging genre combining rich, panoramic live video with extensive interactive control over what to view within the covered scene. Extensive technical development in image capture and interaction modalities makes these high-resolution interactive TV images possible.

The services in this new genre build on current interactive television in two significant ways. First, they invite the viewer to *look into* a panoramic image space, as opposed to *looking at* framed images of a scene. Second, and closely linked to the above, they aim to provide the viewer with *production-like tools* for looking, i.e. for managing this increased control of view selection. These tools by necessity go beyond typical system controls such as audio level, colour and screen settings, since they extend the control into the image content itself. We are thus making the distinction between regular *system interaction* and what we may call *content interaction* – looking into the panoramic image space, using production-like tools for navigating and interacting directly with the image content as the broadcast or program unfolds.

Increased resolution and image size introduce numerous possibilities for viewer interaction as well as for integrating several screens into the viewing experience. With real-time content interaction, television viewers are slowly developing new ways of consuming shows, particularly live events. A sports match may be viewed on television while details are being looked up on a tablet device, or that device might even be showing a different version of the event. Individual viewers may immerse themselves in details of their choice within the broadcast, together or separately on "second screen" devices. At the same time, larger screens in the home serve as enablers for panoramic images.

In the last decade, increased attention has been paid to interactive TV (iTV), leading to development of novel systems offering advanced viewing features [1]. Consequently, a range of interactive functionalities has been suggested for viewers enabling more possibilities than those of standard remote controls. Much work is describing these technical approaches, where there are fewer studies that address how viewers would actually interact with content offering a multitude of possible views.

We here focus on *content interaction,* the possible interactivity that enables the viewers to compose a personalized television experience for themselves. We explore two main strains of interaction as well as look into more subjective perspectives from potential users. For the purpose of exploring these issues, we present three studies; a set of focus group studies conducted to explore basic potential interaction practices among television viewers, and two user studies of two different methods of interacting with the content: using a touch screen on a "second screen" tablet and using wide hand gestures for interacting with a large screen. In both cases interaction with panoramic video includes zooming in and out, panning and tilting, for navigating in the image space. The results of these studies lead to first a set of design challenges, which we subsequently frame as more specific design implications.

Figure 1: Example of a panoramic scene used

2. BACKGROUND AND RELATED WORK

We start this section by providing general observations on qualities of a panoramic image, and then concentrate on state of the art work focusing on systems enabling viewers to control a view selection and experiences around such. The two sections represent the two features that distinguish the new interactive TV genre we investigate; *panoramic view* and *real-time content interaction.*

2.1 Panoramic View

Why do we appreciate panoramic views? On a more general level Lansford and Jones [2] argue that there are values incorporated in a panoramic or a scenic view, which are the reason why properties with panoramic views, e.g. waterfronts are being highly valued. Meitner [3] suggests that a scenic beauty is "*determined jointly by the location of the viewpoint (observer) and the features of the view shed that is represented (experienced). In other words, scenic beauty ratings require a location/place, a landscape to appreciate from that location and a person to do the appreciating*". The experience of the panorama is thus dependent on the viewers' location. In the case of panoramic TV, deciding on what to look at is in this sense just as important as the "*location of the viewpoint*". Latour and Hermant [5] in their text Paris: Invisible City, are critical of what they see as the panorama's claims on capturing the entirety of a scene "*at at glance*", and suggest that "*it's time we updated our panoramas*". In their view, the panoramic view is too constructed to be a useful representation, in that it is dependent on a fixed viewpoint where the illusion is mastered. The above quotes all highlight the importance of the spectator's point of view in watching a panorama. In mediated panoramic events, such as the ones seen in the type of television investigated here, selecting a good point of view is an important consideration.

Wide formats for television and film in the home have seen a rise with the proliferation of large displays, but have their roots in cinematic formats. When the movie format Cinemascope became popular in Hollywood there were concerns with the width of the images [4]: "(...) *Hollywood's creative personnel feared that the wide screen would immobilize the camera and lead to long takes. Some editors were afraid to cut quickly, worrying* ***that viewers would not know where to look*** *in a rapid series of wide compositions*" (our bold). These concerns are brought to the fore again with increased interactive control. However, the action and content in the scene play an important role in guiding the viewer. For example, if the covered scene is a football game then it is probably most common to follow the action around the ball. The viewing is therefore generally guided by the structure of the event, but designing for panoramic TV demands an understanding of how this is performed.

High-resolution panoramic video

Here, we are especially interested in the use of ultrahigh-resolution panoramic video. Panoramic video is generally obtained by stitching streams from several standard high definition cameras, as in [7][10]. Importantly, stitching multiple high-definition images into a panoramic view produces an ultra-high definition image space with enough resolution to frame "virtual cameras" within the panorama, i.e. to enable interactive viewing of arbitrary region-of-interests (RoIs), produced either by the system or manually by using pan/tilt/zoom (PTZ). A single panoramic video represents the view of the scene from a given viewpoint, and the same viewpoint remains during the entire event (by panning, tilting or zooming the viewer is able to choose the viewing direction). In the case of several panoramic videos covering the same event from different points, the viewpoint might change giving users the possibility to also choose the viewing angle of the specific action.

Some systems, although providing a panoramic view of the event, are intended for a specific use, such as large screens in a cinema, as in [10], and do not offer any interaction for the users. The imLIVE [8] demo offers live streaming 360-degree video with interactivity for audiences. The panoramic camera has "only" 2400x1200 pixels, which makes it problematic to use for deep zooming. The Camargus [6], a production oriented system, offers panoramic video of the sport event created by combining an array of high-definition video feeds into one feed. The operator is given the possibility to control a virtual camera, and the recorded content is mainly used for replays.

2.2 Content interaction

When a panoramic view is combined with the tools for "looking into" it interactively, the TV viewers are given the possibility to choose their own view of the event. Emerging panoramic TV services enable navigation and interaction within the image content, in contrast to more familiar control of program features such as audio level, image controls and channel selection. Such content interaction bares a resemblance to actual production work, and accordingly it calls for extended interactive tools beyond the familiar remote control.

Flexible TV production and consumption

Recent technological advances in camera development and image processing have enabled changes in both production and viewing practices. Compared to traditional TV production where a production team produces a program – a single sequence of images selected to guide the viewer through the covered event – there is now the possibility to capture the whole event. Consequently, there are a number of systems, both academic and industrial, that enable arbitrary views of events, and flexibility for TV production and viewing. In all these systems, different types and levels of interactivity are enabled, often being very system specific. Since our concern is with panoramic television, our aim is not to cover all available systems, but to provide a brief look at relevant existing approaches focusing on key features that those systems provide. Beside panoramic television, approaches that have been developed include free viewpoint video and various forms of multi-view video.

In free viewpoint videos, researched extensively in the past decade, viewers can interactively change their viewpoint in the scene; i.e., viewers are able to freely navigate within real-world visual scenes, as known from virtual worlds in 3D computer graphics [24]. In contrast, in traditional videos, the viewpoint is chosen by the producer or director. Generally, in order to get imagery from arbitrary viewpoints, several cameras are placed around the scene, and views from real camera images near the chosen viewpoints are interpolated. Such systems enable new production as well as new possibilities for users, and are often used for sporting events, e.g. to show replays from any angle, as in [12][13].

The LIVE project [9] attempts to give viewers their own personal and interactive broadcast by providing a production support system. The idea is to produce a parallel multistream coverage of a live event, including a backchannel for the consumers to be able to influence the broadcaster through voting [11][14]. Similarly, the My-eDirector project [10] aims to provide an interactive broadcasting service enabling end users to direct their own coverage of large athletics events and thus to take a role of a virtual director adapting the broadcast to their own viewing preferences [16][17].

User interaction with rich video content

As the above brief overview showed there are various approaches that aim to give TV viewers the possibility to choose their own view of an event, typically live sport. While there are many works describing technical approaches, there are still significantly fewer user studies that aim to show how viewers would interact with content offering various views on the event.

Olsen et al. in [15] study a prototype that offers interactive TV sport over the Internet and gives viewers both the possibility to choose the view and to control replays. Experiments showed that sport fans could easily learn the interactive controls and that they would use the interaction such as switching between cameras and "moving in time", rather than passively watch the broadcast. An evaluation of the My-e-Director 2012 prototype implementation of the service, offering personalization capabilities based on user profiles and recommendations, is presented in [17]. Results showed that users may become annoyed when getting continuous annotations and recommendations on switching channels.

Concerning viewers' interaction with panoramic TV, there are only a few relevant works. Neng and Chambel in [18] explore 360° hypervideos focusing on navigation and visualization mechanisms. Their "360° Hypervideo Player" provides features that assist users in orientating themselves in the panorama. The first actual user study of wide-format video is given in [19]. Authors present a study with the focus on omnidirectional video (ODV), a video that enables users to look around in 360°. The focus was on finding characteristics that make a TV-program suitable for enhancement with ODV. Interaction with panoramic video consisted of changing of the viewing angle and zooming in and out. Their findings show that ODV has the potential, yet challenges remain on a technological, content and user levels.

Bearing in mind that there are already specific services for live broadcasting where interactive panoramic video is used, there is a lack of empirical findings to inform the design of such and similar services. Today these services are not specifically requiring particular devices, which leads to a need to explore interaction with content itself. Our goal with this research is to explore how potential viewers of panoramic video can interact with the content to gain a richer viewing experience. The wide possibilities for presenting different parts of the content to viewers point to the need for thorough design guidelines in the relation to live television watching experiences. In order to take a closer look at how panoramic video content might be approached, we first conducted a preliminary focus group study. These focus groups were organized to get ideas and early input from users. Before describing the studies of interaction methods, we therefore briefly go through the method and results of the focus group study.

3. FOCUS GROUP STUDY

The goal of our focus group study was to explore potential views and preferences for interacting with the panoramic format of live television experiences. The study served as a pre-study to our further investigations of interaction methods with panoramic video content in relation to live experiences and shows. We chose a football match as the live event because the potential for following the play closely through following the ball presented a fairly simple example. Live sports were also chosen since the viewing of this type of live event requires real-time interaction, and quick decision making. In addition it was easy to elicit excitement among potential viewers and participants who were interested in football.

We organized three focus groups lasting approximately 1,5 hours each. Participants were 16 regular TV viewers, recruited from within our research organisation. The focus groups began with a general open discussion on TV viewing such as the experience of live content, followed by more concrete tasks. The material used and tasks given varied for each focus group. The participants were given a small set of tasks connected to the different materials. With the material as a helpful asset, users shared stories about their TV watching practices, which worked as a way to help these ordinary TV viewers talk about the content and presentation. All discussions were video- and/or audio recorded and the material was later analysed and structured under different key topics.

Figure 2: Group discussion around mock-ups

The first group was shown still images from a traditionally produced live broadcast of a football game. In total, 14 pictures were shown, representing various view selections like an overview of the game, an overview centered on the ball, a detail frame of a player, selected tackles, and selected audience reactions. Participants were asked to discuss the images and produce their own view by choosing the preferred views and time aligning them.

The second group was presented with keywords connected to live TV. Keywords were used to structure and stimulate the discussion regarding the concepts associated with novel TV services. We initially presented the words mobile, home, and public in order to focus on TV viewing that happens everywhere, in any environment. The words interactivity, content, multiple displays, social connectivity, and immersion represented different aspects of changing TV watching practices. Participants were instructed to choose one or two key words from each stack, and then brainstorm around the live TV topic.

The second and third group looked at mock-ups of user interfaces. The images were similar to the first focus group but presented in panoramic view in the latter sessions. Three pictures of a panoramic view in combination with additional small windows showing regions of interest and various close-ups were displayed to the participants. Figure 2 shows a detail from one of the focus groups.

Overall, focus group participants were positive towards panoramic images of live experiences. We identified two themes through the focus groups that lead to our further development and studies of interaction approaches of the panoramic images; understanding context and social viewing.

Panorama as Understanding Context

One male participant expressed a salient need for a broader view of sports matches specifically: "*What you miss when you see sport on TV is this view (pointing on panoramic image). This is almost never shown and that is what you see when you watch it for real*". Participants expressed how the panoramic view would be able to give them a better understanding of the overall context of the live experience. They generalized across several types of content but particularly the sports events were seen as benefiting from this. One participant expressed: "*If we talk about game play, this is where we see how the game is going (pointing at panoramic image). This doesn't say anything (pointing at a close-up) about what is happening. It is like playing Monopoly and just looking at the dice*". One participant stated that he would like to have "more description" of the panorama so that he would know what he was looking at, which resonated well with others who emphasized an appropriate separation between watching detailed parts of the picture and the overview: There should be "*a balance between detail and 'where are we'*". In this case the panoramic view is then supporting an understanding for "where" we are.

Most of our focus group participants emphasized that they would appreciate to have an easily accessible panorama picture so that they could "*go back to it in one click*", an insight that went into the design of the interaction approaches that we prototyped.

Social Viewing

Another prominent aspect that came up during the focus groups was sociality. During the first focus group, the participants expressed interest in a way to coordinate with friends before the game, for example to choose the same content to watch if they were not in the same place. Doing this before the game would reduce the risk of losing focus once the live game had started. "*My feeling is that maybe it is important, this shared ... that you have the same viewpoint so that you can discuss it with your friends after watching – like 'did you see that tackle' or 'did you see that moment' – that you have shared this specific particular moment and then you have the slow motion replay of that. And you have the same kind of experience of watching*", one participant commented.

Both earlier observations at sports bars [22] and Esbjörnsson et al.'s studies [23] of sport spectators at rallies show that the gathering around a live event (watched in real life or on a screen) involves constant discussion about what is going on. In addition, our participants expressed a desire to have the possibility to share the content in other ways, for example through a web page, where they could see what others were watching and be able to switch between broadcast streams.

Based on the focus groups, we concluded that the emphasis in terms of designing for interaction with panoramic content would be to make sure dynamic viewing is supported, as well as easy access to the panoramic view.

4. INTERACTION APPROACHES

It has been suggested that rich video content calls for interaction techniques beyond remote controls. Gesture interaction and touch interaction on so called second screen devices are two broadly recognized approaches to this end. The following section presents user tests on developed prototypes for these two interaction techniques.

In the second stage of our project, two systems for interacting with such video were developed, and used in laboratory user studies to explore interactive viewing of panoramic video. The first system enables interactive panoramic TV viewing on a mobile device, either a tablet or a mobile phone, and the second one uses hand and arm gestures for interactive panoramic TV viewing in the home environment, also allowing small group interaction. These systems were developed as part of the broader project and were as such designed as probes into how viewers would interact with content produced within the framework of a panoramic, high-resolution video image.

4.1 Tablet Based Second Screen Interaction

The underlying system has been developed to record and transmit high-resolution panoramic live video in a way that allows for multiple interaction techniques on the end user side. One potential interaction method is a secondary screen that can be used to view a selected region of the panoramic image. The prototype uses an Android based application allowing for basic navigation within the panoramic picture, running on the tablet and a server [20]. Initially the client displays the full panorama. By using interactive commands the user can choose the region of interest for viewing. Interactive commands that are supported are panning, tilting and zooming (PTZ). Zooming can be controlled using pinching or using a small invisible slider bar at the left of the image. Panning/tilting is controlled by touching the screen and moving one finger around on the screen in a swiping movement.

One technical limitation is that large panoramic content (7K) cannot be shown on a mobile client, unless it is downscaled to a resolution that can be handled (e.g. 1.2K). Downscaling results in a resolution that is not high enough to see details. Thus, in the system we use here, the server crops and rescales to the resolution of the client screen (while respecting the aspect ratio of the content) and re-encodes the panoramic content based on the interactive commands (requests) from the client within a reasonable small delay, so that it is perceived as real-time.

User Study of Tablet Interaction

We conducted a study of the tablet interaction prototype in October 2012. The content used for the study was a 10 minutes panoramic video (running in the loop) of a football match that was played between Chelsea and Wolverhampton in October 2010, which was available to us (see Figure 1).

The study was conducted as a lab-based test with 16 participants between the ages of 22 to 46, with a median age of 28 years. 9 of them were not interested in football and the rest indicated an interest with one being an avid football fan. Almost half of the participants stated that they had watched panoramic video before in different contexts including large-scale movie experiences.

After a short introduction to the system, the participants were instructed to interact with the content; to try to follow the ball, look at details of their choice and enjoy the game. We recorded the interaction with two cameras, from the front and from the back. After the participant had been interacting with the system for about five minutes the researcher asked them a set of questions about their immediate experiences and impressions. Figure 4 shows a detail from the gesture study.

Figure 3: Interaction with panoramic video on the tablet

4.2 Gestures and Large Screen Interaction

In a second test, we tested a prototype for a large screen where interaction takes place through hand and arm gestures in front of the screen, similarly to game interaction with Microsoft's Kinect [21].

Figure 4: Gesture interaction with panoramic video

The gesture system allows the user to perform interactions with panoramic video displayed on a high definition TV screen or projected on the wall in a home environment. It works in a device-less and marker-less manner and the user is able to control the content using only their hands, either standing or sitting in a chair or sofa. The system is multi-user, i.e. a small group of users can ask for control of the system and interact with it while the others are still present in the scene.

The current implementation of the gesture system is written in C++ and runs on a single laptop needing at least 6 CPUs. The system has been split into multiple connected components where each is responsible for a single task (e.g. head location, hand tracker, gesture classification, etc.). Each component is based on SmartFlow, a piece of software developed by NIST (http://www.nist.gov/smartspace/) to facilitate the communication of software modules. The system uses the color and depth video provided by a single Kinect camera as the principal sensor. No other data provided by the Kinect is employed. The Kinect camera should be placed at heights between 120cm and 250cm and approximate at the middle of the TV screen. A free space in front of the sensor of 2-4m is also recommended.

The functionality in terms of user control of the content includes interactions such as selecting menus presented on the screen, navigating through high resolution panoramic views of the scene by panning, tilting and zooming, and control of the audio by changing the volume, muting or selecting the speaker. Figure 4 shows a detail from the gesture study.

User Study of Gestures Interaction

We conducted a lab-based study of gesture interaction with panoramic TV in a simulated home environment in November 2012. This test included two types of content (the second type of content was not available for us in the first user study). We included both the panoramic video of the aforementioned football match between but additionally, we used 5 minutes of panoramic video of a dance show where the Berlin Philharmonic Orchestra accompanied youth dancers from Sascha Waltz in the show Carmen, at the Arena Berlin in May 2012 (see Figure 5).

The study was conducted with 20 people, between the ages 20 and 39. Most of them had had experiences with panoramic videos before, either on big cinema screens or on tablets or mobile devices. 3 of them had some experience using gesture controls in this context while the other 7 were new users to this type of interaction. Where 14 of the participants interacted as pairs, six of them interacted alone. This enabled us to see any differences in dynamics when users watched and interacted with the system alone as opposed to within a social situation. All the pair users knew each other.

The participants were instructed to interact with the system, using the described gestures to zoom, pan, tilt and turn the volume up and down and mute. We video recorded their interactions from two directions, front and back, where it was possible to see both the screen and the gestures simultaneously. The participants were interacting with the system for about 10 minutes for each of two available contents, though in some cases even longer. Afterwards, the researcher asked them a set of questions about their immediate experiences and impressions, and their responses were audio recorded and transcribed.

We now continue describing the overall findings from the two studies. We keep the findings together rather than reporting individually from each study because the studies revealed comparable issues that are relevant to consider in combination.

5. FINDINGS

Both studies revealed insights into challenges and benefits of interacting with panoramic content, as well as issues regarding each method of interaction. Where the tablet study was mostly focused on detailed zooming and panning we were able to explore social interaction around live event watching in the study of gesture interaction where we observed seven couples. We first discuss four themes around our findings before presenting more specific design challenges and design implications.

5.1 Controlling the View

The main advantage to navigating (e.g. zoom and pan) within a large image is of course to be able to control the view. One of the reasons why participants want more control is that it will give them a personal view of the live show/event: "*It is always the wrong pictures they are showing, showing something completely different, and then you see something like 'that was really weird' or if you think the referee made some mistake and you want to watch it and form your own opinion,*" said one participant from the tablet study after talking approvingly about the navigation possibility on the smaller screen. Most sports viewers know where to look on the screen, which was also observed during our earlier studies in sports bars [22] and essentially supports our observation

Figure 5: Screenshot from dance panoramic video

that participants knew where they wanted to pan the zoomed-in section of the image. They acknowledged that it can be useful to control their own view, as one of the gesture study participants expressed: "*For this it makes sense if you have a lot of things going on and you want to follow*". However, there was clearly a limit to the desire to control and constantly having to actively choose the view: "*I think it's nice to be able to zoom, but I was thinking about constantly doing that. I'd love that sometimes somebody does it for me*", said another participant from the same study.

When we discussed how interactive panoramic video would change television watching, one participant said that it would give her a "*responsibility to take action*". Still, one main concern was the stress of being given too many options. "*If you are your own producer you know that you will miss something and when you have the choice of being very individual in picking some specific scene then I think it's easy for you to feel worried, like 'am I missing something? Is everyone else watching something else?'*". According to these users, having too many options can also affect the experience of the content in a negative way.

In terms of using the larger screen where the full panoramic content was running during the tablet test, participants found it useful to get an overview; one participant explains: "*If I cannot see the ball when I move the picture [on the tablet] then I look up [at the big panoramic image] to see where the ball is. Even if the ball is here and I want to see the whole image, so it quite depends. I think it is quite important to have this overview. You can see the players, where they stand. And the two teams. (...) I guess the position of the players is quite important*" This illustrates well how part of controlling the view was also about being able to view the show "from above". The panorama was thereby giving viewers the visual experience of seeing the game from the grandstand, "(…) *I like it, it is really cool to see the whole field, it is like I'm sitting there.*", said one of the tablet study participants.

Controlling the view by zooming in and out was highly appreciated. It was used by all participants in the interaction user studies and it quickly became natural for them to zoom in and out, although it was not trivial to master technically. However, some participants were worried about content being lost. "*There is a danger with all this zooming in because you lose the overview and then it just takes a few seconds and 'oh where are they now'*". The issue here is how to get the best information, and the zooming is seen as a delicate instrument for getting more details suiting personal needs. Constantly zooming in and out creates a distraction in the interaction with the panoramic content. "*It would be good to be able to move between regions of interest without having to zoom out to the panorama again*", one participant commented, suggesting that a feature similar to a cut between cameras would be useful.

5.2 Social viewing

As we learned in our focus group study, potential viewers are keen on social functionality when it comes to interacting with panoramic content and televised live events. Although only the gestures study was conducted with pairs, it was clear that such interaction with the content enabled social interaction and could be used in a setting of two or more viewers.

However, social interaction was not always viewed in a positive light. Another aspect that emerged from our studies is that interacting too much with the content reduces possibility to communicate with your friends or family when watching happens at the same location. One of the participants in the tablet study said: "*The system is not social, when watching with friends interaction takes too much attention so you are not able to interact with friends.*" On the other hand, it can also be a reason to start discussion: "*Watching with friends (...) if panoramic content could be time shifted, rewound, on the tablet, to show friends and comment, and have a main broadcast on the TV.*"

We found that the relation between what people said they prefer in terms of social viewing and how they interacted together was in some cases contradictory; where they explicitly asked for social functionality such as embedded communication, they were also demonstrating a need to simply watch the content without having to be social around it. This brings us to the next theme, the aspect of shifting viewing between active and passive.

5.3 Active-passive viewing

One of the salient characteristics of television watching is the way viewers shift between passively watching and actively interacting (through channel surfing or more complex interaction with iTV systems); similarly we observed a laid back approach in the gestures study. All participants did, at one point or another, settle on the current view and simply observed the show, if not also leaning back in the chair. They expressed in the post-viewing interviews that such lean back behaviour is part of their television habits, that it is necessary to be able to do that to enjoy the show at certain times.

When talking about potential interaction with the system, our gesture study participants often mentioned less advanced features such as simple volume up and down and channel shifting as well as being able to rewind. This was surprising in that the system afforded a high level of detail in viewing the video content and the possibility to navigate within the larger panoramic image, than what is possible in traditional television broadcast. However, many participants in fact mentioned the downsides of this detailed interaction: they were worried they would miss out important parts of the live show, particularly in a game like football. One participant provided the example that he had lost out on a goal when he had flicked from one screen (television screen) to another (tablet screen) at home the previous night.

In the study on the tablet participants generally expressed positive attitude in interacting with the panoramic video on the tablet. People seemed more willing to interact with the panorama on the tablet than on the TV set: "*I like the panorama on the tablet. I like to zoom/pan because you are used to such interaction from using mobiles in everyday life...so it is not really for a big screen.*" Another participant added: "*Having a panorama overview on the tablet was not problematic, indeed it was calling for interaction...people are used to interaction with fingers.*" Finally interestingly, it was not considered a problem watching on a handheld device: "*People already watch different stuff on mobiles (...) I actually watched handball during Olympics with my friends*". In all, users displayed great familiarity with second screen interaction, both in terms of dividing their attention and in specific interaction techniques.

5.4 Technical Expectations

Our studies also highlighted the harsh requirements that users have on technical workings of these types of systems today.

Expectations on technical quality

Even using ultra-high definition capture, there are technical limits to the resolution that can be provided in a detailed subsection of the larger image. This most critically became evident when test participants zoomed deep into details of the image. "*The resolution is the biggest killer of the application, if I zoom I want a proper resolution, otherwise it doesn't make sense. If I could get detailed view I would zoom in more*", one of them commented. Several participants reported that they sometimes zoomed in past what they perceived as acceptable image quality. "*I wanted to look at the people but then it's out of focus, to see what they are doing.*", one viewer stated, giving a concrete example of a situation where they steered their view into a very narrow section of the scene without getting more detail in the image. The sensitivity and speed of the zoom were other factors that limited the interaction. "*I was able to follow the ball, except for the long shot, when the ball would disappear. But it is also the lag of the system, if zoom was faster, it would help.*" Hence, the allowances of the zoom control – responsiveness, speed and zoom level – are important both individually and in combination, in the experience of interactive navigation in panoramic images.

Properties of the panoramic image

In the tablet study the participants pointed out specifically how a panoramic view is lacking different angles of the camera, like "*It is quite a freedom, but I'm missing different angles like in real broadcast*". Because of the camera viewpoint being fixed, it was problematic in some situations to get a good view despite the possibility to zoom, compared to a normal broadcast of a football match, where there would be several cameras at different locations in the stadium.

Participants did not typically perceive the panorama as a scene that is "always there", when they are watching more detailed regions of the image. This may be due to the relative novelty of panoramic video in the TV context, as compared to standard-framed content, and has the consequence that viewers do not take full advantage of the panoramic scene, for attaining an overview and when selecting regions of interest.

6. DISCUSSION AND IMPLICATION FOR DESIGN

Our research goal has been to understand how to look at panoramic video, i.e. how to do content interaction focusing on the specific qualities of a panorama no matter the underlying devices and interaction techniques. Here we expand on these findings picking out features of relevance for the design of panoramic TV.

6.1 Design challenges

In our studies, we have identified four categories of challenges that emerge when interacting with panoramic video content, and that need to be attended to and supported in systems design.

Balancing active and passive viewing

One of the qualities of interactive panoramic TV content is that it offers many different viewing options within the larger image of the scene, and being a spectator is to some extent a matter of finding material you are interested in. With the possibility to choose, viewers can satisfy their own interests, which is generally appreciated. Yet, there is typically a produced program or live broadcast to return to at will. As seen in our studies, users sometimes want to take it easy and sometimes want to actively control their view of the TV content; even if most of our participants expressed an interest in having their own view of the event, they were also worried that such interaction would involve too much effort and stress, and would cause them to miss important action. Interacting with and looking into high-resolution panoramic video of live broadcasts can be affected by the viewer's specific choices of:

- **Replays.** User decisions on what and when to replay,
- **Choosing region of interest by:**
    - *Zooming.* Typically choosing close-ups or regions of interest to see, for example, facial expressions or details of occurrences,
    - *Selecting pre-defined regions of interest,* provided by the system or a producer,
    - *Selecting overview.* Having the panoramic view, e.g. an overview of the whole field, and
    - *Using navigation.* Doing virtual camerawork in the high-resolution panorama picture.

From a production point of view, all these actions mean overriding the editing decisions of the producer, in the case of a live broadcast. The two modes exist in parallel and produce two alternative views of the scene. Accordingly, managing the transitions between the main broadcast and active control of the image content is a challenge. The tolerance for errors was high among participants while controlling the view themselves, but introducing interactive control at any time may also cause disruptive breaks between the passive and active modes.

Designing interaction with live broadcast content requires detailed attention to the workings of the interaction where at the same time viewers want to have the opportunity to lean back and passively watch the action. It is imperative to support both types of behaviour within interactive television systems, though the balance can be complicated to facilitate with systems like the ones suggested.

Enabling orientation

Interaction with the panoramic picture is connected to orientation. The panorama gives the users a possibility to interact within an event that is dispersed. In the same fashion as tourist signs at scenic spots can give the visitors a more informative experience, panoramic video can work as a guiding image. At the same time, the fixed view point was seen as a limitation by users used to the fast cuts and multiple perspectives of broadcast television.

Users were explaining the use of the panorama as a mechanism for orientation and something you could "go back to" if you loose yourself in the navigation. In that sense the panorama could be understood as an interactive map that gives you a greater understanding of the scene. One user's explanation of the panorama being a menu is enforcing that there is a certain value of organization within the bigger context.

Having the panoramic picture readily available as an overview is highly appreciated, and an important feature for relaxed exploration in panoramic ultra-high definition television formats.

Individual exploration

Knowing what kind of panorama you are looking into also sets the standard for what you can understand, for example a panoramic view of a football stadium is more easily understood by frequent visitors. Sport viewers in traditional television have knowledge on what they want to see but not the equipment to interact and change their view of the whole event. Availability of panoramic video and the freedom to choose the view of the event brings out a specific dimension of TV viewing, namely, the possibility to have different selected views of the same event. This issue is in some respect already handled today with fans at an arena having separate viewpoints. This is although something new among TV viewers, where watching the same thing means the exact same images. Giving the TV viewers a possibility to interact with a panoramic view of a live event is a way of supporting individuals to create their own broadcast.

The tablet use is enabling a scenario where the individual views of the game could be even more personalized. There is no need for negotiation with other family members or friends, whereas experience becomes social and intimate in the case of the invitation by the device owner [25], or in the case of the tablet used as a second screen.

Social aspects

One conclusion that emerges from our studies is that people like being a part of a larger context, which usually involves communication and content sharing. However, our participants also expressed an interest in having their own view of the event, clearly showing that they often know what they want to see.

The social interaction that develops around watching TV can be seen from several perspectives:

- **Collocated viewing.** A group, family or friends, watching the same content at the same place (e.g. at home or at public place such as a sports bar).
- **Remote viewing.** The already well-known problem of coordinating with friends/other viewers watching the same content but located at different places.
- **Free viewing.** A newly formulated problem of watching the same program but from possibly diverse viewpoints showing different contents.

Watching TV, and in particular sports, has always been a social experience, as seen in the cases where family or friends gather to watch TV together or when discussing TV shows 'by the water cooler' (*Did you see the game yesterday?*). With advances in communication technologies, the social component has become more complex; social TV watching is possible for remote viewers too. This is already an ongoing subject of academic research and commercial development. In [27], the role of broadcasting is redefined in sense that it is not more a plain consumable TV but it serves as a source for social interaction where socializing around the content is becoming more important than the content itself. In the same work it is also emphasized that "*Social TV aims to provide multiple remote viewers with a joint watching experience*".

A taxonomy of the social aspects of television based on the presence and type of communication is proposed in [26] suggesting that there are two dimensions of the social aspects of TV. The first dimension concerns the presence of the viewers – collocated or remote; and the second dimension concerns the type of communication between viewers – synchronous or asynchronous. High-resolution panoramic video and the freedom to choose the view of the event bring out a third dimension of co-viewing. This third dimension is concerned with the possibility that viewers will have different views of the same event, thus making social interaction more problematic.

Following the sociability heuristics proposed in [28], specifically the following one: "*Support remote as well as collocated interaction*", we add that the design of interactive panoramic TV should also provide support for social conversation for viewers watching different images of the same event.

6.2 Design Implications

Here we summarize our findings and identified design challenges into design implications. The recommendations refer to how to support viewing panoramic content and to provide the live TV viewers with a better experience, and they apply to the design of high-resolution panoramic TV applications supporting content interaction.

Supporting navigating in a panoramic scene. Individual viewing and exploration should be supported in a way that makes it intuitive, yet takes advantage of production-like tools for content interaction. Careful attention should be given to designing the trade-off between responsiveness, accuracy and speed in navigation, so that viewers can follow action within the scene without getting lost in the image. Making the comparison to manual camerawork, the user-controlled view selection through *panning* and *tilting* (moving the view selection horizontally and vertically) should be fast enough to cover relevant movement in the scene, but respond in ways that do not allow jarring and disorienting virtual camera movements. As a concrete recommendation, for most scenes on a horizontal plane, e.g. a stage or a football field, horizontal movements should be more responsive than vertical ones to produce stable moving view selections. Additionally, pre-configuration through user profiles before an event would allow users to take advantage of system-defined views in a personalized manner. If the viewers can customize their viewing preferences in advance, the interaction will be less of a burden during the event. Manual interaction would then be more balanced to occasions of special interest and would draw less attention away from the content.

Enhanced point of view. Both the literature and our observations point to the importance of the spectator's point of view in watching a panorama. In mediated panoramic events, such as the ones seen in the type of interactive television investigated here, selecting a good point of view is an important consideration. If done well, it is a powerful means of conveying the experience of "being there", as a spectator at the scene of the event. This is a quality that panoramic television is well equipped to convey, compared to other emerging formats offering e.g. free form 3D navigation and other features. Designs of panoramic television should emphasize this quality as a key feature. On the consumption side, which is the main focus here, this would imply establishing the central viewpoint early on to novice users. The affordances of the navigation tools should be explained based on that central perspective, e.g. through the metaphor of being there on the stands of a sport event, but with a set of tools for going into details of the action.

Supporting overview and detailed view selection. To fulfill their full potential for representing a scene, panoramic TV applications should include tools to provide better overviews and relevant detailed views, as well as tools for intuitively combining and shifting between the two. Access to the panoramic overview should be quick and easy. A mechanism like that would enable viewers not to get lost, and this is important since advanced content interaction also increases the risk of getting lost. Overview shots are the default image in most live broadcasts, and serve an additional purpose in panoramic television in that they provide a reference to the image content outside of regions currently framed, whether by the producer, the system or the viewer. Our study shows that viewers do not typically perceive the panorama as a scene that is "always there", when they are watching more detailed regions of the image, suggesting unfamiliarity with the format. Having the full panoramic image readily available would thus also serve the purpose of enabling orientation in the wider scene and reminding viewers of the image space they have at their disposal. Besides navigating in the image using virtual equivalents of pan and tilt movements, *zooming* is a highly appreciated and indivisible part of interactive navigation in panoramic images. As such it should be designed in a way that enables precise and quick viewing of specific regions of interests. Analogously with navigation movements, the responsiveness, speed and smoothness of the zoom need to be carefully balanced in order to meet the expectations users have on the interaction. Zoom level (or magnification) should only go as far as the resolution of the image safely allows. Practically for most current systems in the range of 2K vertically, this would mean view selections that are significantly zoomed in but not close-ups of e.g. individual people at a large distance. Our observations suggest that the ability to perform production-like operations like zooming at all is a very attractive feature. Restricting it to a smaller range is a far better option than risking exposing viewers to low resolution image content. For parts of the scene that are outside of the main action, e.g. the ceiling of an arena, possibility to zoom in should be reduced in order to stop viewers from getting lost in unrecognizable details. Furthermore, shortcuts to specific regions, which are known to be of high interest for majority of people, like the goal area, should be provided.

Support for active/passive viewing shift. Support for lean-back viewing should be enabled by providing easy access to the professionally produced broadcast, or broadcasts by other acknowledged producers, e.g. a local event producer. We suggest that, for example, local producers could take advantage of new technology and produce content for a group of viewers with similar interests. In this case an amateur producer would be equipped with a director-like functionality, making the narrative decisions and sharing the chosen view of the event with people who expressed an interest in it (multicasting). The shared content could, for example, be viewed in a public environment, such as supporter pubs, where the produced material would be oriented toward following a specific team. This would enable group of fans to get an alternative view closer to their preferences and interests with little or no interactivity. Active control of the image content should be offered to viewers, but in non-intrusive manner emphasizing voluntary transition between the passive and active modes.

Designing for co-located and remote social viewing. Social conversation for viewers watching different images of the same event should be enabled. This can include mechanisms for, e.g., following a friend or the most popular view of the event, or sharing preferences. The former can be based on friend rating and/or voting for the most popular view. The latter can include information about whether the user wants to share the chosen view and with whom. An example is adding a possibility to share interesting views; instead of asking, as is the case now, "did you see this or that situation?" users could show it, as to some extent happens now through, e.g., YouTube. To enable the proposed social connectivity, the solution should support any of the existing communication tools.

7. CONCLUSION

Due to extensive technical developments in recent years, a new type of TV content is becoming available to viewers – rich, high-definition panoramic video. In this paper we aimed to provide an understanding of how panoramic video as the interactive material can be used as TV content, whereas we are focusing on the specific panoramic picture values and experiences around it. Interaction with panoramic video assumes looking into the panoramic image space, using production-like tools for navigating (zooming, panning, tilting). Such mass capture of TV content and its delivery to viewers in the form of high-resolution panoramic video generates a vast amount of data. It is hardly possible for all the available content to be consumed at once; hence the viewers need to find their own ways of looking into the panoramic picture. In order to understand how this available mass data can be viewed, we have investigated how future users could benefit from the available technology, but also what potential problems they might face.

Based on the user studies we have identified challenges that come with this new interactive content – balancing active and passive viewing, enabling orientation in the panoramic image space, supporting both individual exploration and social conversations around it. Designs of new services based on interactive panoramic video need to resolve this challenges and emphasize panoramic qualities not present in the traditional TV content. As a start we offer the recommendations on how to support viewing panoramic content and to provide the live TV viewers with a better experience in the form of design implications.

Future studies will orient towards the design of the second iteration of the prototypes based on the findings from this work, primarily concentrating on the orientation mechanism as a support for the individual panorama exploration, but also on providing the balance between relaxation and control.

8. ACKNOWLEDGMENTS

The authors would like to thank their colleagues in the FascinatE project for their contributions to the work reported here. We especially thank our project partners UPC (*Department of Signal Theory and Communications, Universitat Politecnica de Catalunya, Barcelona, Spain*) for providing the prototype for gesture interaction and ALU (*Bell Labs - Multimedia Technologies Domain, Alcatel-Lucent Bell NV, Antwerp, Belgium*) for providing the prototype for tablet interaction. The research leading to these results has received funding from the European Union's Seventh Framework Programme (FP7/2007-2013) under grant agreement no. 248138. We also thank all the study volunteers.

9. REFERENCES

[1] P. Cesar and K. Chorianopoulos. 2009. The Evolution of TV Systems, Content, and Users Toward Interactivity. Found. Trends Hum.-Comput. Interact. 2, 4 (April 2009), 373-95.

[2] N. H. Lansford, Jr. and L. L. Jones. Marginal Price of Lake Recreation and Aesthetics: A Hedonic Approach, 1995. J. Agr. and Applied Econ. 27 (1), July 1995:212-223.

[3] M. J. Meitner. Scenic beauty of river views in the Grand Canyon: relating perceptual judgements to locations, 2004. Landscape and Urban Planning 68 (2004), 3-13.

[4] K. Thompson and D. Bordwell. Film History: An Introduction (third edition), 2010. McGraw Hill International Edition. New York.

[5] B. Latour and E. Hermant. Paris: Invisible City, 2006.

[6] Camargus system. www.camargus.com/ (Accessed 18.11.2010.).

[7] Hego OB1 system. www.ob1.hegogroup.com/ (Accessed 17.11.2011.).

[8] imLIVE demo. www.immersivemedia.com/markets/imLIVE/index.html (Accessed 18.11.2011.).

[9] LIVE project. http://www.ist-live.org/ (Accessed 07.10.2011.).

[10] My-edirector project. http://www.myedirector2012.eu/ (Accessed 07.10.2011.).

[11] S. Grünvogel, R. Wages, T. Bürger, and J. Zaletelj. A novel system for interactive live tv. In L. Ma, M. Rauterberg, and R. Nakatsu, editors, Entertainment Computing - ICEC 2007, volume 4740 of Lecture Notes in Computer Science, pages 193–204. Springer Berlin / Heidelberg, 2007.

[12] A. Hilton, J.-Y. Guillemaut, J. Kilner, O. Grau, and G. Thomas. Free-viewpoint video for TV sport production. Image and Geometry Processing for 3-D Cinematography, volume 5 of Geometry and Computing, pages 77–106. Springer Berlin Heidelberg, 2010.

[13] N. Inamoto and H. Saito. Virtual viewpoint replay for a soccer match by view interpolation from multiple cameras. IEEE Transactions on Multimedia, 9(6):1155–1166, 2007.

[14] J. Jiang, J. Köhler, C. M. Williams, J. Zaletelj, G. Güntner, H. Horstmann, J. Ren, J. Löffler, and Y. Weng. Live: An integrated production and feedback system for intelligent and interactive tv broadcasting. IEEE transactions on broadcasting, first, 2011.

[15] D. R. Olsen, B. Partridge, and S. Lynn. Time warp sports for internet television. ACM Trans.Comput.-Hum. Interact., 17:16:1–16:37, December 2010.

[16] C. Patrikakis, A. Pnevmatikakis, P. Chippendale, M. Nunes, R. Santos Cruz, S. Poslad, W. Zhenchen, N. Papaoulakis, and P. Papageorgiou. Direct your personal coverage of large athletic events. Multimedia, IEEE, November 2010.

[17] C. Z. Patrikakis, N. Papaoulakis, C. Stefanoudaki, A. Voulodimos, and E. Sardis. Handling multiple channel video data for personalized multimedia services: A case study on soccer games viewing. In PerCom Workshops'11, pages 561–566, 2011.

[18] Luís A. R. Neng and Teresa Chambel, Get around 360° hypervideo. In Proceedings of MindTrek '10. ACM, New York, NY, USA, 119-122, 2010.

[19] Lizzy Bleumers, Wendy Van den Broeck, Bram Lievens, and Jo Pierson, Seeing the bigger picture: a user perspective on 360° TV, In Proceedings of EuroiTV '12, ACM, New York, NY, USA, 115-124, 2012.

[20] Jean-François Macq, Nico Verzijp, Maarten Aerts, Frederik Vandeputte, Erwin Six, Demo: Omnidirectional video navigation on a tablet PC using a camera-based orientation tracker", in Proceedings of the Fifth ACM/IEEE International Conference on Distributed Smart Cameras (ICDSC 2011), Ghent, Belgium, August 2011.

[21] X. Suau, J. Ruiz-Hidalgo, and J. Casas, "Real-time head and hand tracking based on 2.5D data", IEEE Transactions on Multimedia , vol. 14, no. 3, pp. 575-585 , 2012.

[22] Oskar Juhlin and Elin Önnevall, On the relation of ordinary gestures to TV screens: General lessons for the design of collaborative interactive techniques, Accepted for publication In Proceedings CHI'13.

[23] A. Engström, M. Esbjörnsson, O. Juhlin, and C. Norlin. More TV! - support for local and collaborative production and consumption of mobile tv. In Proceedings of EuroITV, pages 173–177, 2007.

[24] A. Smolic, K. Mueller, P. Merkle, C. Fehn, P. Kauff, P. Eisert, and T. Wiegand. 3d video and free viewpoint video - technologies, applications and mpeg standards. Multimedia and Expo, IEEE International Conference on, 0:2161–2164, 2006.

[25] Y. Cui, J. Chipchase, and Y. Jung. Personal TV: A Qualitative Study of Mobile TV Users. In Cesar, P., Chorianopoulos, K. and Jensen, J. F. (eds.) Interactive TV: A Shared Experience. Proceedings of 5th European Conference, EuroITV 2007. Springer, 195--204.

[26] C. Konstantinos. Content-enriched communication supporting the social uses of TV. Communications Network, 6:23–30, 2007.

[27] R. Schatz, S. Wagner, S. Egger, and N. Jordan. Mobile tv becomes social – integrating content with communications. In Proceeding of International Conference on Information Technology Interfaces, ITI 2007, pages 263–270, 2007.

[28] D. Geerts and D. De Grooff. Supporting the social uses of television: sociability heuristics for social TV. In Proceedings of the 27th international conference on Human factors in computing systems, CHI '09, pages 595–604, New York, NY, USA, 2009.

Usability Design for Video Lectures

Konstantinos Chorianopoulos
Ionian University, Corfu, Greece
choko@ionio.gr

Michail N. Giannakos
NTNU, Trondheim, Norway
michail.giannakos@idi.ntnu.no

ABSTRACT
There is a growing number and variety of educational video lectures online, but there is limited understanding of their effectiveness in terms of learning and usability. Although there is significant research literature within the individual domains of usability and of video learning, there is limited understanding of their integrated design. In particular, there is limited research on guidelines for usable video lecture design, such as the presence of humans in the video and navigation support through the video. For example, it is established that learners benefit from highly structured learning material, but the manual editing of video is not feasible for most learning organizations and instructors. In order to accommodate this emerging instruction medium we are drawing design principles and models from the research literature on educational technology and video interaction. Moreover, we provide a comprehensive approach to the design of usable video lecture systems and content. Finally, we suggest that learning organizations and instructors should invest additional effort in video systems that support an integrated approach to editing, sharing, and controlling of video lectures.

Categories and Subject Descriptors
K.3.1 [**Computer Uses in Education**] Computer-assisted instruction; J.1 [**Administrative Data Processing**] Education

Keywords
Video Lectures, MOOCs, Learning, Usability Design.

1. VIDEO LECTURES
The usage of web-videos for learning purposes has been increasing over the last years. The number of institutions and business organizations providing their content using web-videos is increasing rapidly. Furthermore, there are many video search engines where someone can find educational videos and there are learning-oriented search engines for educational videos (e.g., youtube.com/edu, my learningtube.com). In addition, most of the universities offering lectures on iTunes and elsewhere. On the top of that, Massive Online Open Courses (MOOCs) are becoming an increasingly important part of education. For instance, students access academic content via digital libraries, discuss with tutors by email and attend courses from their home.

Our focus is on web-videos because this type of advanced Internet-based technology enhances the teaching and learning experience. Learners can access and download teaching materials directly. Like other information objects such as books, notes, and documentaries, web-videos may be reused to support both formal and informal learning. Whether recorded for distribution to students or for use internally by employees, these recordings have the potential to be useful for browsing and searching at a later date. Web-videos can also be integrated in an online learning system (portal, e-class, etc) and combine several services. For instance, learners can use in parallel of web-video an online chat room, video conferencing to communicate with their instructors.

There is no single format for video lectures, so there is no simple choice between the available options. Firstly, the traditional video format consists of a video frame that depicts the instructor and the board. It has been the most popular video lecture format, because it is familiar, as it resembles teaching in a real classroom. Nevertheless, there are other popular formats such as the Khan style of video lecture, which depicts only the writing pen of the instructor, but not his face. Between, these two extremes there are hybrid video formats (figure 1), which combine the main characteristics of the two (instructor face, instructor pen). Moreover, the videos are usually embedded in a distance learning system that may include several other features, such as video navigation, quiz-style questions, discussion forum, peer grading, etc. Although any modern distance instructional consists of many elements, there are no guidelines for aligning those elements in the most effective way for a given audience and course. The most popular and basic video lecture formats are: 1) Video capturing of classroom teaching that includes the whiteboard and the instructor, 2) a close-up of the whiteboard that provides a live view of the main course topics, and 3) simple slides or more elaborate animations with a voice-over, which are also known as webcasts. Besides the above basic formats, there are variations of them, such as the Picture-in-Picture technique that combines a close-up of the instructor, which is overlayed to the slides.

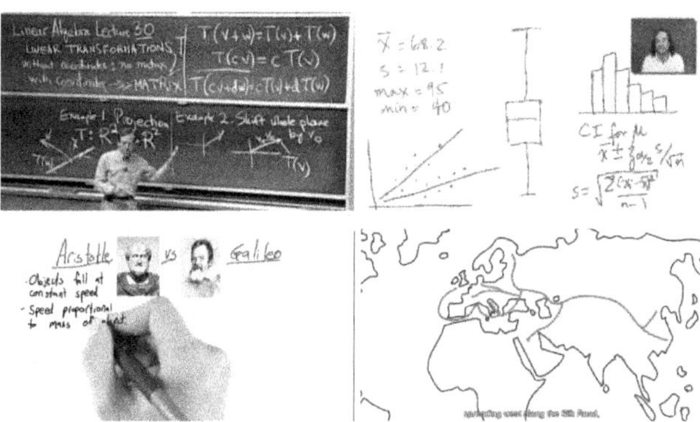

Figure 1. There are many styles of presenting a video lecture, each one with benefits and shortcomings that have to be evaluated against the type of course, the audience, the instructional design, and the technical resources

Web-video technology has many pedagogical affordances and great potential for transforming the teaching and learning environment when it is employed in a suitable way. Thus, the issue is no longer whether instructors should integrate technology in their courses, but how to use technology to transform their courses and create new opportunities for learning. Developing a video lecture is a complex process that requires thorough planning

EuroITV'13, June 24-26, 2013, Como, Italy.
ACM 978-1-4503-1951-5/13/06.

and an implementation procedure. Knowledge of learning theories and instructional implications (e.g., [3]) is a prerequisite for successful realization of the learning content with the most appropriate delivery components. According to Scutter et al., [4], the continuous watching of the lectures has the advantage that information could get "into students heads". Re-listening the courses and taking additional information from the video would also appear to encourage the usability of this medium.

Table 1 Overview of the benefits and shortcoming of the available video lecture presentation styles

Video lecture style	Benefits and shortcomings
Talking head and board	**Simple to capture and share but less usable by the students** *Example*: iTunes U, MIT Open Courseware
Picture-In-Picture (Hybrid style)	**Provides usable cuts between the instructor video feed and the slides or drawing board, but it requires elaborate post-production** *Example*: Coursera
Drawing board	**Video capture of the drawing board with instructor voice over simulates private tutoring** *Example*: Udacity, Khan Academy
Slides and animations	**Includes voice-over and screen cast, so it is simple to capture but might be less friendly without the video feed of the instructor** *Example:* TED Ed, Webcasts, How-to videos

2. USABILITY DESIGN

Besides the video style, there is a need to provide navigation support through the content and connect it to the rest of the instructional design, such as quiz questions, and peer-support in the forum. Thus, the navigation support for video depends mostly on the instructional design and on the video capturing system. Most video lecture systems provide basic video navigation, such as play, pause, and random seek. Depending on the availability of video segments, annotations, extra video (viewing angles, instructor, slides) a video lecturing system might also provide an additional user interface for connecting the video to the rest of the instructional design. For example, Udacity and Khan Academy employ a highly segmented video capturing approach.

Figure 2. The combination of the video lecture with a user interface has been performed according to the resources of each organization, but there are no guidelines for the integration and automation across video editing, sharing, and controlling

In order to support video learning, various technological tools have been developed. For example, matterhorn, centra, edx are just few of them. These tools provide an easy way for a learner who has missed a lecture, the opportunity to catch up, but also enable other, especially slow learners, to review difficult concepts. Despite the widely usage of video lectures and the amount of studies have been conducted upon them (e.g., [1, 2]), usability design principles have not developed.

Although the design of user interfaces for video navigation have been extensively explored in research communities such as interactive TV (EuroITV) and multimedia (MM), the educational context of video lectures raises novel research issues. In addition, it is also not clear how findings from EuroITV and MM communities can be used to improve video lectures for learning. On the one hand, it is obvious that a carefully authored video lecture that includes multiple annotated video segments is beneficial for the learner, but, on the other hand, the majority of instructors and organizations who choose to upload their video online, they do not have the resources to go through this elaborate post-editing.

Table 2. Overview of the properties for each one of the three basic components of a video lecture system

Video lecture system components	Description
Controlling	**Students control video lectures at their pace and the system collects information for video analytics** *Example*: Video topics and quiz index, player buttons (e.g., speed, camera, seek, pause)
Sharing	**Instructors and students are empowered to remix and create mash-ups of video lectures based on existing material** *Example*: TED Ed
Editing	**Capturing and post-processing systems are employed by instructors and editors to produce new content** *Example*: iTunes U Course Manager, Opencast Matterhorn

In conclusion, the basic guidelines for usable video lectures are:

- Video style and video navigation should be designed according to the topic and the learner.
- Video editing, sharing, and controlling should be integrated with instructional design.
- Video analytics should be employed for assessing the effectiveness of the video lecture and the navigation system.

3. REFERENCES

[1] Gkonela, C. and Chorianopoulos, K. 2012. VideoSkip: event detection in social web videos with an implicit user heuristic. Multimedia Tools and Applications, 1-14.

[2] Giannakos, M. N., et al., 2013. Analytics on video-based learning. In Proc. of LAK '13, ACM, NY, USA, 283-284.

[3] Passerini, K., Granger, M. 2000. A developmental model for distance learning using the Internet. Computers and Education, 34(1), 1–15.

[4] Scutter, S., et al. 2010. How do students use podcasts to support learning? Australas J Educ Technol, 26(2), 180-191.

Out of the Box Selection and Application of UX Evaluation Methods and Practical Cases

Marianna Obrist
School of Computing Science
Newcastle University
King's Walk
NE1 7RU Newcastle upon Tyne, UK
marianna.obrist@ncl.ac.uk

Hendrik Knoche
Dept. of Media Technology
Aalborg University
Sofiendalsvey 11
9200 Aalborg SV, DK
hk@create.aau.dk

Santosh Basapur
Motorola Mobility, Inc.
600 N US Highway 45
Libertyville, IL 60048 USA
sbasapur@motorola.com

ABSTRACT
The scope of user experience supersedes the concept of usability and other performance oriented measures by including for example users' emotions, motivations and a strong focus on the context of use. The purpose of this tutorial is to motivate researchers and practitioners to think about the challenging questions around how to select and apply UX evaluation methods for different usage contexts, in particular for the 'home' and 'mobile' context, relevant for TV-based services. Next to a general understanding of UX evaluation and available methods, we will provide concrete UX evaluation case examples and hands-on exercises. Participants will be engaged in a reflection on the practical use of methods, connected problems and challenges.

Categories and Subject Descriptors
H5.m. Information interfaces and presentation (e.g., HCI): Misc.

General Terms
Measurement, Human Factors, Design.

Keywords
User Experience, Evaluation Methods, Case Examples, Practice.

1. INTRODUCTION
User experience (UX) is a complex concept that has gained large popularity both in academia and industry in the last decade (e.g., [1],[2],[5]). It contains various contributing factors such as "*users' emotions, beliefs, preferences, perceptions, physical and psychological responses, behaviors and accomplishments that occur before, during and after use* [of a product, system or service]" for example in the ISO's definition [4]. Designers have to arrive at a synthesis of these possibly competing factors and face hard decisions in trading them off for another. Researchers are interested in measuring and modeling these contributing factors and resulting overall UX of prototypes or actual implementations of systems and services. Equipped with metrics embedded in a method these factors can be evaluated and hopefully will help informing the trade-offs required in designs.

Developing applications that are explicitly intended to improve the UX can be seen as an important step towards a "positive HCI" [3]. UX frees the discourse from the old obsession of HCI to avoid poor designs, which was not geared at providing guidance to arrive at good designs. At the same time we need to keep in mind that we cannot *design the UX* but only *design for UX* because people appropriate technology individually especially in discretionary use contexts as McCarthy and Wright [6] point out.

In order to compare design solution and inform better designs we need to understand how to best measure the individual factors and the overall UX. Developing these metrics is a current challenge in UX research. Partly this is due to the fact that certain parts like immersion or enchantment are hard to measure without affecting the experience. The current ambitious definitions of UX ([4],[5]), however, do not include the operationalizations such that the UX and its contributing factors can be measured and quantified. Requirements for these measures include amongst others: validity and reliability, speed, and cost efficiency, applicability for concepts, ideas, and prototypes [3]. The goal of this tutorial is:

- to provide participants an overview on evaluation methods for studying UX,
- to have the participants apply selected methods on concrete examples (e.g., home and mobile context),
- to discuss strengths and limitations of selected methods (used as hands-on examples) in an UX design and evaluation approach.

As tutorials should follow an interactive and participative approach, we will provide the participants with some initial background based on our expertise, followed by an interactive session. Based on applied examples from the organizers' research expertise (e.g., mobile TV, 3DTV, second screen) we will prepare some provocative problem statements on studying UX to elicit views and responses from the participants. Thereby the tutorial ensures a practical focus based on a common theoretical and empirical basis, confronting the participants with the challenges of selecting and applying a method 'out of the box' of methods (such as provided on http://www.allaboutux.org/).

2. AUDIENCE
The tutorial aims to bring together researchers on UX from a scientific and industrial background. Target participants are in particular designers of TV-based systems and services, as well as researcher concerned with requirements elicitation and evaluation of UX qualities of such services. Beside an interactive program throughout the tutorial we will also engage participants in before the

EuroITV'13, June 24-26, 2013, Como, Italy.
ACM 978-1-4503-1951-5/13/06.

conference, by for instance, providing them access to practical examples or video material for selected UX evaluation cases.

The tutorial aims to attract both experienced designers and researchers around the field on UX, but also affords newcomers a clear overview at what is going with regard to UX evaluation and design approaches. In the first part of the tutorial we will present a background on the topic (particularly refereeing to the collection of UX evaluation methods http://www.allaboutux.org/) to inspire and stimulate the discussion in the practical session applying selected methods on a concrete case examples. The participants will have the opportunity to apply and internalize this knowledge and share their own expertise in the second part of the tutorial. The overall goal of this combination is to learn from each others' experiences based on clearly defined topics and guiding questions.

3. SCHEDULE

The tutorial is foreseen as a half-day event and brings together practitioners and researchers on UX.

09:00 – 09:30: **Introduction**. The scope of UX research and approaches, qualities of UX.

09:30 – 10:00: **Practical Case Examples**: example from 'home' and 'mobile' context are presented as short videos, showing the main situations we are confronted with when exploring users experiences. Group building around each example.

10:00 – 10:30: Coffee Break

10:30 – 11:30: **Selected Method Presentations**: 6 to 8 selected UX methods will be presented in detail by the organizers (key characteristics, advantages/disadvantages, success/failure stories).

11:30 – 12:00: **Group work**: each group selects a method for the case examples (following the previous presentations of selected methods and from the website http://www.allaboutux.org/).

12:00 – 13:00: **Plenary**: each group presents their methodological approach for their cases. Organizers will provide feedback.

12:30 – 13:00: **Conclusions**: Presentation of further guidelines on the usage of UX methods, A&Q round.

The schedule provides a framework to transfer and interactively discuss different approaches for investigating relevant qualities of UX around iTV. The tutorial will be a forum to share individual knowledge with other participants from research and industry.

4. INSTRUCTORS

Marianna Obrist is a Marie Curie post-doc Fellow at the Culture Lab at the Newcastle University, UK. The focal point of her research is on user experience (UX), UX studies, methods and evaluation. Over the last years, Marianna was Assistant Professor in the HCI & Usability Unit at the University of Salzburg, Austria. She was involved in the organization of diverse workshops, SIGs, and tutorials at HCI-related conferences (e.g., CHI, NordiCHI, MobileHCI, EuroITV).

Hendrik Knoche is an assistant professor at Aalborg University's department of Architecture, Design and Media Technology. He has extensive experience in interactive, TV centric services. His PhD focused on adapting video content to mobile devices and he was involved in various user studies relating to mobile social networking, automated content adaptation, biometric systems, video quality visual experience and Quality of Experience.

Santosh Basapur is a Principal Staff Researcher at Motorola Mobility Inc. and has been practicing UX Design and Research for over 8 years. His expertise is in using methods from various domains like HCI, Anthropology, and Human Factors and applying them to UX research. His current research is in iTV UX using second screens and Mobile Wellness Applications. He has published at conferences like CHI, HFES, Euro-iTV, IASDR, Design & Emotion, HCII, and Pervasive Health.

5. Related Experiences & Publications

The instructors of this tutorial have been actively involved in the previous EuroiTV conferences, as organizers and presenters. Some selected publications are listed below outlining the instructors' field specific experience:

Basapur, S., Mandalia, H., Chaysinh, S., Lee, Y., Venkitaraman, N., and Metcalf, C. (2012). FANFEEDS: evaluation of socially generated information feed on second screen as a TV show companion. In *Proc. EuroiTV '12.* ACM, NY, USA, 87-96.

Knoche, H., Sasse, M. A. (2008) Getting the big picture on small screens: Quality of Experience in mobile TV. In Ahmad, A. M .A. & Ibrahim, I.K. (eds.) *Multimedia Transcoding in Mobile and Wireless Networks*, Ch.3, 31-46, Information Science Reference.

Obrist, M., Wurhofer, D., Meneweger, T., Grill, T., and Tscheligi, M. (2012). Viewing Experience of 3DTV: An Exploration of the Feeling of Sickness and Presence in a Shopping Mall. In: *Entert. Comp. J.* (online 21 March 2012).

6. Acknowledgments

This work is supported partly by the Marie Curie IEF Action (FP7-PEOPLE-2010-IEF).

7. REFERENCES

[1] Beauregard, R., Younkin, A., Corriveau, P., Doherty, R., and Salskov, E. (2007). Assessing the Quality of User Experience. *Intel Technology Journal: Designing Technology with People in Mind*, 11.

[2] Hassenzahl, M. (2008). User experience (ux): towards an experiential perspective on product quality. In *Proc. IHM'08*, ACM Press, NY, USA, 11–15,

[3] Hassenzahl, M., and Tractinsky, N., (2006). User Experience - a research agenda. In *BIT Journal*, 25(2), 91-97.

[4] ISO DIS 9241-210:2010. *Ergonomics of human system interaction - Part 210: Human-centred design for interactive systems*. ISO. Switzerland.

[5] Law, E., Roto, V., Vermeeren, A., Kort, J., Hassenzahl, M. (2008). Towards a shared definition of user experience. In *Proc CHI'08*, 2395-2398

[6] McCarthy, J., and Wright, P. (2004). *Technology as experience*. Cambridge, Massachusetts: MIT Press.

Designing Gestural Interfaces for the Interactive TV

Radu-Daniel Vatavu
University Stefan cel Mare of Suceava
13, Universitatii, 720229 Suceava, Romania
vatavu@eed.usv.ro

ABSTRACT
Home entertainment and interactive TV have known fundamental technological advances in the last years by incorporating developments from sensing, processing, and communications technologies. As such systems become more complex and present more options to their users, careful interaction design is needed in order to deliver the expected level of user experience. This tutorial explores novel interfaces for home entertainment scenarios, presents gesture acquisition and recognition techniques, and sets the focus on designing gestural interfaces for today's interactive smart TVs. Case studies will be discussed and a practical design session reusing the Wii Remote controller for designing interactive TV scenarios will be included.

Categories and Subject Descriptors
H.5.2 [Information interfaces and presentation]: User Interfaces. Input devices and strategies.

Keywords
Gesture-based interfaces, home entertainment, interactive TV, gesture recognition, novel interfaces, smart TV, Wii, Kinect.

1. INTRODUCTION
Today's home entertainment systems possess enormous potential to engage their users into complex enriching activities, all taking place in the comfort and familiarity of the home environment [1]. Home entertainment is therefore transforming from a standalone electrical appliance (the TV set) into an experience shared throughout the environment (the interactive TV [4]). Digital and real mix in complex engaging scenarios for the design of which engineers and architects must work together to deliver new standards of multisensory experiences [3,10]. In this vision of future home entertainment, remote controls cannot find their place any longer. As such standard input devices cannot handle the ever increasing functionalities of smart TVs without ruining and diminishing user experience [2], new interaction paradigms must be explored. Among these, gestural interfaces [5,13-16,18] represent a potentially viable candidate, with an increased interest manifested recently from the TV industry [21-23]. This tutorial explores gestures for the interactive TV by focusing on practical technology, examples, and case studies from research and industry, and offers a practical design session.

2. TUTORIAL DESCRIPTION
Presentation format: PowerPoint slides, video demonstrations, and interactive content. The web page of the course is hosted on the instructor's web site (http://www.eed.usv.ro/~vatavu).

Tutorial level: between introductory and intermediate.

EuroITV'13, June 24-26, 2013, Como, Italy.
ACM 978-1-4503-1951-5/13/06.

Detailed outline and schedule:

1. **Introduction: Why do we need gestures in the interface?** [20 minutes]
   An introduction will be provided to participants on the existing gestural interface research by highlighting benefits and pitfalls of using gesture commands [8,9,15].
2. **Gesture acquisition: How to choose the right technology?** [30 minutes]
   The main technologies available today to acquire human gestures will be presented: worn and hand-held equipments (e.g., motion sensing game controllers, gloves, mobile phones and devices) and non-invasive solutions (video cameras). Advantages of each approach will be discussed in the context of the home entertainment environment [15,18].
3. **How to recognize gestures? Popular and simple gesture recognition techniques** [30 minutes]
   A simple, efficient, and widely-used technique for recognizing gesture motions will be presented: the nearest-neighbor classification approach employing Dynamic Time Warping and Euclidean cost functions (employed by popular gesture recognizers [6,7,20]). Specific notions such as recognition accuracy, training set, false positives, and recognizer invariance will be clarified for participants. The new $P gesture recognizer [17], able to recognize gestures articulated in many different ways, will be presented to participants as part of a discussion highlighting how to handle user variability in gesture input.
4. **Gesture set design** [20 minutes]
   Guidelines on how to design and evaluate gesture commands will be provided. Design criteria for gestures, such as ease of execution [12], learnability, and fit-to-function [19], will be explained and exemplified to participants.
5. **Gestures for the interactive TV** [40 minutes]
   Practical applications and case studies of how gestures have been integrated into interactive TV and home entertainment scenarios will be discussed. Examples from research [5,13,14,16,18] and industry [21-23] will be analyzed.
6. **Practice session: Reusing the Wii Remote for interactive TV scenarios** [30 minutes]
   An interactive session will be provided in which participants will be introduced to a low-cost, widely-available motion sensing device (the Wii Remote, http://wii.com) that can be used to rapidly prototype and evaluate gestural interface designs. The device will be employed to create a simple gesture-based interface for controlling the TV set. After a short introduction of the device, attendees will be involved into a participatory design exercise of a gestural TV interface.

Attendees: Researchers and practitioners working in HCI, user interfaces, and design of home entertainment experiences that plan to include gestures into their designs. A mixture of participants from research and industry will be ideal. No previous knowledge of gesture-based interaction is required. Participants will be introduced into today's widely used gesture acquisition and recognition techniques (with focus on how to select the right technology and recognition algorithm), and case studies will be discussed (focus on the guidelines to consider when mapping gestures to functions). A practical design session involving the Wii Remote controller is included in the tutorial.

Tutorial history: The tutorial was delivered at the European Conference on Interactive TV and Video (EuroITV'12) in Berlin, and was well received by all participants. For example, one participant witnessed that the tutorial *"was quite comprehensive and included a lot of background information. This is how a tutorial should be."* The 2013 version of the tutorial will contain updated academic references and industry case studies of TV gestural interfaces.

3. INSTRUCTOR

Radu-Daniel Vatavu works in Human-Computer Interaction with focus on novel interactions and gestural interfaces. He has developed gesture recognition algorithms [11,17], designed computer vision systems [14,15], and investigated human performance aspects of gesture production [12]. He is also interested in designing novel interfaces for home entertainment scenarios such as interactive TV systems, with a specific concern towards human aspects and how technology helps to enhance everyday human-human interactions.

Research expertise relevant to the iTV community includes the TV coffee table [14] and the interactive TV wall [13]. The interactive coffee table [14] allows viewers to control their TV set with simple hand gestures, while reducing fatigue problems and privacy issues for video-based home entertainment scenarios. Design guidelines for novel interfaces and future TV environments are discussed in [15]. The interactive TV wall [13] represents a system for controlling a large video projection display by using the Wii Remote. The project introduces the concept of multiple media screens controlled with standard WIMP interaction techniques. In terms of user gesture preferences, studies of user elicited gestures revealed high level of consensus for frequent home entertainment control tasks [16,18].

Teaching experience: R.-D. Vatavu is a Lecturer at the Computer Science Department of the University Stefan cel Mare of Suceava where he teaches Algorithms Design, Pattern Recognition, and Advanced Programming for undergraduate and master students.

4. REFERENCES

[1] Baillie, L., Benyon, D. 2008. Place and Technology in the Home. Computer Supported Cooperative Work, 17, 2-3, 227-256

[2] Bernhaupt, R., Obrist, M., Weiss, A., Beck, E., Tscheligi, M. 2008. Trends in the living room and beyond: results from ethnographic studies using creative and playful probing. Comput. Entertain. 6, 1

[3] Canizares, A. G. (2006) Fun Rooms: Home Theaters, Music Studios, Game Rooms, and More. Harper Design, 2006

[4] Cesar, P., Chorianopoulos, K. 2009. The Evolution of TV Systems, Content, and Users Toward Interactivity. Found. Trends Hum.-Comput. Interact. 2, 4 (April 2009), 373-95

[5] Dezfuli, N., Khalilbeigi, M., Huber, J., Müller, F., Mühlhäuser, M. 2012. PalmRC: imaginary palm-based remote control for eyes-free television interaction. In Proc. of the 10th European conference on Interactive TV and video (EuroITV '12). ACM, New York, NY, USA, 27-34

[6] Kristensson, P.O., Zhai, S. 2004. SHARK^2: a large vocabulary shorthand writing system for pen-based computers. In Proc. of ACM Symp. on User Interface Software and Technology (UIST '04), 43-52

[7] Liu, J., Zhong, L., Wickramasuriya, J., Vasudevan, V. 2009. uWave: Accelerometer-based personalized gesture recognition and its applications. Pervasive Mob. Comput. 5 (6), 657-675

[8] Norman, D.A. 2010. Natural user interfaces are not natural. interactions 17, 3 (May 2010), 6-10

[9] Norman, D.A., Nielsen, J. 2010. Gestural interfaces: a step backward in usability. interactions 17, 5 (September 2010), 46-49

[10] Rushing, K. (2006) Home Theater Design: Planning and Decorating Media-Savvy Interiors. Quarry Books, 2006

[11] Vatavu, R.D., Grisoni, L., Pentiuc, S.G. 2009. Gesture Recognition Based on Elastic Deformation Energies. In Gesture-Based Human-Computer Interaction and Simulation. LNCS 5085. Springer, 1-12

[12] Vatavu, R.D., Vogel, D., Casiez, C., Grisoni, L. 2011. Estimating the perceived difficulty of pen gestures. In Proc. of INTERACT'11, Springer, 89-106

[13] Vatavu, R.D. 2011. Point & click mediated interactions for large home entertainment displays. Multimedia Tools and Applications, 59 (1), Springer Netherlands, 113-128

[14] Vatavu, R.D., Pentiuc, S.G. 2008. Interactive Coffee Tables: Interfacing TV within an Intuitive, Fun and Shared Experience. In Proc. of the 6th European Interactive TV Conf., LNCS 5066, 183-187

[15] Vatavu, R.D. 2010. Creativity in Interactive TV: Personalize, Share, and Invent Interfaces. In A. Marcus, A. Cereijo Roibas, R. Sala (Eds.) Mobile TV: Customizing Content and Experience. Springer, 121-139

[16] Vatavu, R.D. 2012. User-defined gestures for free-hand TV control. In Proc. of the 10th European conference on Interactive TV and video (EuroITV '12). ACM, New York, NY, USA, 45-48

[17] Vatavu, R.D., Anthony, L., Wobbrock, J.O. 2012. Gestures as Point Clouds: A $P Recognizer for User Interface Prototypes. In Proc. of the 14th ACM Int. Conf. on Multimodal Interaction, ICMI'12, 273-280

[18] Vatavu, R.D. 2013. A Comparative Study of User-Defined Handheld vs. Freehand Gestures for Home Entertainment Environments. Journal of Ambient Intelligence and Smart Environments, 5(2). IOS Press, 187-211

[19] Wobbrock, J.O., Morris, M.R., Wilson, A.W. 2009. User-defined gestures for surface computing. In Proc. of CHI '09, 1083-1092

[20] Wobbrock, J.O., Wilson, A.W., Li, Y. 2007. Gestures without libraries, toolkits or training: a $1 recognizer for user interface prototypes. In Proc. of UIST '07, 159-168

[21] Samsung Smart TV, http://www.samsung.com/us/video/tvs

[22] LG Magic Remote, http://www.lg.com/global/magicremote/

[23] Philips uWand, http://www.uwand.com/

4th International Workshop on Future Television: Focus on Multi Screen Applications *MultiScreen 2013 / FutureTV 2013*

Jean-Claude Dufourd
Telecom ParisTech
France

Stephan Steglich
Fraunhofer FOKUS
Germany

Lyndon Nixon
STI International
Austria

Raphaël Troncy
EURECOM
France

Vasileios Mezaris
CERTH-ITI
Greece

ABSTRACT
This paper describes a full day workshop taking place at the 11th European Interactive TV conference (EuroITV 2013), Como, Italy, on June 24th, 2013, in conjunction with the workshop on Multi-User Services for Social TV, MUSST.

Categories and Subject Descriptors
H.5.2 [**Information Systems**]: INFORMATION INTERFACES AND PRESENTATION (I.7) – *User Interfaces (D.2.2, H.1.2, I.3.6);* H.5.4 [**Information Systems**]: INFORMATION INTERFACES AND PRESENTATION (I.7) – *Hypertext/Hypermedia (Architectures)*

General Terms
Algorithms, Economics, Experimentation, Human Factors, Legal Aspects, Security, Standardization.

Keywords
distributed interfaces, home network, multiscreen, television.

1. INTRODUCTION
This paper describes a full day workshop taking place at the 11th European Interactive TV conference (EuroITV 2013), Como, Italy, on June 24th, 2013, in conjunction with the workshop on Multi-User Services for Social TV, MUSST.

2. MULTISCREEN
"Multiscreen"/"Second Screen"/"Companion Screen" are both buzzwords and a holy grail: there is a lot of hype, and behind it, tremendous unrealized potential. Beyond silos that penalize the domain of multiscreen with extreme business restrictions, there is a need for open standards and platforms, as well as matching production tools. But even in the W3C, the picture is unclear about which set of standards is the best bet. So this workshop intends to bring some issues in focus:

- Issues and bottlenecks: usability, synchronization (media and/or application), content creation including debug, content adaptation to device size and input methods...
- Standards: W3C (WebIntents, Network Service Discovery, Sysapps), HbbTV, ETSI, ...
- Platforms: (closed/native) iOS, Android, (open/web based) webinos, Coltram, ...
- Authoring/production tools: few existing tools if any, what is needed ?
- Deployments and real-life experiences, including new innovative usages of the second screen.

This workshop aims at facilitating exchanges around Multiscreen applications, creating a community of interested parties, from the TV industry, from the mobile world, from the Internet, and is relevant to application developers and designers, human computer interface specialists, platform developers, TV channels, standard gurus, etc, for the purpose of developing common interests, for example to create a W3C Community Group around MultiScreen Apps.

Attendance is not limited to contribution presenters, but submitting a written contribution is highly recommended, in one of the forms described below.

3. TOPICS
- Multiscreen issues and bottlenecks;
- Multiscreen-related standards;
- Multiscreen platforms;
- Multiscreen apps development and production tools;
- Multiscreen deployments and real-life experiences;
- Business models for multiscreen applications;
- Security issues with multiscreen apps.

EuroITV'13, June 24–26, 2013, Como, Italy.
ACM 978-1-4503-1951-5/13/06.

MUSST: Workshop on Multi-User Services for Social TV

David Geerts
iMinds / KU Leuven
Parkstraat 45 Bus 3605
3000 Leuven, Belgium
+32 486 63 32 61
david.geerts@soc.kuleuven.be

Oskar van Deventer
TNO
Brassersplein 2
2612 CT Delft, Netherlands
oskar.vandeventer@tno.nl

Christian Köbel
University of Applied Sciences THM
Wilhelm-Leuschner-Str. 13
61169 Friedberg, Germany
christian.koebel@iem.thm.de

Bettina Heidkamp-Tchegloff
RBB
Marlene-Dietrich-Allee 20
14482 Potsdam, Germany
Bettina.Heidkamp@rbb-online.de

ABSTRACT
Social TV has been an active area of research for more than a decade now, but most research has focused on interaction between remote participants, either through direct communication or more indirect e.g. in the form of social recommendations. However, an issue that has been mostly neglected in social TV research is how to deal with multiple co-located users in the home. This workshop addresses that gap by investigating possible social TV services for multiple users in the home. The objective is to bring together researchers and practitioners working on different aspects of multi-user services for social television including multiple devices: user research, prototype design, technical implementation, system scalability, business models, etc.

Categories and Subject Descriptors
H.5.1 [**Information Interfaces and Presentation**]: Multimedia Information Systems, J.7 [**Computers in Other Systems**]: Consumer Products, H.4.3 [**Information Systems Applications**]: Communications Applications

General Terms
Measurement, Design, Economics, Experimentation, Human Factors, Standardization, Theory, Legal Aspects, Verification.

Keywords
Interactive Television, IPTV, Social TV, Multi-User Services

1. INTRODUCTION
The world of consumer devices has undergone drastic changes in the last couple of years, which has also had an impact on the interactive television landscape. While some research is already taking this multi-device context into account, little research has actually addressed the multiple user context. Currently, personalisation and recommendation algorithms do not work across devices and do not support a group of users. Group dynamics, such as new users entering or leaving the room are not considered either. As watching TV is often a group experience, this lack impedes service and content providers to introduce personalisation and tailored content recommendations in a Hybrid Broadcast Internet environment.

EuroITV'13, June 24–26, 2013, Como, Italy.
ACM 978-1-4503-1951-5/13/06.

Several systems for multi-user interaction are currently being developed in the world of gaming, such as the Nintendo Wii [2] and the multimodal interface of the Microsoft Xbox Kinect [1]. These developments are not yet taken up in interactive television applications: there is no identification of individual users, nor is there personalisation of services based on such identification or context awareness.

Currently, the media industry is developing technologies for advanced personalised media services, like personalised channels, personal service composition, targeted advertising, presence and content recommendations [3]. An analysis of these use cases showed all of them assume personalisation aimed at a single individual user [4]. In research however, the multi-user context is being explored somewhat more. An example is an EPG implementation that featured multi-user policies and showed the combined channel-order preferences of users detected with RFID cards, with the most preferred channel by the group showing on top [5].

The most widely used approaches to content recommendations for single users are content based and collaborative filtering approaches, as well as combinations of these two. Hybrid recommender systems often combine collaborative and content based filtering to take advantage of the benefits of both techniques [6]. In the area of content recommendations for multiple users, the recommender system proposed by Shin & Woo generates recommendations for users by merging user profiles and combining their common interests [7]. More research is needed into the best solutions for multi-user recommendations.

Finally, the inclusion of so many varied services into the social TV space requires clever and scalable division of work between parts of the service implemented on the end-devices and parts implemented in the cloud.

2. WORKSHOP TOPICS

The MUSST2013 workshop invites both researchers and practitioners from diverse backgrounds (user research, technical, business, ...) that work in the area of social and/or personalized interactive television services to submit position papers that address the following topics:

- Social TV applications supporting multiple co-located users
- Multi-user interaction with the TV using the remote control, second-screen devices, gestures or voice control
- Personalization and recommendation for multi-user households and groups
- Identification of multiple users in the living room (e.g. with voice or face recognition)
- System scalability for offering multi-user services to a large audience
- Division of these services between end devices and the cloud

We will invite the authors of the best papers of the workshop to extend their paper and submit it to a special section in the Elsevier Entertainment Computing Journal.

3. ORGANIZERS

David Geerts holds a PhD in Social Sciences at the KU Leuven, where he is research manager of the Centre for User Experience Research (CUO) at the faculty of Social Sciences. David is specialized in user-centered design and evaluation of ICT applications, with a special interest in social interactive television. He organized several workshops, special interest groups, and tutorials at international conferences on this topic (mainly at EuroITV2008-2012 and CHI2006-2012). David Geerts is member of IFIP TC14 Entertainment Computing and its WG6 on interactive television, is co-founder of the Belgian ACM SIGCHI chapter (CHI Belgium), is part of the EuroITV steering committee and was program chair of the EuroITV2009 conference.

Dr. M. Oskar van Deventer is TNO's senior scientist on media networking. His focus is on the realization of international standards and - research project. He is work package leader in the European HBB-Next project on next-generation Hybrid Broadcast Broadband (HBB). He is active contributor and editor of several ITU-T, ETSI and IETF standards, most recently on IPTV, media synchronisation and Content Delivery Networks (CDN) interconnection. He has won several international awards in the area of Mobile Gaming and he has achieved a Guinness World Record. His research current focus is on HBB and CDN. He is author of one book, more than 100 publications, over 50 patents applications, over 500 standardization contributions and several standards.

Christian Köbel is currently a PhD. student at Polytechnic University ISPJAE in Havana, Cuba. He received his diploma degree in Telecommunications from the University of Applied Sciences - Technische Hochschule Mittelhessen in Friedberg, Germany in 2008. Christian's research is centered on dynamic resource management in wireless ad-hoc networks. Currently he is working at the University of Applied Sciences as a research assistant on multi-device video/audio synchronization solutions within the HBB-NEXT framework project. Christian's research has a strong focus on innovative cross-layer systems in distributed network technologies, incorporating a global view on all involved network assets.

Bettina Heidkamp-Tchegloff coordinates multimedia research and development in EC and internal projects for RBB's Department of Production & Operation. She is currently Consortium Leader of HBB-NEXT and monitors RBB's activities in the EC co-funded projects 3D Vivant, Linked TV, FI-CONTENT. To date she has led eight EC projects in the area of ICT. Bettina Heidkamp-Tchegloff holds a master's degree in Media Studies, English and Modern History from Freie Universität Berlin. On completion of her professional journalistic training, she worked in journalism/PR and project management for a number of broadcast related EC co-funded projects before she joined ORB, RBB's predecessor.

4. ACKNOWLEDGMENTS

This workshop is organized by members of the EU HBB-NEXT project and has received funding from the EC Seventh Framework Programme (FP7/2007-2013) under Grant Agreement n°287848.

5. REFERENCES

[1] Microsoft Xbox 360, "You are the controller. No gadgets, no gizmos, just you", http://www.xbox.com/en-US/kinect, introduced in 2010

[2] Nintendo Wii, "A revolution of motion controlled gaming to people of all ages and families everywhere", http://www.nintendo.com/wii/console, 2006

[3] ETSI TISPAN, "Service Layer Requirements to Integrate NGN Services and IPTV", TS 181 016 V3.1.1, July 2008.

[4] Ray van Brandenburg, Oskar van Deventer, Mike Schenk (TNO), Georgios Karagiannis (University of Twente), "Multi-User Interactive TV: the Next Step in Personalisation", EuroITV 2010, Tampere, Finland, 9th-11th June 2010.

[5] Ray van Brandenburg, "Towards multi-user personalized TV services, introducing combined RFID Digest authentication", Graduate Thesis TNO + University Twente, 10 December 2009.

[6] J.J. de Wit, "Evaluating Recommender Systems -- An evaluation frame-work to predict user satisfaction for recommender systems in an electronic programme guide context", Master's thesis University of Twente, May 2008.

[7] C. Shin and W. Woo, "Socially Aware TV Program Recommender for Multiple Viewers", IEEE Transactions on Consumer Electronics, Vol. 55, No. 2, pp.927-932, May 2009

1st International Workshop on Interactive Content Consumption at EuroITV 2013

Rene Kaiser
JOANNEUM RESEARCH
Graz, Austria
rene.kaiser@joanneum.at

Omar A. Niamut
TNO
Delft, The Netherlands
omar.niamut@tno.nl

Goranka Zoric
The Interactive Institute
Stockholm, Sweden
goga@mobilelifecentre.org

Graham Thomas
BBC
London, UK
graham.thomas@bbc.co.uk

Categories and Subject Descriptors

H.5 [**Information Interfaces and Presentation**]: [Multimedia Information Systems]

General Terms

Human Factors

Keywords

content consumption, interaction, multimedia content

1. WORKSHOP AIM

This workshop focuses on novel forms of interactive content consumption. It will explore the shifting balance between lean-back passive TV consumption and lean-forward interactivity; this shift is especially relevant considering the explosion of companion screen interaction, multi-modal gesture/voice control and advanced audiovisual content interaction such as free viewpoint video and user-controlled shot selection. New media formats and consumption paradigms have emerged that allow for new types of interactivity.

The workshop's objective is to provide a highly interactive discussion forum that will allow to capture a comprehensive view on this research area. During the workshop, an overview on new content interaction concepts, research activities and future challenges in this area will be concluded and documented. An interdisciplinary view on the topic shall be compiled by contributions from technical research, user-centric studies, and industry developments. Part of the discussions is fueled by technical demonstrations of interactive content consumption forms. The workshop aims to examine and evaluate new forms of content interaction by discussing the field along three axes:

- Recent technological advances that provide new forms of audiovisual content interaction;
- User studies that evaluate new types of audiovisual content interaction;
- Technologies that supports the user in finding the balance between passive consumption and lean-forward interaction.

EuroITV'13, June 24–26, 2013, Como, Italy.
ACM 978-1-4503-1951-5/13/06.

The following will sketch areas and aspects that are considered within the scope of the workshop. We seek technological research on more active interaction with audiovisual content, e.g. immersive pan-tilt-zoom (PTZ) navigation, game-like interfaces to media, or intelligent storytelling and narrative engines. The workshop will deal with both recorded and live media access. Both mobile and domestic consumption may be investigated. Beyond entertainment, other domains of interactive TV are sought. Topics are not limited to these examples, but should remain within the overall scope of the workshop. Below, some of the questions that the workshop aims to answer are listed:

- How can content personalization be enhanced through interactivity, and at which abstraction level do users want to interact?
- Which new requirements for content capture and production, user-generated, professional and prosumer, do new forms of interactive media imply?
- Do trends in content consumption behavior influence technical research by revealing new challenges?
- What do studies on interaction with content in the realm of social media sharing reveal?
- How can forms of (inter-)active media access be designed to be interwoven with passive consumption modes?
- How can interactivity enhance the experience of people watching together, even when they are in disjoint locations?
- Can 3D free viewpoint video lead to an interactivity revolution in the media industry?
- Which technical advances are needed to allow the industry to offer more interactive media services?
- How does the balance between active and passive consumption affect the Quality of Experience?

The workshop is seeking 3 types of submissions: full research papers for presentations, short papers for poster oral presentations, and technical demos. One keynote speaker will be invited.

2. WORKSHOP FORMAT

We have developed a workshop format to stimulate networking and knowledge transfer among the participants. The full day workshop will be an active forum to discuss research challenges, methodologies and results in a field that is gaining attention based on quickly changing content consumption needs and habits. More than half the time will be reserved for discussion. The chairs will establish an informal atmosphere, inspired by Barcamps. In an active moderating role, they will make sure the workshop's questions will be answered and documented, yet will allow some flexibility where appropriate to meet the interest of the audience. Results will be collected on flipcharts along multiple questions which emerge throughout the day, e.g. what are the latest innovations in that field? Which research activities exist to tackle unsolved challenges? How could we combine different interaction technologies to the benefit of the user? Throughout the day, the audience will be encouraged to contribute, and especially to comment existing inputs (*I'd love to collaborate on this! ... This has already been solved in my project!*). Live comments by both the local and remote audience through Twitter will be promoted and collected. The workshop will consist of the following sessions:

- Welcome and presentation of workshop aims.
- Interactive participant introduction in Barcamp style (name, affiliation, role, 3 #hashtags).
- Invited short keynote. Shall give an overview on research and industry trends, and provide a common ground for the subsequent workshop exercises.
- Pitches (á 2min) to kickstart the poster/demo session, where, in parallel, posters and demos are exhibited and discussed. This shall establish an understanding of each other's work, approach, and focus.
- — *lunch break* —
- A fishbowl discussion format, based on the aspects raised beforehand. There is a limited number of *active* seats. If you want to say something, you have to take an empty seat or wait for one. This format of a dynamically changing working panel proved to work well for discussions among experts on concrete questions.
- Some 4-6 research paper based talks. Questions will be allowed during the talks. Chairs will moderate, and collect statements relevant to the overall workshop aim.
- Concluding session. The group will revisit what has been collected throughout the day. Conclusions will be summarized.

3. ORGANIZING COMMITTEE

The organizing committee consists of 4 members of the FascinatE project who complement each other well, since they cover different areas of technical research, HCI research and industry.

Rene Kaiser is a key researcher for JOANNEUM RESEARCH and has been involved in a number of European projects dealing with automation of content production such as NM2, Aposdle, TA2 and Vconect. His research focus is on *Virtual Director* software, on automating shot selection through cinematographic behavior models. Further he is interested in automating non-linear video production, enabling the user to interactively influence the narrative path while watching. Rene was responsible for the organization of the Interactive and Immersive Entertainment and Communication Special Session at MMM'12. He is part of a group hosting the annual PhD cooperation workshop at the i-KNOW and i-SEMANTICS conference, and has been organizing the Barcamp Graz, a yearly 3-day *unconference* which is an interactive and open discussion format.

Omar Aziz Niamut is a senior research scientist at TNO. He has a background in the coding and transport of digital signals. His PhD research included the development of universal audio coding algorithms. He advised the European Parliament on the harmonization of mobile TV and was editor of the technology report on mobile TV in the European Commission — assigned European Mobile Broadcast Council. He participated in ETSI TISPAN standardization with over 200 contributions on interactive services for next-generation IPTV and advised the Singapore government on the use of IPTV standards. His current focus is on social and interactive media services over next-generation networks and immersive media delivery. He is the author of academic papers and articles in the fields of audio coding, interactive IPTV services and immersive media. Omar holds over 15 patents and has contributed a chapter to a book on Social TV. He has been a member of the EuroITV TPC, for the Systems and Enabling Technologies track.

Goranak Zoric is a senior researcher at the Mobile Life Center, employed by the Interactive Institute, since January 2010. She holds a M.Sc. (2005) and a Ph.D (2010) in electrical engineering with a major in telecommunications and information science at the University of Zagreb, Croatia. Goranka's current research includes interdisciplinary human centric approach in the field of live video interaction with the focus on mobile technologies. Goranka's previous work was in the field of automatic face animation of virtual characters and their applications in networked and mobile environments. She is a guest editor on the special issue in Personal and Ubiquitous Computing journal on Video interaction. Goranka previously organized the eNTERFACE'06 workshop on multimodal interfaces as part of SIMILAR NoE project.

Graham Thomas is the Section Lead for Production Magic at BBC R&D. His PhD from Oxford University included the development of motion estimation methods for standards conversion, which led to an Emmy award and a Queen's Award. Since 1995 he has been leading a team of engineers developing 3D image processing and graphics techniques for TV production. His work has led to many commercial products including the free-d camera tracking system, Chromatte retroreflective chromakey cloth, and the Piero sports graphics system. Graham holds over 20 patents and has contributed chapters to several books on broadcast applications of computer vision. He is a chartered engineer, member of the IET, and a Visiting Professor at the University of Surrey. Graham previously organized a tutorial on AR in broadcasting at ISMAR, and is on the CVMP organizing committee.

4. ACKNOWLEDGMENTS

The workshop is supported by the European FP7 research project 'FascinatE[1] – Format-Agnostic SCript-based INterAcTive Experience" under grant agreement no.248138.

[1] `http://www.fascinate-project.eu`

4th EuroITV 2013 Workshop on Interactive Digital TV in Emergent Economies – Thinking Outside the TV Box

Vicente F. de Lucena Jr
University of Amazonas -- Ceteli
Campus Universitário
Manaus – AM, 69077000, Brazil
+55 92 3305 4680
vicente@ufam.edu.br

Artur Lugmayr
Tampere Univ. of Technology (TUT)
EMMi - Entertainment and Media Management Lab.
P.O.Box 541, Korkeakoulunkatu 8, Tampere - Finland
+358 40 849 0773
artur.lugmayr@tut.fi

Zhiwen Yu
Northwestern Polytechnical University
Institute of Ubiquitous and Intelligent Computing
Xi'an, Shaanxi, 710129, P. R. China
+86 29 8849 1544
zhiwenyu@nwpu.edu.cn

Arpan Pal
Tata Consultancy Services
Innovation Lab, Kolkata
Bengal Intelligent Park, 3rd & 4th Floor, India
+91 33 6636 7295
arpan.pal@tcs.com

ABSTRACT
The popularity of TV worldwide and particularly in emergent markets opened a window of opportunities for many people working with Interactive Digital TV and related topics. New solutions and new business approaches are supposed to rise from this scenario, creating many new job opportunities. The main goal of this workshop is to bring together researchers, educators, and industry related people working in the deployment of solutions for DTV Systems and associated Applications in emergent countries. Along this workshop, discussion about the adopted policies, technologies, state-of-the-art, middleware solutions, content models, interactivity systems, and future applications in the researchers' countries are going to be the main action. Indeed, it is going to be a global forum where the participants will be able to identify common issues and exchange experiences, sharing the adopted solutions for common known problems. The organizers hope to keep building and maintaining a research network that will help people working in developing countries to increase their technical contributions in Interactive Digital TV related issues.

Categories and Subject Descriptors
A.0 [**General**]: Conference proceedings.

A.1 [**Introductory and Survey**]

General Terms
Documentation, Economics, Reliability, Experimentation, Human Factors, Standardization, Languages, Theory, Legal Aspects.

Keywords
Interactive Digital TV Systems; Digital TV Standards; New Applications for Interactive Digital TV

EuroITV'13, June 24–26, 2013, Como, Italy.
ACM 978-1-4503-1951-5/13/06.

1. INTRODUCTION
More than only an entertainment media, Interactive Digital TV has been used as an efficient tool to improve people's lives. Digital social networks, E-government, E-bank, and many other digital applications are been moved to the TV world and are already accessible in many communities in Europe, USA and Asia. In fact this movement is the proof that the information age is a reality and it is possible to witness several actual examples of how people's lives have been transformed by these instruments. In fact, access to information has become easier and people work cooperatively despite being scattered around the globe, even in extremely remote locations. Overall, wireless communication has revolutionized the lives and the way of working of many people and, with the ever growing miniaturization in electronics, many new devices have been developed and inserted in every-day life.

With the deployment of Interactive Digital Television in emerging economies in South America like Brazil and Argentina, in India, China, Russia, in Africa and in West Europe countries, something alike happened. There is a growing demand for new services, new devices and even more content to be provided, as well as many ways of integrating these new features with existing technologies [1]. Reports about the changes obtained after the introduction of Digital TV in developing countries will help us to understand its real importance and to plan how to contribute for its future [2].Within the scope of this workshop we would like to focus on 'thinking outside the box' of traditional Digital TV platforms – we aim at contributions in the wider field of audio-visual services in emerging countries as well.

Additionally to its natural potentiality for digital inclusion, the Interactive Digital Television provides a great incentive to its adoption, for it means a cheaper way to supply access to technology and knowledge in emerging economies [3]. It also provides scenarios for integrating Television with other areas of knowledge such as smart environments and ubiquitous systems, allowing access to services until now not easily achievable [4]. These and other related topics are the main discussion object of this workshop. We would like to learn from developers and scientists coming from emergent countries the adopted policies, technologies, middleware, content models, and about the future applications they are currently working on.

2. WORKSHOP

The immediate goals of the workshop are to discuss the state of the art of Interactive Digital TV in developing countries; to identify new applications that may benefit the life of people living in these countries; to define common characteristics of the adopted technologies; to discuss middleware definitions and interactive software development for Digital TV; to discuss content production and new copyright policies for digital production; to discuss policies and regulations in emergent countries; and to identify topics of mutual interest leading to future cooperation partnerships.

Having the workshop goals in mind, a list of subjects that could find resonance in the community involved was defined. The main topics of interest include: the state of the art of interactive digital TV in emergent economies; interactive applications for Digital TV; educational applications; healthcare applications; T Government and T Commerce proposals; description of the characteristics of the adopted technologies; new middleware definitions; interactive software development for Digital TV; integrating semantic technologies with interactive DTV; content production and copyright policies for digital production; and service deployments.

In fact, the accepted papers will cover many of the items listed above and they have been selected on the basis of their relevance to the workshop and on the quality of the presented work. In order to facilitate the discussion of the presented topics, all selected papers were listed in the workshop web site. Participants were able to read them in advance and were asked to bring their notes and thoughts to the discussion table.

The workshop itself was planned to have two major parts. In the first one, authors should present their works in a 30 minutes slot of time, this time was divided in 20 minutes for the presentation itself and 10 minutes for particular questions about the work presented. The second part of the workshop consists of a round table discussion about the main subjects presented. The discussion focus shall be about the issues faced by the participants themselves in order to consolidate Interactive Digital TV in their own countries. We planned also to summarize the state of this technology in the countries represented in the workshop.

An illustrative poster containing the results of the discussions is going to be released and presented to the other participants of the conference in the poster section giving them the opportunity to learn about our conclusion.

We believe that Interactive Digital TV is a very interesting topic all over the world and that is a great opportunity to develop new media, new ways of interacting with consumers and new job opportunities for scientists, developers, artists and other related professionals. Emerging countries like the ones that participated in this workshop have a significant role in the development of this technology.

This workshop will have reached its goals if the participants succeed in building a permanent discussion forum, capable of contributing to each other works and facilitating the communication among people with interests in developing Interactive Digital TV Systems.

3. ORGANIZERS

Prof. Dr.-Ing. Vicente Ferreira de Lucena Jr: is professor of Digital Systems in the University of Amazonas in Manaus, Brazil. He pursued his Ph.D. (Dr.-Ing.) at the University of Stuttgart in Germany where he worked with software engineering for automation systems. Since 2002 he has been working with software development for interactive digital TV and convergent systems in Brazil.

Prof. Dr. Artur Lugmayr describes himself as a creative thinker and his scientific work is situated between art and science. Starting from July 2009 he is full-professor for entertainment and media production management at the Department of Business Information Management and Logistics at the Tampere University of Technology (TUT): EMMi – Entertainment and Media Production Management. His vision is expressed as to create media experiences on emerging media technology platforms.

Prof. Dr. Zhiwen Yu is a professor and vice dean in the School of Computer Science at Northwestern Polytechnical University, China. He received his B.Eng., M.Eng., and Ph.D. degree of Engineering in computer science and technology from Northwestern Polytechnical University. He has worked as an Alexander Von Humboldt Fellow at Mannheim University, Germany, and as a research fellow at Kyoto University, Japan, and a post-doctoral researcher at Nagoya University, Japan. His research interests cover pervasive computing, context-aware systems, human-computer interaction, and mobile Internet.

Dr. Arpan Pal has more than 20 years of experience in the area of Signal Processing, Communication and Real-time Embedded Systems. He is with Tata Consultancy Services (TCS), where he is heading research at Innovation Lab. He is also a member of Systems Research Council of TCS. His main responsibility is in conceptualizing and guiding R&D in the area of cyber-physical systems and ubiquitous computing with focus on applying the R&D outcome in the area Intelligent Infrastructure. His research interests include Home Infotainment, Mobile phone and Camera based Sensing and Analytics, Physiological Sensing, M2M communications and Internet-of-Things based Applications. He is a B.Tech and M.Tech from Indian Institute of Technology, Kharagpur, India in Electronics and Telecommunications.

4. ACKNOWLEDGMENTS

Our thanks to the Euro ITV Steering Committee, to the Tutorial and Workshop Chairs, and to our Hosts from the Politecnico di Milano University for their support and kindness.

5. REFERENCES

[1] Standards – A Guide to MHP, OCAP, and JavaTV. Elsevier Oxford, UK

[2] Srivastava, H. O. 2002. Interactive TV – Technology and Markets. Artech House, Inc. Norwood, USA.

[3] Lugmayr, A.; Niiranen, S. and Kalli, S. 2004. Digital Interactive TV and Metadata – Future Broadcast Multimedia. Springer-Verlag, New York, USA.

[4] Pagani, M. 2003. Multimedia and Interactive Digital TV: Managing the Oportunities Created by Digital Convergence. IRM Press, London, UK.

TV's Future is Linked: Web and Television across Screens

4th International Workshop on Future Television at EuroITV 2013

Raphaël Troncy
EURECOM
Sophia Antipolis
France
raphael.troncy@eurecom.fr

Lyndon J B Nixon
STI International Consulting and Research GmbH
Vienna, Austria
lyndon.nixon@sti2.org

Vasileios Mezaris
Information Technologies Institute - Centre for Research and Technology Hellas (CERTH)
Thessaloniki, Greece
bmezaris@iti.gr

Categories and Subject Descriptors

E.2. [**Data Storage representations**]: Linked representations
H.5.1 [**Multimedia information systems**]
H.5.4 [**Hypertext/hypermedia**]

Keywords

Web multimedia; online media; media metadata; media semantics; media descriptions; Linked Media; hypermedia; hypervideo; interactive video.

INTRODUCTION

This workshop follows from the successful 1st and 2nd Workshops on Future Television held at EuroITV 2010 and 2011 and organized by the EU project NoTube, and last year's 3rd Workshop sponsored by the EU project LinkedTV (http://www.linkedtv.eu). Previously, we have focused on how Web-TV integration could make much more possible than simply pasting Web content over the TV program, or giving the TV viewer a hard-to-use Web browser on their screen. Rather:

> "The growing amount of data available on the social and semantic Webs, APIs to access and manipulate that data, and trends towards common identifiers and metadata schema allow future television new opportunities to link TV viewing with both the viewer's personal Web profiles on social networks, and with social activities across the Web also from friends and similar people."

This initial focus on Web data sharing and the social TV experience has, interestingly, been caught up by the TV market itself. Second screen TV experiences with Web content tie-in, or social network integration, are being supported by an emerging ecosystem of mobile device applications as well as applications on the TV platform itself. This original vision of *future* television is rapidly becoming the present reality. However this shift in how TV is experienced uncovers new barriers and limitations in the present day technologies and networks.

This workshop is now supported by the new EU project LinkedTV (started October 2011). LinkedTV considers how TV and the Web will, in the future, become seamlessly interwoven, independent of the content or device. Rather than switch between "TV" and the "Web" they will co-exist, content on screen being intuitively mashed-up together, and interaction by the user shifting between content and data from the different sources effortlessly. Last year's workshop focused on such mashups and interweaving solely on the main screen of the dedicated TV device, but both within the project and outside it in the Web and TV community the existence of **multiple screens** is becoming the new challenge to address. This is not only smartphones and tablets, but also dedicated device controllers (WiiU, Xbox), as well as corollary screens around the home (digital radios, fridges, Internet alarm clocks, etc.).

We will explore in this 4th Future Television workshop the research and development necessary for linking the worlds of TV and the Web, with special emphasis on moving the integrated Web-TV experience which is currently partially being provided on single screens (the TV, but also the mobile) into **second screens and multiple screens**, and on examining how the use of more than one screen, combined with various interaction and sensor possibilities (touch, gesture, speech, movement), could contribute new ideas and more intuitive usage possibilities for the future vision of "Linked Television".

EuroITV'13, June 24–26, 2013, Como, Italy.
ACM 978-1-4503-1951-5/13/06.

Author Index

www.ingramcontent.com/pod-product-compliance
Lightning Source LLC
LaVergne TN
LVHW061245100826
845148LV00008B/1027
* 9 7 8 1 4 5 0 3 2 4 4 1 0 *